Fluorescence microscopy is used for studying the distribution of substances which are present in very small amounts. The high sensitivity of the method makes it ideal for studying the distribution of substances in living cells. Its techniques are used mainly in biology and medicine, but are also valuable in coal petrology and elsewhere. The best-known application is in immunofluorescence.

This new work provides comprehensive coverage of all aspects of fluorescence microscopy. It covers instrumentation, applications to a wide variety of fields, and the history of the technique. There is a chapter on quantitative techniques, including scanning: this aspect is dealt with in more detail in a companion volume, *Quantitative fluorescence microscopy*.

Volume I deals with instrumentation and techniques for fluorescence microscopy, and includes a chapter on quantification and scanning. It gives practical information on how to use fluorescence microscopes, as well as explaining the underlying theory. The technique of photographing fluorescence is explained in detail.

Volume II deals with the applications of fluorescence microscopy in many fields. It includes information on fluorochromes and on autofluorescence. An invaluable appendix provides an alphabetical list of fluorochromes, giving information concerning chemical structure, fluorescence properties, applications and suitable filter combinations.

Fluorescence microscopy

Fluorescence microscopy

F.W.D. ROST
Professor of Anatomy, University of New South Wales

VOLUME I

The right of the University of Cambridge to print and sell all manner of books was granted by Henry VIII in 1534. The University has printed and published continuously since 1584.

CAMBRIDGE UNIVERSITY PRESS
Cambridge
New York, Port Chester, Melbourne, Sydney

Published by the Press Syndicate of the University of Cambridge
The Pitt Building, Trumpington Street, Cambridge CB2 1RP
40 West 20th Street, New York, NY 10011-4211, USA
10 Stamford Road, Oakleigh, Victoria 3166, Australia

First published 1992
Printed in Great Britain at the University Press, Cambridge

British Library cataloguing in publication data
Rost, F. W. D.
Fluorescence microscopy. Vol. 1
I. Title
502.823

Library of Congress cataloguing in publication data

Rost, F. W. D.
Fluorescence microscopy / F.W.D. Rost.
p. cm.
Includes bibliographical references and index.
ISBN 0-521-23641-X (v. 1)
1. Fluorescence microscopy. I. Title.
QH212.F55R66 1992
502′.8′2 – dc20 91-19168 CIP

ISBN 0 521 23641 X vol. I hardback
ISBN 0 521 41088 6 vol. II hardback
ISBN 0 521 42277 9 set of two vols. hardback

SE

Contents of Volume I

Contents of Volume II

Preface

My goals in preparing this book have been to describe, firstly, the structure and use of fluorescence microscopes; secondly, the basic physical and chemical principles involved, so that the fluorescence microscopist will understand what is happening; and thirdly, some of the various applications of fluorescence microscopy, in broad overview, to alert users to techniques which might be useful to them. I have assumed that the reader consulting this book will usually have a practical problem and desire a pragmatic answer; accordingly, the main part of the text describes current knowledge and techniques; and therefore almost all purely historical considerations are dealt with separately in Volume II, Chapter 16.

I originally intended to deal with quantitative aspects in more detail, spread over some 15 chapters, but later it was decided to publish the quantitative chapters as a separate book (*Quantitative fluorescence microscopy*) and to summarize quantitative aspects in a single chapter in the present volume. *Quantitative fluorescence microscopy* also discusses, in greater detail, scanning microscopy (including confocal microscopy) and some other topics which are best studied quantitatively, such as photobleaching (fading).

Similarly, we decided to print *Fluorescence microscopy* as two volumes. The first volume deals with the microscope and the technique of using it; the second volume discusses applications of fluorescence microscopy and (in the last chapter) the history of fluorescence microscopy.

Acknowledgements

I am very much indebted to those friends and colleagues who kindly read and commented on chapters or sections of the book. They are (in alphabetical order): Associate Professor A. E. Ashford (Chapter 10), Professor D. St. C. Black (the structural formulae in Appendix 7, many of which he very kindly redrew in accordance with modern conventions), Miss E. Buss (Appendix 1), Professor A.C. Cook (Chapter 11), Mr A.C. Dickes (Chapter 6), Mr B. J. Eglinton of Wild-Leitz Pty Ltd (Chapter 4), Dr B. Freeman (Chapter 17), Mr R. H. Freere of Wild-Leitz Pty Ltd (Chapter 4), Professor J. B. Furness (Chapter 12), Dr R. C. Hiller (chlorophylls), Dr J.-P. Hornung (neurobiology), Professor F. H. Kasten (Chapters 13, 18), Professor J. C. Kelly (Chapter 2), Mr R. Kerr of FSE Pty Ltd (Chapter 4), Mr W. Loeb of Carl Zeiss Pty Ltd

(Chapters 3, 4, 5), Mr B. Miller of Kodak (Australasia) Pty Ltd (Chapter 6, Appendices 4A and 4B), Dr C. J. F. van Noorden (Chapter 15), Mr R. J. Oldfield (Chapters 1, 2, 3, 5, 6, 7, 8), Professor A. G. Everson Pearse (Chapter 12), Professor M. van der Ploeg (Chapters 14, 16, Appendix 1), Professor J. S. Ploem (Chapter 3), Dr Laura Powers (Chapter 11), Dr D. H. Rapaport (neuronal tracing), Mr D. Regel of HD Scientific Pty Ltd (Chapter 4), Professor K. G. Rienits (Chapters 9, 10), Mr G. D. Schneider (Chapter 17), Professor P. J. Stoward (Chapter 15), Associate Professor I. Törk (Chapter 6), Associate Professor D. J. Tracey (neuronal tracing), Mr J. Unvala of Selby-Anax (Chapter 4), Mr P. Vardy (Chapter 17), and Dr N. G. M. Wreford (Chapter 8). Of course, I alone must accept responsibility for all errors of commission and omission.

Special thanks are due to Professor A. G. Everson Pearse (Royal Postgraduate Medical School, London) for many things. I had the great privilege of working under him for almost a decade (1965–1974), and it was as a direct consequence of this that I became involved in fluorescence microscopy in general and microspectrofluorometry in particular.

I am very grateful to numerous friends and colleagues who have given me the hospitality of their laboratories and provided much valuable information. I particularly wish to thank (in alphabetical order): Professor G. Bottirolli, of the University of Pavia; Professor T. Caspersson of the Karolinska Institut, Stockholm; Mr Karl-Heinz Hormel of Ernst Leitz Wetzlar GmbH, Wetzlar, formerly of E. Leitz (Instruments) Ltd, London; Professor M. van der Ploeg and Professor J.S. Ploem of the University of Leiden; Dr Andreas Thaer, of Helmut Hund Kg, Wetzlar, formerly of E. Leitz GmbH and of the Battelle Institut, Frankfurt; Dr F. Walter of Ernst Leitz Wetzlar GmbH, Wetzlar.

Professor Frederick H. Kasten (Louisiana State University, USA), besides reading chapters, generously sent me material, including advance copies of typescripts of his monographs on fluorochrome staining techniques and the history of fluorescence microscopy. I am grateful to numerous people for supplying photographs; acknowledgements are made in the captions.

Preparation of the book was commenced while I was employed at the Royal Postgraduate Medical School, London, and completed at the University of New South Wales. I am indebted to Library staff of the University for much assistance. Mr Patrick de Permentier, Ms Jenny Flux and Ms Stacey McClelland provided technical assistance and help with bibliographic work. A substantial amount of the text was typed by Mrs Lorraine Brooks. For conversion of word-processor text from an older system to WORD on a Macintosh, I am very grateful to the late Mr Peter Hughes of Macquarie University and to Mr Paul Halasz. I am also indebted to Mrs Gillian Rankin of Macquarie University and to Ms Alicia Fritchle for redrawing diagrams, to Miss Jenny Norman (Macquarie University) and Mr Collin Yeo for assistance in photography, and to Mrs Mary Armstrong for secretarial assistance.

Last but not least I wish to express my appreciation of the collaboration of Dr Alan Crowden and Mrs Sandi Irvine of CUP.

Fred Rost
Sydney, 1991

Abbreviations and symbols

A	adrenaline
AC	alternating current
ASA	American Standards Association (film speed)
B&W	black-and-white
BChl	bacteriochlorophyll
c	concentration of fluorophore
°C	degrees Celsius
CA	catecholamine
cd	candela
CIF	carbonyl- or carboxylic acid induced fluorescence
Chl	chlorophyll
d	optical path length (depth)
DIN	Deutsches Industrie Norm
DA	dopamine
DIC	differential interference contrast
DC	direct current
DNA	deoxyribonucleic acid
E	(1) quantum energy of photon
	(2) optical extinction (optical density)
EI	exposure index
EM	electron microscopy
Em	emission maximum
Ex	excitation maximum
eV	electron volt
F	relative fluorescence intensity
FDFV	freeze-dried formalin-vapour-fixed
FIF	formaldehyde-induced fluorescence
FPR	fluorescence photobleaching recovery
FRAP	fluorescence redistribution after photobleaching

FRP	final reaction product
h	Planck's constant
Hz	Hertz (cycles per second)
I	intensity of light
I_a	intensity of absorbed light
I_f	intensity of fluorescence
I_t	intensity of transmitted light
I_0	intensity of incident light
ISO	International Standards Organization (film speed)
J	joule
K	(degrees) Kelvin
lm	lumen
LP/mm	line pairs per mm
MDy	microdensitometry
MFy	microfluorometry
mm	millimetre
ms	millisecond (10^{-3} s)
MSF	microspectrrofluorometer
MSFy	microspectrofluorometry
MW	molecular weight
NA	(1) numerical aperture
	(2) noradrenaline
NADH	nicotinamide adenine dinucleotide
nm	nanometre (10^{-9} m)
ns	nanosecond (10^{-9} s)
P	work function
PRP	primary reaction product
Q	quantum efficiency
QI	quartz-iodine (tungsten-halogen)
QVIM	quantitative video intensification microscopy
r	a constant
RNA	ribonucleic acid
s	second
S_0	singlet ground state of molecule
S_1, S_2	1st, 2nd, excited singlet states
sb	stilb
t	time
T_0	triplet ground state of molecule

T_1	1st excited triplet state of molecule
THz	terahertz (10^{12} Hz)
UV	ultraviolet
V	volts
VIM	video intensification microscopy
W	watts
WL	wavelength
x	amount of unknown substance
γ (gamma)	contrast factor
ϵ (epsilon)	extinction coefficient
ϕ (phi)	quantum efficiency
λ (lambda)	wavelength
μ (mu)	refractive index
μm	micrometre (10^{-6} m)
μs	microsecond (10^{-6} s)
ν (nu)	frequency
τ (tau)	lifetime of fluorescence

1

Principles of fluorescence microscopy

Fluorescence microscopy is a technique of microscopy whereby fluorescent substances are examined. Because many substances are fluorescent, or can be made so, fluorescence microscopy is widely applicable. It has a number of advantages over other forms of microscopy, offering high sensitivity and specificity. Because fluorescence is observed as luminosity on a dark background, particular constituents of the specimen can be seen even in in extremely small amounts.

Development of fluorescence microscopy

It was the German microscopist Köhler, best known for the development of a system of illumination, who invented fluorescence microscopy. Köhler was experimenting with ultraviolet microscopy, which, because of its shorter wavelength, can give a finer resolution than microscopy with visible light. Köhler (1904) discovered that tissue is fluorescent when irradiated with ultraviolet light. This phenomenon is known as *autofluorescence*.

Soon after, it was found that stained sections of tissue could be examined in a fluorescence microscope if the dyes used in staining were fluorescent. Such dyes were termed *fluorochromes* to distinguish them from *diachromes*, which are coloured dyes made visible because of their absorption of light.

A period of study of fluorochromes, almost exclusively by German microscopists, culminated in the application of the metachromatic dye Acridine Orange to vital staining. Acridine Orange fluoresces either green or red, depending upon the nature of the stain–substrate interaction; it was found that, in general, DNA in nuclei of living cells stained green, while nuclei of dead cells (containing denatured DNA) and cytoplasmic RNA both fluoresced red. This discovery, with its obvious applications to biology and medicine, led to further acceptance of the value of fluorescence microscopy.

However, the major factor in the extension of use of fluorescence microscopy was the development of the fluorescent antibody technique by Coons and co-workers in 1941. Fluorescence microscopy provided the high sensitivity needed for the demonstration of antigen–antibody reactions *in situ*. The Coons technique, although commonly known as the fluorescent antibody technique, is applicable to the labelling of antigens and complement as well, and is better referred to as the immunofluorescence technique. These techniques brought about a revolution in immunological research, and for the first time made fluorescence microscopy important enough for it to become widely used.

As a result, fluorescence microscopes became more readily available, leading to developments in other fields. Of these, possibly the most significant was the formaldehyde-induced fluorescence technique for the demonstration of neurotransmitter amines. Some substances, themselves non-fluorescent, can be converted to fluorescent compounds by a chemical reaction with another substance which is itself also non-fluorescent. This kind of reaction is referred to as induced fluorescence. The most important histochemical methods for inducing fluorescence are those which demonstrate neurotransmitter amines and related substances, by condensation reactions with formaldehyde, glutaraldehyde or glyoxylic acid, which result in a fluorescent product. These methods have been extensively applied in studies of the distribution of adrenergic, dopaminergic and serotoninergic nerves. These methods are described further in Volume II, Chapter 12. The history of fluorescence microscopy is described in more detail in Volume II, Chapter 18.

The basic principles of fluorescence microscopy

The principles of optical microscopy apply as much to fluorescence microscopy as to any other technique of optical microscopy. There are of course some additional complications. It will be assumed that the reader is familiar with ordinary transmission microscopy; however, some information on basic microscopy is given in Appendix 3A.

In a fluorescence microscope, the specimen is illuminated with light of a short wavelength, e.g. ultraviolet or blue. Part of this light is absorbed by the specimen, and re-emitted as fluorescence. The wavelength of the re-emitted light is longer than that of the incident light. To enable the comparatively weak fluorescence to be seen, despite the strong illumination, the light used for excitation is filtered out by a secondary (barrier) filter placed between the specimen and the eye. This filter, in principle, should be fully opaque at the wavelength used for excitation, and fully transparent at longer wavelengths so as to transmit the fluorescence. The fluorescent object is therefore seen as a bright image against a dark background (see Plates).

It follows that a fluorescence microscope differs from a microscope used for conventional absorption microscopy mainly in that it has a special light source and a pair of complementary filters. The basic arrangement is shown in Fig. 1.1. The lamp should be a powerful light source, rich in short wavelengths: high-pressure mercury arc lamps are the most common. A *primary* or *excitation filter* is placed somewhere between the lamp and the specimen. The filter, in combination with the lamp, should provide light over a comparatively narrow band of wavelengths corresponding to the absorption maximum of the fluorescent substance (fluorophore). The *secondary, barrier or suppression filter* prevents the excitation light from reaching the observer's eye, and is placed anywhere between the specimen and the eye (preferably in the body tube). Its transmission should be as low as possible in the range of the light used for excitation, and as high as possible within the spectral range of the emission from the specimen.

The functions of the excitation and barrier filters may be more clearly demonstrated with the aid of Fig. 1.2. The figure relates to a hypothetical fluorophore (fluorescent object) which absorbs in the blue and fluoresces in the green. The absorption spectrum of the fluorophore is shown on the left; this indicates that the substance absorbs predominantly in the blue, hardly at all in the violet, and the spectrum has a long tail with a secondary peak in the ultraviolet. The emission curve is an approximate mirror image of the longest-wavelength peak of the excitation curve.

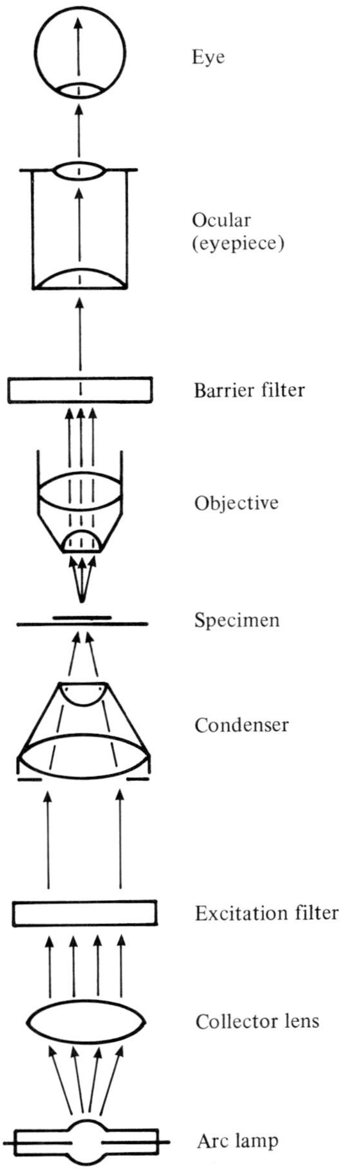

Fig. 1.1. Optical diagram of a fluorescence microscope using dia-illumination (illumination through the specimen). A brightfield substage condenser is shown, although a darkground condenser is more usual. More detailed diagrams are given in Chapter 3, and illustrations of fluorescence microscopes in Chapter 4.

The broken lines show ideal curves for the transmission of the two filters. The excitation filter should ideally have complete (100%) transmission in the region of the absorption peak of the fluorophore, while having zero (0%) transmission (complete opacity) at the wavelengths of fluorescence. Conversely, the barrier filter should have 100% transmission at the wavelengths of fluorescence, with 0% transmission in the region used for excitation. The two filters are therefore complementary: they allow the passage of short-wavelength light as far as the specimen, blocking its passage to the eye, and transmitting the fluorescence.

A more typical fluorophore will have several absorption peaks. For example, FITC (Fig. 1.3) has a peak in the ultraviolet as well as one in the blue. The emission curve is a mirror image of the longer-wavelength (lower quantum energy) peak. In principle, either peak can be used for excitation; broad-band excitation, utilizing both peaks, is also possible. Light used for exciting at the short-wavelength peak has a quantum energy higher than that required to excite at the longer-wavelength peak; therefore, the fluorophore molecules are raised to a higher energy level when excited by the shorter-wavelength light. This increases the probability of a photochemical reaction rather than fluorescence, both because of the higher energy level and because the additional absorbed energy must be released as heat. The increased tendency to photochemical reaction, and increased local heating, both tend to manifest themselves as fading. In principle, therefore, *the light used for excitation should correspond in wavelength to the longest-wavelength absorption peak of the fluorophore.*

Obviously, light at a wavelength not corresponding to an absorption peak of the fluorophore is not absorbed by the fluorophore (or absorbed very little) and cannot excite fluorescence. Such light can, however, excite other fluorophores which may

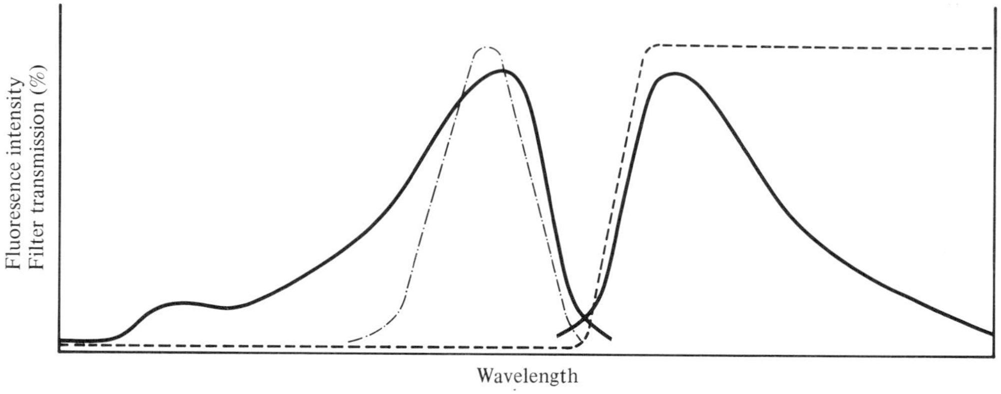

Fig. 1.2. Excitation and barrier filters in relation to the spectra of a fluorophore. Solid lines, the excitation (left) and emission (right) spectra of a hypothetical fluorophore. This fluorophore has peak absorption in the blue, with a secondary peak in the UV; the emission spectrum has a peak in the green, with a long tail extending into the red. An excitation filter with a narrow bandpass in the blue (-·-·-·-) provides optimal excitation. A barrier filter (-----) is opaque at the excitation wavelengths, while transmitting the major part of the emission. Diagram redrawn after Rost (1980).

happen to be present. This means that *the specificity of the fluorescence observation is increased by confining the wavelengths used for excitation to a narrow band corresponding to an absorption peak of the fluorophore*. Also, such additional irradiation may be absorbed by other substances, with the consequent generation of heat.

Opaque or very thick objects can be examined using epi-illumination, a technique whereby the light for excitation is reflected downwards through or around the objective onto the specimen. Although this technique is, in principle, only essential for examination of opaque objects such as intact organs, it has some advantages also for the more conventional transparent specimens, as will be explained in Chapter 3.

Because the image seen in the microscope may consist of only a few small fluorescent areas in an otherwise black field, fluorescence microscopy is sometimes supplemented with other forms of microscopy, e.g. phase-contrast, to enable the specimen as a whole to be visualized and to show the position of fluorescent areas in relation to the rest of the specimen. These combined techniques will be described in Chapter 5.

The position of fluorescence microscopy in relation to other techniques is summarized in Table 1.1. Conventional, transmitted-light, absorption microscopy is appropriate for coloured objects of resolvable size, and instrumentally is the simplest form of

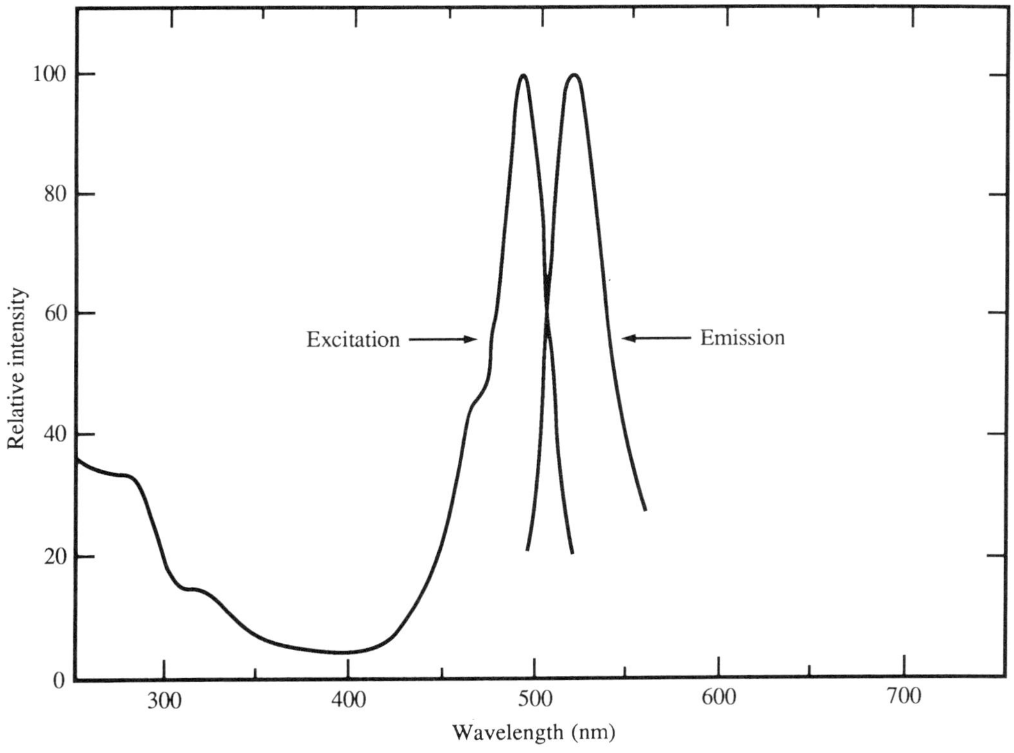

Fig. 1.3. Excitation (left) and emission (right) spectra of Fluorescein isothiocyanate (FITC). Note that the emission spectrum has a major peak in the blue, and a secondary peak in the UV. The emission curve is a mirror-image of the main excitation peak. Re-drawn after Hansen (1964).

Table 1.1. *The relationship of fluorescence microscopy to other microscope techniques, and suitability of various microscope techniques*

Object	Absorption	Retardation	Reflection	Fluorescence
Coloured	Suitable	Usually unsuitable	Usually unsuitable	Suitable
Transparent	Impossible	Suitable	Impossible	Impossible[a]
Opaque	Impossible	Suitable	Suitable	Suitable[b]
Small particles: detection below limit of resolution	Limited	Limited	Unlimited[c]	Unlimited[c]

[a] Unless substance absorbs in the ultraviolet region.
[b] Using epi-illumination.
[c] Detection of isolated small particles possible below limit of resolution, limited only by sensitivity.

microscopy. Colourless, transparent objects can be studied only by retardation techniques (polarization, phase-contrast, interference); these techniques depend upon conversion of phase retardation (by materials of high refractive index) to changes in intensity which can be seen by the eye. An exception is darkground illumination, which may reveal colourless transparent objects by refraction at interfaces of differing refractive indices. Darkground microscopy is otherwise suitable mainly for particulate matter, and (like fluorescence microscopy) can reveal the positions of particles too small to be resolved. Fluorescence microscopy is closely allied to transmission (absorption) microscopy in its range of application, but possesses particular advantages: great sensitivity for detecting small amounts of (fluorescent) substances or small particles, and the possibility of application to opaque objects. Since fluorescence involves two wavelengths, optical specificity can be substantially increased by careful selection of filter combinations to favour the excitation and emission of some particular fluorophore.

Fluorescence microscopy has two particular problems, apart from the purely technical one of having the necessary equipment. The fluorescence image as seen in the microscope is weak compared to that obtained by almost all other kinds of microscopy. This makes particular demands on the efficiency of the system, so as not to waste light, and may lead to difficulties in interpretation due to difficulty in seeing properly at low light levels (see Chapter 7) as well as difficulties in photography (Chapter 6). To make matters worse, the specimen usually fades more or less rapidly under irradiation, and may fade too quickly to be photographed.

As in other forms of microscopy, there are three basic kinds of fluorescence microscopy: qualitative, quantitative, and analytical. The first is concerned with morphology, or with whether or not something (e.g. an immunological reaction) is present. Quantitative fluorescence microscopy is concerned with finding out how much of a specific substance is present in a specified region of the specimen. Analytical fluorescence microscopy is the characterization of a fluorophore by measurement of excitation and emission spectra or other characteristics such as polarization or decay time.

Combination with other viewing modes

Fluorescence microscopy can easily be combined with simultaneous or alternate darkground illumination or phase-contrast microscopy, to facilitate viewing of non-fluorescent parts of the specimen (see Chapter 5). Combinations with interference contrast and polarization microscopy are also possible but a bit more difficult.

Special techniques

Total internal reflection microscopy (Axelrod *et al.*, 1983; Gaub *et al.*, 1986) is a technique adapted to the study of tissue surfaces adjacent to the glass slide. Its principle is that the excitation light is totally internally reflected at the water–glass interface, but penetrates a short distance into the liquid and can thereby selectively excite fluorescent molecules in the liquid near the interface. Standing-wave fluorescence microscopy (Lanni, 1986) is a technique for obtaining optical cross-sections, to enable three-dimensional reconstructions.

Applications

Fluorescence techniques can be applied to all kinds of material. Because fluorescence microscopy requires more complex and expensive instrumentation than conventional transmitted-light microscopy, the former is usually reserved for those applications in which its high sensitivity is of importance: that is, to examine substances present in low concentrations. Fluorescence microscopy can also be applied to detect particles below the resolution of a light microscope, and in histochemistry to visualize substances which cannot be seen by conventional microscopy – e.g. neurotransmitter amines. Biological material is commonly stained in some manner with a fluorescent stain: the procedures which may be applied are summarized in Table 1.2.

Autofluorescence is the fluorescence of naturally occurring substances, such as chlorophyll, collagen, and fluorite. Autofluorescence is also sometimes called *primary fluorescence*. Most plant and animal tissues show some autofluorescence when excited with ultraviolet light of reasonably short wavelength (e.g. 365 nm). Sometimes this autofluorescence is a nuisance, as it may conceal or be confused with a specific fluorescence. The term autofluorescence is sometimes extended to cover the fluorescence of drugs and other exogenous substances which may be present in tissues, e.g. tetracycline antibiotics which accumulate in growing bone. Autofluorescence is treated in Volume II, Chapters 9, 10 and 11.

Induced fluorescence is fluorescence which has been induced by a chemical reaction which converts a non-fluorescent substance, already present in tissue, to a fluorescent substance. An example is formaldehyde-induced fluorescence: neurotransmitter amines such as noradrenaline can be converted to fluorescent substances by treatment with formaldehyde. Induced fluorescence methods are commonly applied to

Table 1.2. *Major fluorescence methods in biological fluorescence microscopy*

1.	***Autofluorescence***	
	a.	Chemically defined substances
		e.g. NADH, vitamin A, chlorophylls
	b.	Chemically ill-defined substances
		e.g. collagenous and elastic fibres, amyloid, lipofuscins
2.	***Induced fluorescence***	
	a.	Bischler–Napieralski reactions for arylethylamines
		Formaldehyde-induced fluorescence (FIF)
	b.	*o*-Phthalaldehyde-induced fluorescence
2.	***Direct fluorochromy***	
	a.	Without pretreatment
		(sometimes supravital)
	b.	With pretreatment
		Feulgen
		Periodic-acid–Schiff (PAS)
		Masked metachromasia
4.	***Indirect fluorochromy (affinity reactions)***	
	a.	Immunofluorescence
	b.	Nucleic acid hybridization
	c.	Lectin binding
	d.	Analogue cytochemistry
5.	***Enzymatically induced fluorescence***	
	a.	Fluorescent enzymatic product
	b.	Enzymatic product made fluorescent by further reaction

noradrenaline, adrenaline, serotonin, and histamine. Induced fluorescence is treated in Volume II, Chapter 12.

Fluorochromy is the process in which tissues are stained with fluorescent dyes, as compared to diachromy in which tissues are stained with coloured dyes. Many coloured dyes are also fluorescent, and can be used as either fluorochromes or diachromes. Hence, many conventionally stained preparations can be examined with either a conventional microscope or a fluorescence microscope. However, the use of fluorescence microscopy extends the range of dyes which can be used, because of the additional possibility of using dyes which are colourless or nearly so but fluoresce in the visible region after excitation with ultraviolet light. Importantly, fluorescence microscopy will reveal the presence of very tiny amounts of dye; this high sensitivity makes the method suitable for supravital staining and neuronal tracing. Fluorochromes can be used as fluorescent tracers, revealing information about their environment in the tissues by changes in fluorescence spectrum or other characteristics. Microfluorometry is sometimes more convenient than microdensitometry for purposes of quantification, even with a coloured dye. Fluorochromy is treated in Volume II, Chapters 13, 14 and 16 and Appendix 7.

Immunofluorescence is a special form of fluorochromy, in which the site of an antigen–antibody reaction is revealed by labelling one of the components of the reaction (the antigen, the antibody, or complement) with a fluorescent dye. Similar techniques are used for the demonstration of other cytochemical affinity reactions, such as in nucleic acid hybridization and in the binding of lectins, hormones and other substances. I refer to these techniques as *indirect fluorochromy*, since the fluorochrome is bound indirectly (through at least one intermediate molecule) to the substrate. Immunofluorescence and other methods of indirect fluorochromy are treated in Volume II, Chapter 14.

Enzymatically induced fluorescence is obtained by incubating tissue with a substance which is converted to a fluorophore by the action of enzymes in the tissue: see Volume II, Chapter 15.

Quantification by microfluorometry

For quantification, microfluorometry has four advantages as compared to microdensitometric methods: distributional error does not occur, so that scanning or two-wavelength techniques are not required; optical specificity can be obtained by selection of appropriate wavelengths both for excitation and for measurement of emission; great sensitivity can be obtained; and thick or opaque objects can be examined. Disadvantages of microfluorometry include the necessity for artificial standards, and relatively rapid fading of the specimen. It is important to ensure that the background fluorescence is kept to a minimum and that an adequate standard is available.

Quantitative methods of fluorescence microscopy have been developed for the most part comparatively recently. There still remain a number of instrumental and other difficulties to be solved, the most important of which is probably the provision of adequate standards. However, several authors have already reported successful applications of quantitative fluorescence microscopy particularly to the determination of DNA by a fluorescence Feulgen technique. Quantification is reviewed in Chapter 8, and treated in detail in *Quantitative fluorescence microscopy* (Rost, 1991).

Microspectrofluorometry is a quantitative technique in which the specimen is irradiated with nominally monochromatic light, and the amount of light emitted from a specified region of the specimen and within a specified wavelength interval is measured in relation to the intensity of the irradiation.

Analytical fluorescence microscopy using a microspectrofluorometer is perhaps one of the most thrilling of the modern applications of fluorescence microscopy. With such an instrument it is possible to measure excitation and emission spectra from specified regions of microscope preparations. This can be applied either for the identification of unknown fluorophores or to the investigation of the conditions of fluorescence of known fluorophores. It has been possible to identify catecholamines by their formaldehyde-induced fluorescence, to confirm that the formaldehyde-induced fluorescences of normal melanocytes, benign naevus cells and malignant melanoma cells are identical; it is also possible to determine the ideal conditions for excitation and examination of specified fluorochromes and to investigate their fading characteristics.

Kinetic studies essentially involve studying the fluorescence parameters described above over a period of time; examples include the study of fading rates, enzyme kinetics, time-resolved fluorescence and phosphorescence.

Scanning enables automatic assessment of specimens, as in cancer diagnosis, and when combined with video recording facilitates the study of fluorescent probes in living cells. An important recent development is confocal fluorescence microscopy, which offers greatly increased resolution and discrimination against background, and the possibility of optical sectioning and three-dimensional reconstruction (see Chapter 8).

Review literature

The other current books on fluorescence microscopy are those of Ploem & Tanke (1987), Becker (1985; superseding an earlier work by Koch, 1972), and of Birk (1984) on instrumentation and techniques; there is also a valuable section on imaging and quantitative fluorescence microscopy in the volume edited by Taylor *et al.* (1986). The only previous general text was that of Haitinger (1938), subsequently reprinted (Haitinger, 1959) with additional matter added by the editors. A general review was given by Price & Schwartz (1956). Early applications to histochemistry were reviewed by de Lerma (1958), who gave a long bibliography; a more recent review is my chapter (Rost, 1980) in a standard textbook. Recent texts on immunofluorescence include those of Kawamura & Aoyama (1983), Wick, Traill & Schauenstein (1982), Nairn (1976), Kawamura (1969), Goldman (1968), and Bals (1966); see also proceedings of conferences related to immunofluorescence (Beutner, Nisengard & Albini, 1983; Knapp, Holubar & Wick, 1978; Hijmans & Schaeffer, 1975; Holborow, 1970). Applications to botany were ably but briefly reviewed by O'Brien & McCully (1981). Current fluorescent staining methods (with fluorescent dyes) are given by Kasten (1981). Additional material of general interest is to be found in conference proceedings edited by Taylor *et al.* (1986) and Thaer & Sernetz (1973), and in the specialist publications of Sani (1964) and Zelenin (1967).

2

Fluorescence: physics and chemistry

Fluorescence is a property of chemical substances, such that, when irradiated with light of a short wavelength, some of the light is absorbed and part of the energy from the absorbed light is re-radiated as light, almost always of a longer wavelength. For example, fluorescein absorbs ultraviolet light and re-emits green light. The incident light is described as the *excitation* or *exciting light*, while the fluorescence is also called the *emission* or *emitted light*.

Fluorescence is one of several types of luminescence, the nature and mechanism of which will now be explained. Essentially, we are considering what happens when molecules are irradiated by light, i.e. bombarded by photons. Molecules may exist at any one of several energy levels, each associated with a particular arrangement of the electronic orbitals. Molecules normally exist at the lowest such energy level, known as the ground state. Electronic states at higher energy levels are referred to as excited states. To proceed from the ground state to an excited state, the molecule must acquire energy, while the return transition from an excited state to the ground state requires a corresponding loss of energy. In the phenomenon of *luminescence*, this loss of energy is accomplished by the emission of a photon, i.e. the production of light. Different types of luminescence are recognized, according to the cause of the excitation and the nature of the electronic transitions involved. In *photoluminescence*, the required energy is obtained by absorption of photons, such as ultraviolet (UV) or blue light. Two forms of photoluminescence are recognized, *fluorescence* and *phosphorescence*; the difference lies in the manner of the emission – in practical terms, fluorescence usually involves a process in which light is re-emitted almost instantaneously after excitation occurs, while phosphorescence takes place after a delay, so that phosphorescent objects continue to glow for some time after the excitation source has been removed. The precise differences will be explained below. If the excitation energy is obtained from the chemical energy of a reaction, the process is called *chemiluminescence*. Other forms of luminescence, not relevant here, are known.

The mechanism of the absorption–emission process was described by Sir George Stokes in 1852. He was also responsible for naming fluorescence after the mineral fluorite (fluorspar), which exhibits a bluish-white fluorescence under ultraviolet irradiation.

It is impossible to discuss fluorescence microscopy without a proper understanding of the physico-chemical principles involved; therefore, this chapter is devoted to a discussion of these, the nature of light and of the interaction of light with matter. It is, however, assumed that the reader is familiar with basic optics. For a detailed monograph on relevant optics, the reader is referred to the text by Slayter (1970).

Recent excellent reviews are those of Guilbault (1985) and Lakowicz (1983). In compiling this chapter, I have relied heavily on information from Barenboim, Domanski & Turoverov (1969), Hercules (1966), Seliger & McElroy (1965), Slayter (1970), Udenfriend (1962, 1969), Guilbault (1973) and Wotherspoon, Oster & Oster (1972).

THE NATURE OF LIGHT

Light is a form of electromagnetic radiation, differing only in quantum energy, wavelength and related parameters from other forms of electromagnetic radiation such as radio waves, X-rays and gamma rays. There has been argument in the past as to whether light consists of waves or of streams of particles; experimental evidence was found to support both theories, and it is now necessary to accept that both are true. It now seems best to regard a ray of light as consisting of streams of entities called *wavicles*, having both a wave structure and a particulate nature.

Optical phenomena may be considered in three ways: *geometrical optics*, which is based on the concept of the 'ray'; *wave optics*, which treats light as a wave motion; and *quantum optics*, which treats light as a stream of particles termed *photons*. In fluorescence microscopy, each of the three points of view is relevant in relation to certain aspects.

Geometrical optics deals with the behaviour of light as directly observed macroscopically; it describes the paths travelled by light beams and predicts the location and magnification of images. The physical nature of light itself is not considered, so that geometrical optics is unable to make predictions concerning the interactions of light beams with matter or with each other. In microscopy, geometrical optics is useful in relation to the focussing action of lenses and lens systems, and the light paths through the instrument from the light source (lamp) to the observer's eye or a camera.

Wave optics treats light as a periodic oscillation of electric and magnetic fields in space. These oscillations are similar to the mechanical waves observed in water and other fluids. Wave optics is required to discuss the phase-contrast, interference and polarization microscopes, the phenomenon of polarized fluorescence, the relationship between microscope images and diffraction patterns, and to explain the limited resolution of the microscope.

Quantum optics treats light as a stream of particles, known as photons, each carrying a definite amount of energy (one quantum) and with zero mass (i.e. zero rest mass). This concept is required to explain the interaction of light with matter. In particular, the quantum approach to optics is required to explain the absorption of light by matter, fluorescence, and the detection of light by photoelectric devices.

The wave nature of light

Following the groundwork laid by C.A. de Coulomb (1736–1806), A.M. Ampère (1775–1806), K.F. Gauss (1777–1855), and Michael Faraday (1791–1867), the modern theory of electromagnetism was created by James Clerk Maxwell (1831–1879) in 1864.

This theory predicted that electromagnetic waves will propagate through vacuum at a constant velocity – the speed of light. An electromagnetic wave consists of an oscillating electric field, together with an oscillating magnetic field, both perpendicular to the direction of motion of the wave. It is a postulate of the special theory of relativity, and a proven experimental fact within current experimental error, that this velocity is invariant; that is, each observer measures exactly the same value for the speed of light *in vacuo* whatever his, her or its own velocity with respect to the source of light.

The wavelength of light is usually measured in nanometres (nm; 10^{-9} m). This replaces the earlier Ångström unit (Å; almost exactly equal to 0·1 nm). The wavelength of visible light varies from 400 nm to 700 nm according to the colour (see Fig. 2.1). Wavelengths of light shorter than 400 nm constitute the ultraviolet; wavelengths longer than 700 nm are infrared. The approximate wavelengths in nm corresponding to pure colours of the visible spectrum are:

Violet	Blue	Green	Yellow	Orange	Red
400–450	450–490	490–580	580–590	590–620	620–700

Wavenumber is the number of cycles of a light wave per centimetre, and is the reciprocal of the wavelength expressed in centimetres. This quantity is sometimes used instead of wavelength or frequency (to which it is proportional); however, it would be better in such cases to refer to the quantum energy (see below).

The quantum theory of light: photons

As early as the beginning of the eighteenth century, Newton proposed a particle or corpuscular theory of light. However, it was not until the end of the nineteenth century that the particle-like manifestations of light were clearly demonstrated: the photoelectric effect, Compton scattering, the spectral distribution of blackbody radiation, and the phenomena associated with the interaction of photons with atoms and molecules (absorption, fluorescence).

The modern quantum theory was initiated by Max Planck in 1900. This arose from consideration of the spectral distribution of radiation from a black body. This had been one of the problems engaging the attention of physicists during the closing years of the century. A black body is one that has a surface that absorbs all radiation incident upon it. When such a body is at a temperature above absolute zero, it radiates heat. The feature which had puzzled physicists was the distribution of wavelengths of this radiation, which depends only on the temperature and not on the composition of the body.

The observed distribution of wavelengths of heat rays differed greatly from that predicted by the calculations of Lord Rayleigh, based on a wave theory of radiated heat. Planck convinced himself that the deductions of Rayleigh, confirmed by J.H. Heans, were sound, and that therefore the premises must be at fault. The essence of Planck's innovation was the premise that the energy of an oscillating system can only have discrete (quantized) values, determined by multiples of a universal constant, h, now called the Planck constant.

Einstein, around 1905, suggested that light is propagated through space in the form of a corpuscular radiation, each corpuscle (quantum, photon) having the energy $h\nu$, where ν is the frequency of the light. A beam of light consisting of photons of one particular quantum energy is said to be monochromatic (i.e. of only one colour). In practice, a truly monochromatic beam is impossible to achieve; however, the term may be applied to a beam which contains photons of only a narrow range of wavelengths, or of which most of the photons have wavelengths close to the mean wavelength. A polychromatic beam of light contains photons of more than one wavelength (or

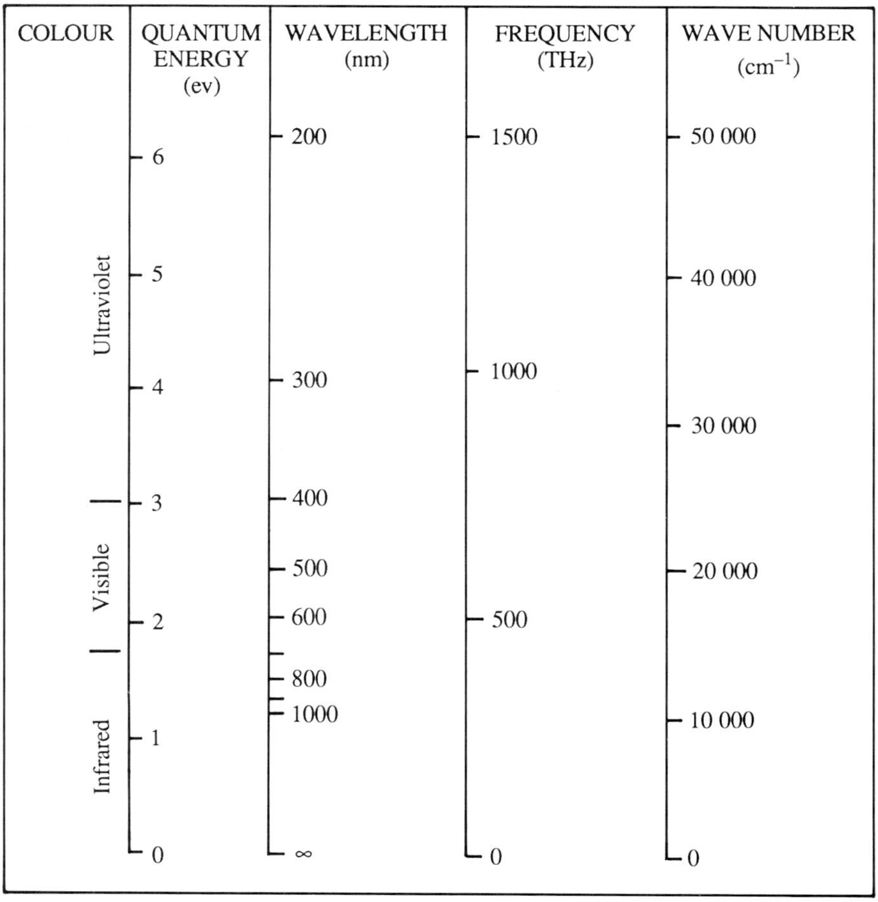

Fig. 2.1. Relationship between quantum energy, wavelength, frequency, wavenumber, and colour of light. 1 terahertz (THz) = 10^{12} Hz.

quantum energy). White light is polychromatic light having a substantial range of wavelengths present. The appearance of white is subjective. The frequency distribution of wavelengths or quantum energies in a beam of light is called the spectral distribution (or spectrum) of the beam.

The photoelectric effect

By the year 1900, it had become apparent that some light radiations, particularly those towards the ultraviolet end of the spectrum, were capable of causing the emission of negative electricity from a conducting surface illuminated by the radiations. The classical experiments of Lenard showed that the energy of the emitted electrons was directly related to the frequency of the incident light and quite independent of the intensity of the incident light over an intensity range of a million to one. Such a behaviour is in complete disagreement with classical electromagnetic theory, for the energy of emission of the electrons should depend on the strength of the electric field which acts upon them, and this field is directly dependent on the magnitude of the electric vector of the electromagnetic wave. On the other hand, experiment showed that the number of electrons emitted was proportional to the intensity of the incident radiation.

In 1905, Einstein suggested his *law of the photoelectric effect*. He postulated that electromagnetic radiation (such as light) is propagated through space in the form of a corpuscular radiation, consisting of small bundles of energy termed *quanta* or *photons*. The energy of each photon is directly proportional to the frequency of the radiation; an increase in the intensity of radiation corresponded simply to an increase in the number of photons per unit time and not to any increase in energy of the individual photons, an increase in the light intensity results in an increase in the number of electrons emitted per unit time without affecting the kinetic energy of the individual electrons.

The results of Lenard's experiments are summarized by Einstein's *equation of the photoelectric effect*, which states that:

$$E=\tfrac{1}{2}mv^2=h\nu-P$$

where E is the energy with which the electron escapes, h is the Planck constant, ν is the frequency of the incident radiation, and P represents the energy required just to liberate the electron with no appreciable velocity.

Looked at another way: when a photon of light falls onto a metal plate, the whole of its energy $h\nu$ may be given up to an electron. Part of this energy (P) may be used to drag the electron away from the metal; the remainder of the photon's energy is retained by the electron as kinetic energy. The quantity P is called the *work function* and is characteristic of the illuminated material. If the incident radiation has a frequency ν_0 corresponding to the quantum energy just sufficient to release the electron, the kinetic energy of the electron will be zero. At frequencies lower than ν_0, no electrons can be ejected.

The photoelectric effect can be used to measure the intensity of light. In a *photoelectric cell* two metal electrodes are placed in a vacuum inside a glass envelope, and electric potential is maintained between them. The electrode connected to the negative side of the power supply is called the cathode. If light be allowed to fall on the cathode, and the photons of light have sufficient quantum energy, they will knock out electrons from the cathode. These electrons will be attracted towards the positively charged anode, with the result that a current will pass between the electrodes and can be measured. The strength of the current is proportional to the intensity of the light. The relationship between the wavelength (quantum energy) of the light and the intensity of the current may be complex. In principle, at any given wavelength, a certain proportion of incident photons may be expected to be successful in releasing an electron. The proportion of successful to total photons is known as the quantum efficiency of the photoelectric cell. This efficiency will vary with the wavelength. Below a certain level of quantum energy, the efficiency will be zero and no current will flow. At the other end of the spectrum, the efficiency of the photoelectric cell will be limited by the transparency of the envelope: ordinary glass will not transmit ultraviolet, while quartz will. The efficiency of the device, particularly at long wavelengths, can be increased by selecting a suitable material for the cathode. In practice, a metallic cathode is coated with a compound of one or more rare earths.

Photomultiplier tubes (photo tubes) are based on a photoelectric cell, but are more sensitive for detecting light because of a built-in amplification system. Photomultipliers are described in more detail in Chapter 8.

THE STRUCTURE OF MOLECULES

In 1911 Sir Ernest Rutherford (1891–1937) proposed the 'nuclear' or 'solar system' model of the atom. This proposed a heavy, positively charged nucleus, around which electrons circle in orbits resembling those of planets around our Sun; the electrons are held, however, by electrostatic attractions rather than by gravitational attraction. A major difficulty in this theory was at first that electrons are charged particles, and charged particles are known to radiate energy when they are accelerated (e.g. in circular orbits). Hence, the very property which permits an electron to be bound to the nucleus would seem to destroy the atom immediately!

In 1913, Niels Bohr (1885–1962), after spending a year with Rutherford, advanced a simple theory which offered a solution to the problem and which led to the modern theory of quantum mechanics. Bohr offered two postulates. First, of the infinite number of orbits possible to an electron in Rutherford's model, Bohr postulates that *only a discrete number of orbits is allowed to an electron, in which orbits the electron cannot radiate.* The permitted orbits are those in which the orbital angular momentum of the electron is an integral multiple of $h/2\pi$ where h is the Planck constant. Bohr's second postulate concerns the absorption and emission of radiation by an atom – the fundamental interaction between matter and electromagnetic radiation. According to the first postulate, an electron cannot radiate while in one of the allowed discrete orbits. The second states that (a) *radiation in the form of a single discrete quantum is emitted or absorbed as an electron jumps from one orbit to another*, and (b) *the energy of this radiation is equal to the energy difference between the orbits.*

In order for an electron to make a *transition* (jump) from a lower orbit to a higher one, the atom must absorb a photon of exactly the appropriate energy. Conversely, when an electron makes a transition from a higher orbit to a lower one, a photon is emitted. The total energy of the system is conserved. The energetics of the processes of absorption and emission may be symbolized by:

$$E_1 + h\nu = E_2 \text{ (absorption)}$$
$$E_2 = E_1 + h\nu \text{ (emission)}$$

Note that only one quantum (photon) is absorbed or emitted, even if the electron jumps over several intermediate orbits during the transition. However, an electron may cascade from a higher orbit to the lowest orbit, emitting a series of single photons of different energy as it jumps from one to another of a series of adjacent orbits. The lowest orbit, in which the atom or molecule has least energy, is called the ground state.

Bohr's theory is an approximation to the actual dynamics of atomic and molecular phenomena. All subatomic particles (e.g. electrons) also exhibit a wave nature, just like photons. Hence, there is an inherent uncertainty in both the exact position and velocity of an electron (the *Heisenberg uncertainty principle*). Electrons may be portrayed as a cloud surrounding the nucleus. The position of the electron can be stated only in terms of probability, the most probable position being one of the Bohr orbits. A region of space in which there is a high probability (99·9%) of finding an electron is called an *orbital*. Any given orbital can have either one or two electrons. If there are two electrons in the same orbital, then to satisfy the *Pauli exclusion principle* they must spin in opposite directions. The Pauli exclusion principle states that no two electrons may have the same set of values for the four quantum numbers which collectively specify the orbital of the electron.

Molecular energy states

Molecules can be regarded as consisting of two or more atomic nuclei, surrounded by a number of electron orbitals. The overall pattern of electrons in an atom or molecule is referred to as the *electronic state* of the atom or molecule. The distribution of negative charge in the molecule, determined by the electronic state, determines in turn the geometry of the molecule. Any given molecule can exist in one of several electronic states, depending on the total electron energy and on the symmetry of the electron spins.

The electronic state with least energy is known as the *ground state*. Under normal conditions at room temperature, the majority of molecules are in the ground state. In states with higher energies, called *excited states*, one or more electrons occupy higher-energy orbitals than in the ground state. The energies of the excited states differ from that of the ground state by fixed amounts which are characteristic of the molecule.

A further characteristic of an electronic state is its *multiplicity*. The spin quantum number (S) of the molecule is the absolute value of the sum of electronic spins within the molecule. The multiplicity of the molecule is defined as the quantity ($2S+1$). This nomenclature derives from atomic spectral considerations, spectral lines of atomic absorption spectra being resolvable into several closely spaced lines if the multiplicity is greater than 1. The multiplicity of an electronic state may be either *singlet* or *triplet*. A singlet state is one in which all the electrons have their spins paired, i.e. there are equal numbers of electrons with positive and with negative spins, so $S=0$ and the multiplicity is 1. A triplet state is one in which there is one unpaired set of electron spins; $S=1$ and the multiplicity is 3. The ground state of an organic molecule is nearly always a singlet state: this must be so if each orbital is either occupied by two electrons or is empty.

In an excited state in which one electron has jumped from a doubly filled orbital to an empty one, this electron may either spin in a direction opposite to that of its former fellow (singlet state) or in the opposite direction (a triplet state). The various energy levels of singlet states are usually referred to as S_0 (ground state), S_1, S_2, etc., and those of the triplet states as T_0, T_1, T_2, etc. A triplet state always has slightly lower energy than the singlet state with the same electronic configuration, e.g. T_1 has lower energy than S_1 (see Fig. 2.2). Therefore, the lowest excited state of an organic molecule is usually a triplet state.

The energy levels of electronic states are indicated diagrammatically in Fig. 2.2. Each energy level is represented by a line, with singlet states shown on the left and triplet states on the right. Superimposed on the energy levels of the electronic states are other levels, more closely spaced, corresponding to levels of vibrational energy: this relates to vibration of the atomic nuclei. Figure 2.3 shows vibrational energy levels superimposed on electronic state energy levels, each vibrational energy level being numbered 0, 1, 2, etc. Further small variations in the energy level of a molecule are due to changes in energy of rotation and flexion of bonds (not shown in diagrams).

Molecules in excited states must be regarded as distinct physical and chemical entities, having properties which may differ considerably from those of the molecules in the ground state. Because of its additional energy, a molecule is more reactive chemically in an excited state than in the ground state. The excited molecule may return to the ground state by emission of a photon (fluorescence or phosphorescence), or by generation of heat; or the molecule may undergo a chemical reaction, such as decomposition, polymerization, or reaction with another molecular species.

THE EXCITATION PROCESS

The absorption of light

A molecule may be excited by absorption of a photon of sufficient energy to produce a transition from the ground state to an excited state. The amount of energy which a photon must possess to produce a particular excited state is equal to the difference in energy level between that state and the previous state of the molecule.

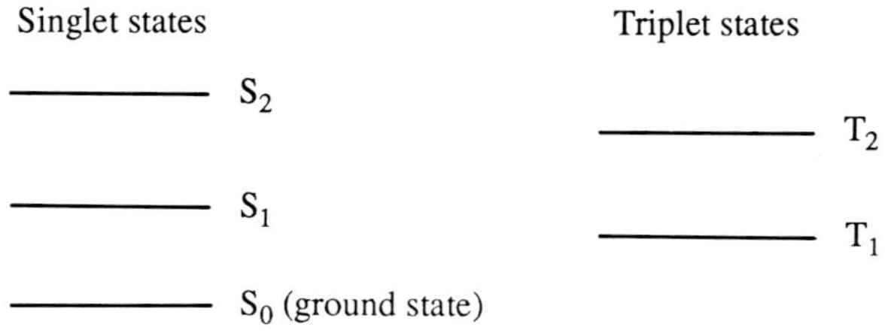

Fig. 2.2. Energy levels of electronic states: singlet states (S_0, etc.) on the left, triplet states (T_1, etc.) on the right.

When a photon is absorbed by a molecule, the energy of the photon is taken up by an electron, which jumps to a higher orbital. The molecule, which we may assume to have been initially in the ground state, is now in some vibrational level of an excited state. No change in multiplicity has occurred. Because of the transfer of an electron to a higher orbital, the distribution of electronic charge in the molecule is altered. This takes place instantaneously, immediately the photon is absorbed. Any changes in the positions of

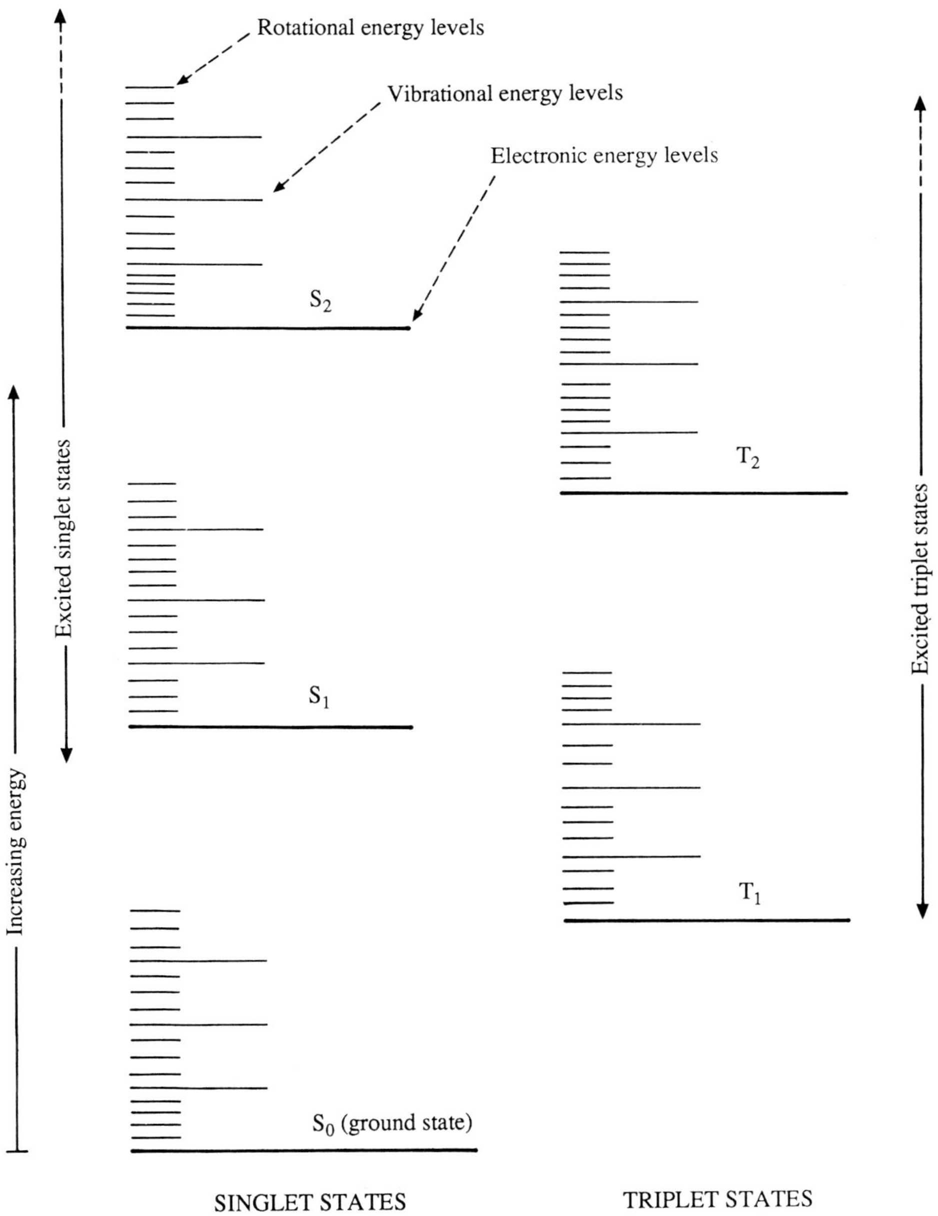

Fig. 2.3. Molecular energy levels: singlet and triplet electronic states with superimposed vibrational and rotational energy levels (Jablonski diagram).

the atoms in the molecule occur later, but within 100 ps (10^{-10} s), usually in about 10 ps (10^{-11} s).

An absorption spectrum is a measure of the probability of photons of given wavelength, i.e. of given quantum energy, being absorbed by the molecule. This probability depends upon the availability of an appropriate transition between the ground state and some level of an excited state, requiring an amount of energy for the transition equal to the energy of the photon. Because a range of photon energies can produce such a transition, electronic absorption spectra consist of bands rather than lines.

Return to the ground state

Under normal conditions, the surplus energy of an excited molecule is invariably lost, with an ultimate return to the lowest vibrational level of the ground state (see Fig. 2.4). Several mechanisms may be involved in this process. In most cases, the surplus energy is converted to heat; alternatively part of the energy may be released as a quantum of radiation (fluorescence or phosphorescence) and the remainder converted to heat. The mechanisms involved in this return to the ground state are vibrational relaxation, radiationless conversion, and luminescence (fluorescence and phosphorescence).

Vibrational relaxation

Under the circumstances which we are considering, we may assume that prior to irradiation all molecules are at their lowest energy level. Therefore, excitation will usually occur from the lowest vibrational energy level of the ground state to some vibrational level of an excited state.

Usually, if the excited molecule is liable to collide with other molecules (e.g. in solution), the excess vibrational energy is converted into heat, and the excited molecule will thereby undergo relaxation to the lowest vibrational energy level of the excited state. This process is usually complete in 10 ps (10^{-11} s). From this point, the molecule may return to the ground state by emission of a photon (fluorescence), or by generation of heat (internal conversion); or may change to an excited triplet state (intersystem crossing) and then return to the ground state by emission of a photon (phosphorescence); or may undergo chemical change. Which of these possibilities actually occurs in a particular case depends upon the relative probability of each process.

Fluorescence

A molecule in an excited singlet state may return to the ground state by emission of the surplus energy as a photon. This process is called fluorescence. That part of the molecule responsible for the fluorescence is known as the *fluorophore*. For reasons to be explained in the next sections, fluorescence is almost always the result of a transition between the lowest energy level of the first excited state (S_1) and some vibrational level of the ground state (S_0). The quantum energy of the emitted photon is equal to the difference in energy between these two levels. It follows that the quantum energy (and therefore wavelength) of the emission are independent of the wavelength of the photon producing excitation. Because the emitted photon has less energy than the excitation photon, the wavelength of the emission is longer than that of the excitation (Stokes' Law). Sometimes, however, if a molecule is already in a higher vibrational energy level before it absorbs the photon, the energy of emission can be higher than that of excitation. The lifetime of an excited singlet state, and therefore the decay time of fluorescence, is approximately 1 to 10 ns (10^{-9} to 10^{-8} s).

The process of fluorescence can therefore be summarized as follows (see Fig. 2.4).

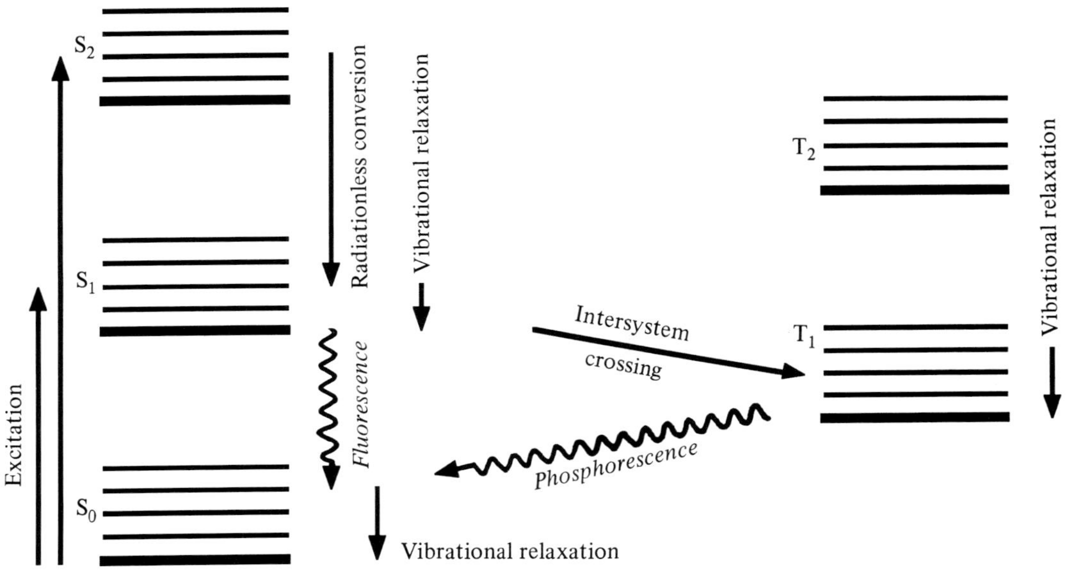

Fig. 2.4. Illustration of the transitions between molecular energy levels during absorption of light, fluorescence and phosphorescence. Only electronic and vibrational energy levels are shown. Absorption of a photon produces excitation (left) from the ground state (S_0) to an excited state (S_1 or S_2). From S_2, S_1 is quickly reached by radiationless conversion. The energy level falls to the lowest vibrational energy level of S_1 by vibrational relaxation. Fluorescence is produced by radiative transition to S_0, with subsequent vibrational relaxation to the lowest vibrational level of the ground state. If intersystem crossing occurs from S_1 (singlet) to T_1 (triplet), phosphorescence may occur by radiative transition to S_0.

A molecule, initially at the lowest vibrational level of the ground state, absorbs a photon and is excited to some vibrational level of a singlet excited state. The surplus vibrational energy is converted to heat, and the molecule relaxed to the lowest vibrational energy level of the first singlet excited state (S_1). A photon is emitted, with a transition of the molecule to some vibrational level of the ground state. The surplus vibrational energy is again converted into heat, and the molecule returns to the lowest vibrational energy level of the ground state.

The fluorescence normally observed is called *Stokes fluorescence*. This is re-emission of less energetic photons, having a longer wavelength. Emission at shorter wavelengths than that of excitation may also occur, and is known as *anti-Stokes fluorescence*. The additional energy required may come from thermal energy or be associated with a molecule with many highly populated vibrational energy levels. The re-emission of photons with the same energy as the absorbed photons is called *resonance fluorescence*: it is never observed in solution because of solvent interactions, but occurs in solids and gases, and is the basis of atomic fluorescence.

The characteristic spectra of fluorescence are the *excitation spectrum* and the *emission spectrum*. The excitation spectrum is a plot of excitation wavelength against the total intensity of the emitted fluorescence. The spectrum is measured by exciting at different wavelengths and measuring for each excitation wavelength the intensity of the fluorescence emission. The emission spectrum is a plot of the relative intensity of the emitted radiation at each wavelength. These spectra may be considered as probability distributions: the excitation spectrum is the distribution of probabilities that a photon of given quantum energy will be absorbed and give rise to fluorescence, and the emission spectrum is the distribution of probabilities that a photon will be emitted with a given quantum energy.

The characteristic skewed shape of fluorescence excitation and emission curves (see Fig. 2.5) can be easily explained in the light of the facts given above. An excitation spectral curve has a peak at the photon energy, or wavelength, corresponding to the

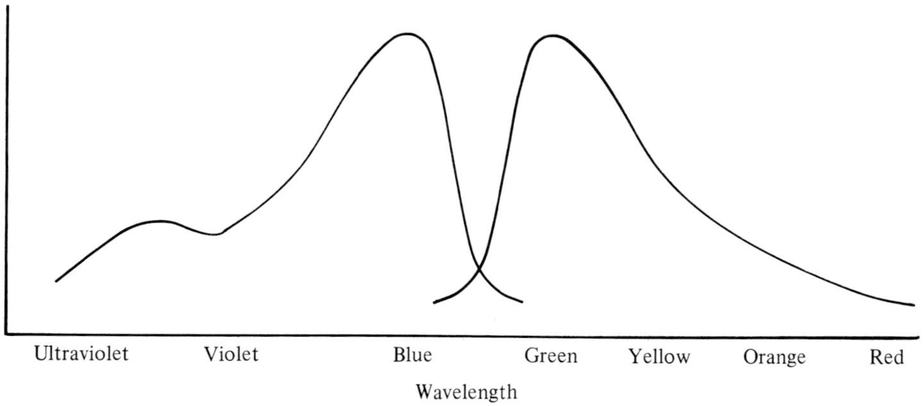

Fig. 2.5. Typical fluorescence excitation spectrum (left) and emission spectrum (right). Note the symmetry of the emission spectrum and the long-wavelength peak of the excitation spectrum; also the overlap of the spectra.

energy difference between the ground state of the fluorophore and some favoured vibrational level of its first excited state (Fig. 2.6). Incident photons with insufficient energy to raise the fluorophore to its first excited state will not be absorbed to produce fluorescence, and therefore the curve falls away rapidly below the peak. On the other hand it matters less if an incident photon has more energy than is necessary, since surplus energy can be converted into heat. Accordingly, the excitation spectrum has a long tail on the high-energy side, with additional peaks corresponding to second and higher excited states.

The shape of the emission spectrum is usually similar to that of the first absorption band (lowest peak of the excitation spectrum), but in 'mirror-image' form. Sometimes there are two peaks. The peak corresponds to the energy lost in the transition from the lowest vibrational level of the first excited singlet state (S_1) to some favoured vibrational level of the ground state (S_0). The curve falls off steeply on the high-energy side because of the improbability of the emission of photons with higher energy than that represented by the difference between the lowest vibrational levels of S_1 and S_0, since fluorescence is virtually always from the lowest vibrational level of S_1 and the energy of the molecule cannot fall below that of the lowest vibrational level of S_0. However, photons of much lower energy can be emitted, corresponding to transitions to higher vibrational levels of S_0. Accordingly, an emission spectrum has a long tail on the low-energy (long-wavelength) side.

These considerations are of some importance in relation to the examination of mixed fluorophores. If two fluorophores (say A and B) are present, one (A) requiring excitation by a higher photon energy than the other (B), both can be excited by light of a sufficient quantum energy to excite A. If light of a low quantum energy is used, just sufficient to excite B, then A will be only very slightly excited (or not at all). Such

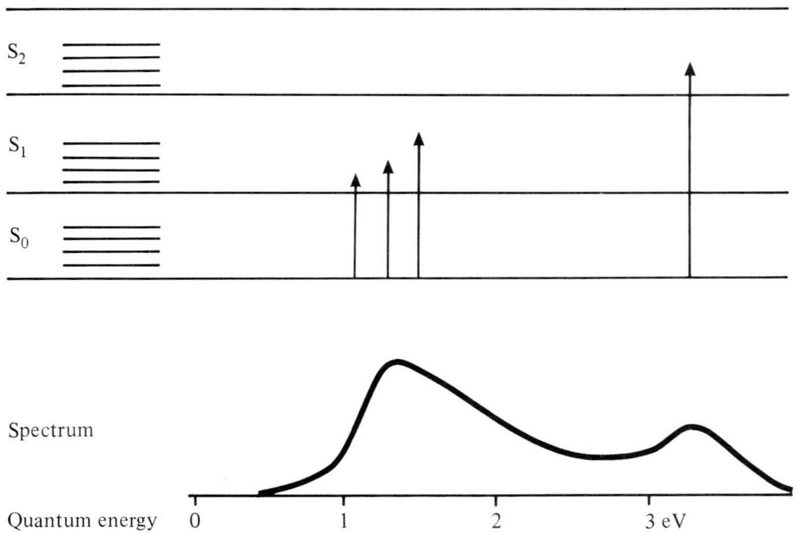

Fig. 2.6. Transitions between electronic states corresponding to various points of an absorption (or excitation) spectrum.

circumstances occur regularly in tissue preparations: autofluorescence (proteins etc.) tends to require a high quantum energy for excitation and can therefore be reduced by using light at a quantum energy only just enough to excite the desired fluorophore. Increasing the quantum energy of the incident photons (decreasing the wavelength) beyond that necessary to excite the desired fluorophore is only deleterious, because of the increased autofluorescence and because the redundant energy can only go to heating the specimen.

The difference between the wavelengths of the excitation and emission maxima is known as the *Stokes shift*. This indicates the energy dissipated during the lifetime of the excited state before fluorescence. In principle it should be measured in energy units (eV), but in practice is measured in wavenumber units, and may be calculated from the corrected maxima (expressed in nm) as:

$$\text{Stokes shift} = \left(\frac{1}{\lambda_{ex}} + \frac{1}{\lambda_{em}}\right) \times 10^7$$

Since fluorescence at room temperature occurs from the first excited state, the wavelength of emission should not be changed by altering the excitation wavelength. If, in practice, such a change is observed, it implies that a mixture of at least two fluorophores is present, each with different excitation and emission characteristics. As the excitation wavelength is changed, so the degree to which each fluorophore is excited varies, and so does the proportion which each contributes to the total emission. Such changes in the emission spectrum of a single substance in solution were until recently thought to indicate the presence of an impurity; however, Chen (1967) reported that quinine and another substance in acid solution show shifts of their fluorescence emission spectrum with changes in wavelength of excitation. Fletcher (1968) named these *B shifts*, and demonstrated them in several fluorescent compounds with rotatable auxochromic groups, in numerous solvents. He postulated that the B shift is due to the fluorescent molecule existing in at least two different 'average' conformations, each with its own distinct energy transitions. These conformations can be considered as being stabilized by the interaction of the solvent with the molecule in both its ground and excited states; the B shift then results from the excitation of different proportions of these two conformations.

The generally accepted relationship between the absorption and emission spectra, based on thermodynamic grounds, is the Kennard–Stepanov formula (Stepanov, 1957):

$$F(\nu) = bA(\nu)\nu^3 \exp(-h\nu/kT)$$

where $F(\nu)$ is the number of fluorescence photons per unit frequency (energy) interval, b a constant, ν frequency, h the Planck constant, k the Boltzmann constant and T absolute temperature. At the time of writing, the most recent practical application of this theory appears to be that of Björn & Björn (1986).

Internal conversion

It is possible for an excited molecule to pass from the lowest vibrational level of an excited state to a high vibrational level of a lower excited state, without the emission of radiation. This process is known as *radiationless conversion*. The transition from S_2 to

S_1 by radiationless conversion, because it occurs between two singlet states with energy levels close together, has a high probability and is very fast, being usually complete in less than 10^{-11} s in nearly all polyatomic molecules. Since this process is much more probable than emission of a photon, a molecule in a higher excited state (above S_1) has a high probability of descending to S_1 by radiationless conversion, before fluorescence has had time to occur. In consequence, fluorescence almost invariably occurs from the lowest excited singlet state and not from a higher state. An exception is the situation at very low temperatures (e.g. 77 K), at which the material is solidly frozen and a proportion of the molecules are unable to reorient themselves before fluorescence takes place; in such case fluorescence may occur from the initial excited state. The same principles apply to the triplet state, phosphorescence occurring from the lowest excited triplet state. Since fluorescence always occurs from S_1 irrespective of the excited state to which the molecule was initially raised, it follows that the spectrum of fluorescence emission is independent of the energy of the photon producing excitation, i.e. of the excitation wavelength.

Phosphorescence

Although it is rare for molecules to enter an excited triplet state directly from the ground state, in many molecules there is an efficient process whereby an excited singlet state may be converted to an excited triplet state. This process is called *intersystem crossing*. It is immediately followed by vibrational relaxation whereby the molecule falls to the lowest vibrational energy level of the triplet excited state. Return to the ground state from the triplet state by emission of a photon is *phosphorescence*. Since it involves a triplet–singlet transition, it is less probable and therefore takes longer than fluorescence. The lifetime of a triplet excited state is in the range of 100 μs to 1 s (10^{-4} to 1 s), and therefore phosphorescence decay is very much longer than that of fluorescence. The energy level of a triplet state is, in general, lower than that of the corresponding singlet state, so that T_1 is at a lower level than S_1. It follows that the wavelength of a phosphorescence is generally longer than that of a fluorescence.

Delayed fluorescence

Delayed fluorescence is a form of luminescence which takes place over a time scale similar to that of phosphorescence but otherwise has the nature of fluorescence. The delay is due to a double intersystem crossing, from singlet to triplet and back to singlet. The spectrum of delayed fluorescence may be different (see Grzywacz, 1967).

Delayed fluorescence and phosphorescence of plant pigments were reviewed by Krasnovsky (1982). Delayed luminescence, with spectra similar to those of the pigment fluorescence, has been observed in both aerobic and anaerobic solutions of chlorophylls and their analogues. However, the mechanism of the afterglow is different in aerobic and anaerobic conditions.

Rayleigh scattering and the Raman effect

If a molecule is excited by an absorbed photon to a higher vibrational level without any electronic transition, energy is entirely conserved and a photon of the same energy is re-emitted within 10^{-15} s as the molecule returns to its original state. The emitted light has therefore the same wavelength as the exciting light. The phenomenon is called *Rayleigh scattering*. Raman bands are satellites of the Rayleigh-scatter peak with a constant energy (frequency) difference from the exciting radiation, and are due to vibrational energy being added to or subtracted from the exciting photon.

Quantum efficiency

The quantum efficiency is defined as the ratio of the number of quanta emitted to the number absorbed. It may be regarded also as the proportion of excited molecules which will luminesce, or in other words the probability that an excited molecule will luminesce (fluoresce and/or phosphoresce). It is measured as the ratio of the number of quanta emitted to the number absorbed, and is usually represented by the Greek letter ϕ (phi). Fluorescence intensity is determined by the absorptivity (which determines the amount of light absorbed) and the quantum efficiency (the proportion re-emitted). The quantum efficiency is determined by competition between the various processes involved. Quantum efficiency is increased by reducing the temperature, which reduces the probability of collisional deactivation.

Lifetime of the excited state

It has previously been indicated that the lifetime of an excited singlet state, and therefore the decay time of fluorescence, is of the order of 1 to 10 ns (10^{-9} to 10^{-8} s). Phosphorescence takes place over a much longer time scale. Experimental procedures for the direct measurement of the lifetime (τ) of the excited state have been available for some time, and is now possible with commercial instrumentation employing a very short (nanosecond) flash for excitation. Determination of decay time can be used for identification by comparison with known samples, and for detection of molecular interactions involving fluorophores. By making observations or measurements at a known period after a brief flash, individual components of mixtures of fluorophores can be distinguished. This technique is probably not applicable to qualitative fluorescence microscopy, but has considerable potential for quantitative studies.

FLUORESCENCE POLARIZATION

Molecules preferentially emit fluorescence of which the electric vector is parallel to the electric field produced by the oscillating electrons. If the molecules are oriented while excited with unpolarized light, the fluorescence will be partially polarized. In biology, this phenomenon is generally known as *difluorescence* by analogy with the term dichroism. Polarization of fluorescence can be detected by placing a piece of Polaroid™ film as an analyser in the eyepiece of the microscope; the analyser or the specimen or both should be free to be rotated, in order that the analyser be set to the correct orientation in relation to the polarized light.

If required, excitation by substantially plane-polarized light is normally obtained by the elimination from the exciting light of light waves whose electric vectors do not all lie in a single chosen plane, using a polarizing filter. The polarized radiation so obtained selectively excites those molecules whose absorption transition moments have a significant component in the plane of the electric vector of the exciting beam, e.g. molecules whose long axes lie essentially in the plane of polarization. If these molecules were firmly fixed, when excited with polarized light they would emit radiation which is polarized in the same direction as the exciting light. However, molecules not firmly fixed may rotate during the interval between absorption and emission, so that the emitted light will tend to be more randomly polarized. The degree to which the emitted fluorescence remains plane polarized will depend inversely upon the amount of Brownian rotation occurring in the interval between absorption and emission of light (see Chen, 1973). The rate of rotation is least for large molecules, such as phycobiliproteins.

For most dyes in solution, the relaxation time of the rotation of the molecules in solution is much shorter than the fluorescence lifetime, and therefore almost completely depolarized fluorescence is to be expected and is indeed observed. On the other hand, fluorescent macromolecules such as proteins have longer rotational relaxation times and the degree of polarization of fluorescence is indicative of the relaxation time.

THE MEASUREMENT OF LIGHT

The strength of a beam of light may be measured as either the amount of energy per unit time or the number of photons per unit time. For monochromatic light, the two measurements are linearly related; for polychromatic beams, they are not, because of the differing energies of the quanta.

The technique of measurement of the intensity of fluorescence is known as *fluorometry* or *fluorimetry*; probably the former spelling is more commonly used. Measurement of fluorescence with a microscope is known as *microfluorometry* or *microfluorimetry*, which will be discussed in Chapter 8. For comments and references on nomenclature of units of measurement, see Mielenz (1976). Fluorescence intensity is usually measured in arbitrary units, related to either the current flowing in a photomultiplier tube or the number of photons detected per second. The former is more commonly used, and is suitable for strong intensities; the latter is more appropriate at very weak intensities.

The *intensity* of a light source (such as a lamp) is the amount of energy emitted per unit time, per unit area of the source, per unit frequency interval, per unit solid angle in a given direction. The *flux* on to a detector surface, or through a diaphragm, is the amount of energy per unit time onto or passing through a unit area of the surface, per unit frequency interval. Hence, the unit of monochromatic light flux is Joules per second per square metre per Hertz ($J\ s^{-1}\ m^{-2}\ Hz^{-1}$). Lamps can be compared objectively in terms of their total light output, spectral distribution of emitted light, intensity of light at a particular wavelength, and the brilliance of the incandescent area. Certain units of measurement are relevant to such comparison: they are as follows.

Candela (cd). Unit of luminous intensity of a light source, defined as the luminous intensity, in a given direction, of a source that emits monochromatic radiation of frequency 540×10^{12} Hz and that has a radiant intensity in that direction of 1/683 watt per steradian (W sr^{-1}). The candela was formerly defined in terms of the emission of freezing platinum.

Lumen (lm). An arbitrary unit by which the flux (rate of flow) of light is evaluated in terms of its visual effect. It is the luminous flux emitted from a point source of uniform intensity of one candela into unit solid angle so that the total flux from one candela is 12–57 lm.

Stilb (sb). Unit of luminance (brightness, luminous intensity) emitted per unit area in a given direction, and equal to 10000 candela per square metre (1 cd cm^{-2}).

Characterization of fluorescence spectra

The characteristic spectra of fluorescence have been defined above: the *excitation spectrum*, which is a plot, for each wavelength of excitation, of the total intensity of the emitted fluorescence; and the *emission spectrum*, which is a plot of the relative intensity of the emitted radiation at each wavelength. Other parameters are the *fluorescence lifetime*, i.e. the distribution of probabilities that a photon will be emitted during a given interval after excitation, and *polarization* of fluorescence.

In graphs of both excitation and emission spectra, the abscissa (*X* axis) is normally a scale of wavelength or of some related quantity; in the present work, a linear scale of photon energy in eV is generally used, a reciprocal scale of wavelength being usually also given. By convention, red is on the right-hand side and ultraviolet on the left, i.e. wavelength increases from left to right. A linear scale of wavenumber, often used, has the same effect (see Fig. 2.1). The *wavenumber* is the reciprocal of the wavelength, usually expressed as the number of waves per centimetre; it is directly proportional to the frequency and quantum energy. Emission spectra should be plotted in terms of relative quanta emitted per unit quantum energy interval, following the recommendations of Chapman *et al.* (1963) except that a scale of eV is suggested rather than frequency or wavenumber. Excitation spectra should be plotted in terms of relative quanta (photons emitted for a constant number of incident photons). Many authors still use a linear scale of wavelength, and plot emission spectra as relative quanta per unit wavelength interval. In the case of moderately sharp peaks in the visible region, this appears to cause only a small shift in the position of the peak.

CHEMISTRY OF FLUORESCENT SUBSTANCES

Structure and fluorescence

The fluorescence (or lack thereof) of a molecule is dependent upon the structure of the molecule and upon the environment in which it is situated. Many organic molecules, and some inorganic ones, are fluorescent. In relation to organic compounds, useful fluorescence is restricted to compounds possessing a fairly large conjugated system, in which electrons, less strongly bound than σ electrons, can be promoted to π^* anti-bonding orbitals by absorption of photon energy without excessive disruption of bonding. In practice, this means that the majority of usefully fluorescent organic compounds are aromatic or heterocyclic.

In liquid solution, most unsubstituted aromatic hydrocarbons exhibit quite strong fluorescence at ultraviolet or visible wavelengths. As the size of the conjugated system increases, so usually do the wavelength of fluorescence and the quantum yield. For example, benzene (I) and naphthalene (II) fluoresce ultraviolet, while anthracene (III), naphthacene (IV), and pentacene (V) fluoresce blue, green, and red, respectively.

I

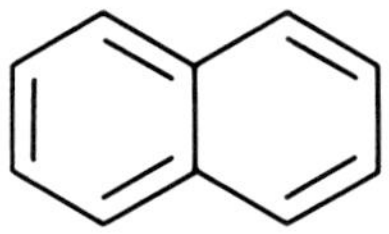

II

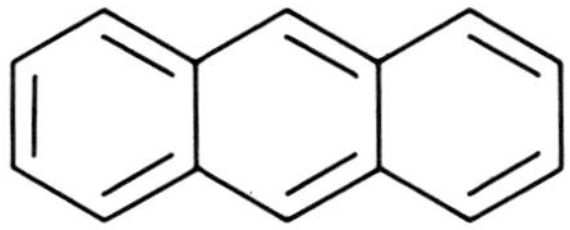

III

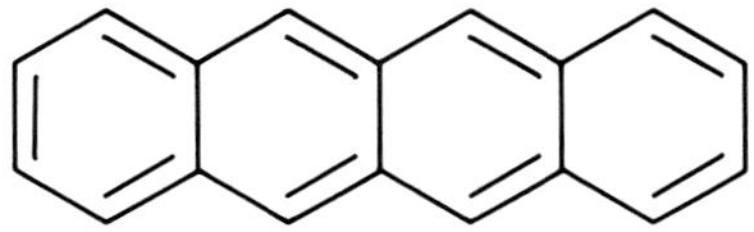

IV

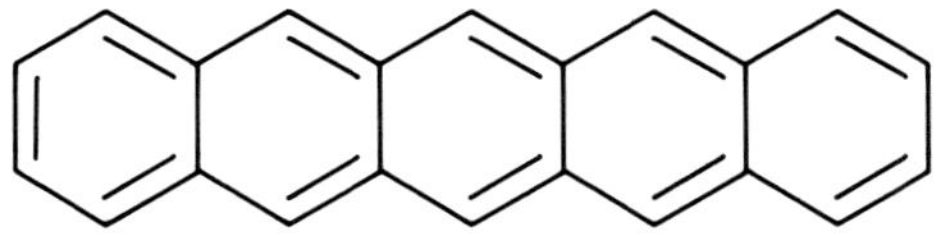

V

The most strongly fluorescent aromatic compounds are usually characterized by rigid, planar structures. For example, fluorescein (VI) has intense green fluorescence, whereas the somewhat similar phenolphthalein (VII) has only weak red fluorescence. The fluorescein molecule has a large, planar, conjugated region, whereas phenolphthalein has three smaller conjugated regions which are not coplanar. Further examples of substances in pairs, having generally similar structure, of which one is fluorescent and the other essentially nonfluorescent are given by Udenfriend (1962). Many examples of fluorescent molecules are described in Volume II: naturally-occurring substances in Chapters 9, 10 and 11, and synthetic compounds mainly in Chapter 13 and Appendix 7. See also Haugland (1989).

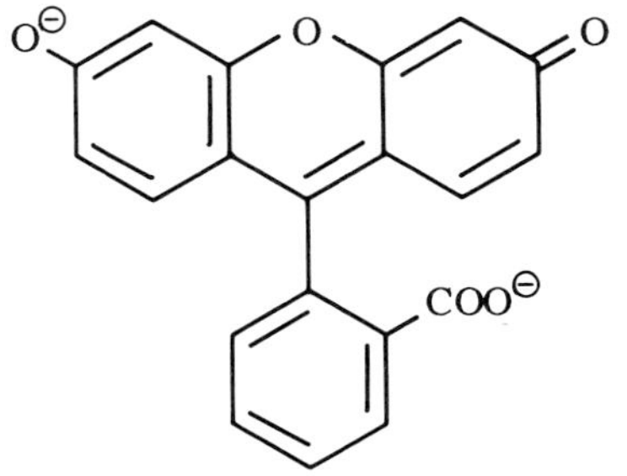

VI

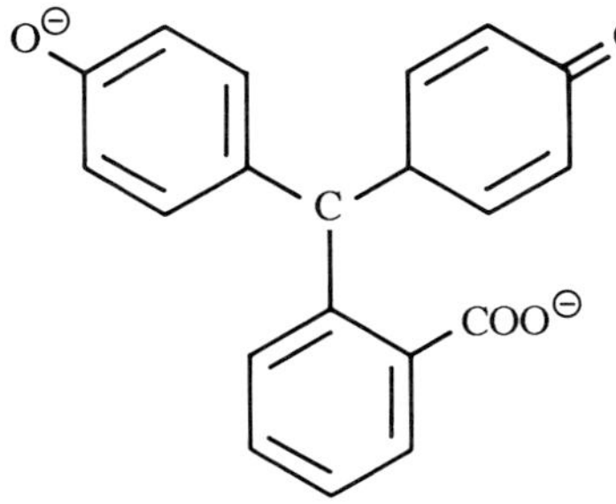

VII

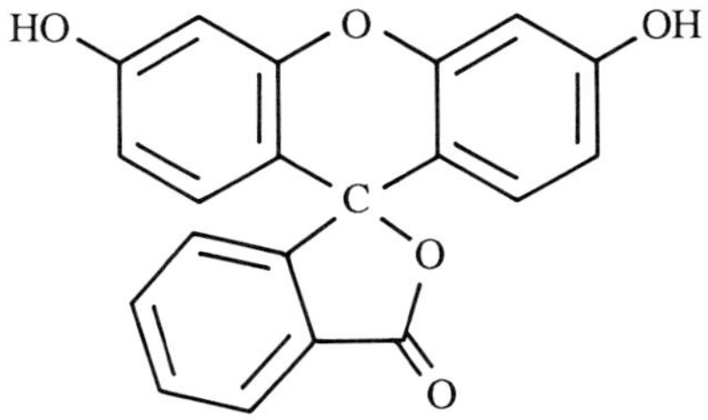

VIII

Environmental effects

The fluorescence of organic molecules is strongly dependent upon environmental factors, such as ionic strength, pH, temperature, viscosity or rigidity of the medium, and binding to macromolecules. Hence, the mounting medium affects the nature of the fluorescence, and the fluorescence of a fluorochrome is affected also by its binding to a substrate.

pH changes, by altering the ionic dissociation, affect both the fluorescence spectra and the quantum efficiency. The effect of pH is particularly strong in some compounds, such as fluorescein, which can therefore be used as pH indicators because of a marked change from strong to weak fluorescence or a change in colour according to pH. The structure of the fluorescein molecule in alkaline or neutral environment (VI) differs markedly from its structure in acid environment (VIII). Such changes can affect the intensity of fluorescence in two ways: changes in absorbance correspondingly affect the the amount of light absorbed (i.e. the number of molecules excited per unit time) and changes in quantum efficiency.

The quantum efficiency of fluorescence depends essentially on competition between the various possible processes, notably fluorescence and deactivation by contact with other molecules or by radiationless conversion. The more the molecule moves around, the more likely it is to interact with another molecule and therefore the less likely it is to fluoresce. Molecular movement can be reduced by lowering the temperature, or by making the material more viscous or rigid. Such procedures are particularly important for the observation of phosphorescence.

Binding to a macromolecule can alter the fluorescent properties of a small molecule (such as a fluorescent dye) by changing the excitation and/or emission maxima, the quantum efficiency, the degree of fluorescence polarization, and the temperature coefficient of fluorescence. This, on the one hand, means that data on the fluorescence of a dye in solution cannot be applied exactly to the same dye in the bound state on stained tissues; and on the other hand, that a small fluorescent molecule with known fluorescence characteristics can be used as a probe, its fluorescence characteristics changing according to its fate within a cell. The high quantum efficiency of phycobiliproteins is believed to be due to the fluorophore being rigidly held by the protein.

Fluorescent dyes in high local concentration may form dimers or polymers, with different fluorescence characteristics. This change of colour is called *metachromasia*. Metachromasia is a phenomenon in which a single dye gives two different colours, called the orthochromatic and metachromatic colours of the dye. The best known metachromatic fluorochrome is Acridine Orange, whose colours are green and red. Metachromasia is the result of the formation of dye dimers and polymers *in situ* on the substrate; this may occur if adjacent dye molecules bind on to the substrate very closely to one another, as occurs (for example) in the case of Acridine Orange binding to RNA. A change from orthochromasia to metachromasia involves a shift in the excitation (absorption) peak towards shorter wavelengths, and a decrease in the molar extinction coefficient at the absorption maximum. There is a corresponding but opposite shift of the emission spectrum towards longer wavelengths, and also a tendency to a reduction in quantum efficiency.

The relation of the structure and environment of molecules to their fluorescence is dealt with at length by Wehry (1973); the relation of structure to fluorescence in certain laser dyes is discussed by Drexhage (1976); see also Barenboim *et al.* (1969), Udenfriend (1962, 1969), Chen (1973), Udenfriend, Guroff & Zaltzman-Nirenberg (1967), and Goodwin (1953).

PHOTOBLEACHING, PHOTOACTIVATION AND QUENCHING

This section deals with factors which influence the brightness of the fluorescence, including the photochemical effects of the irradiating light on the specimen. These topics are dealt with in greater detail in *Quantitative fluorescence microscopy* (Rost, 1991).

Photochemical reactions

Because of their additional energy, molecules in excited states tend to be more chemically reactive than in the corresponding ground state. Irreversible reactions may take place from the lowest vibrational level of a singlet state, or from a triplet excited state of which the decay time is usually longer, giving more opportunity for a chemical reaction. The reaction may be decomposition, polymerization, or reaction with another molecule. Ionization in the excited state may occur reversibly. An excited molecule may polymerize with another molecule in the ground state, producing a polymer known as an *excimer*. The excimer will have an emission spectrum different from that of the monomer.

Fading, photobleaching

It is characteristic of fluorescent preparations that they nearly always fade during irradiation. The terms photobleaching and fading are used roughly interchangeably to refer to progressive loss of fluorescence intensity during irradiation. Photobleaching is the more specific term, as fading can also be taken to refer to the effects of long-term storage. The term photodecomposition is also used, implying (not necessarily correctly) destruction of the fluorescent molecule. Photobleaching must be distinguished from decay, i.e. very rapid progressive reduction of intensity of fluorescence after cessation of irradiation. I therefore distinguish, in order of increasing lifetime, between *decay* after excitation, *photobleaching* during irradiation, and *fading* during long-term storage. *Recovery* is partial or complete restoration, after photobleaching, of the ability to fluoresce. The subject of photobleaching is particularly important in relation to photomicrography (Chapter 6) and quantification (Chapter 8). Photobleaching is usually a considerable nuisance, particularly in quantitative studies, in clinical diagnosis where two or more people may need to examine a preparation, and in photomicrography; on the other hand, fading has been put to use in studies of molecular motion by examination of recovery after photobleaching. An excellent review of photobleaching was given by Picciolo & Kaplan (1984).

The mechanism of photobleaching is still largely unknown. I believe that several factors are involved, and that their relative influence varies from one specimen to the next and even within parts of the same specimen. Photobleaching is due mainly to photochemical reactions induced by the light used for excitation. As explained above, the absorption of light (prerequisite for fluorescence) entails the raising of molecules to the excited state. The excited state is virtually a different chemical species, usually much more reactive than in the corresponding ground-state molecule. A small but significant proportion of the excited molecules, instead of fluorescing, undergo a photochemical reaction with the production of a new molecule which may be non-fluorescent, or at least non-absorbent at the excitation wavelength. It is commonly believed that photobleaching is due to a reaction between the excited fluorophore and oxygen to form a non-fluorescent product. A non-oxidative mechanism is implied by the studies of Johnson *et al.* (1982), while Picciolo & Kaplan (1984) found evidence of both oxidative and non-oxidative mechanisms, on the basis of different fading curves with and without an anti-oxidant and on the assumption that fading in the presence of an anti-oxidant is unlikely to be oxidative.

The rate of photobleaching depends upon several factors, including the nature of the fluorophore, its chemical environment, and the intensity and quantum energy of the excitation. Some fluorophores are particularly liable to rapid photobleaching (e.g. porphyrins), while others are relatively stable (such as lignin and Acriflavine). In metachromatic staining, e.g. by Acridine Orange or Coriphosphine O, the metachromatic (red) form bleaches more rapidly than the orthochromatic (green) form. During prolonged irradiation, the rate of photobleaching varies. During the first few milliseconds of irradiation, photobleaching does not occur. There is then usually a period of rapid bleaching, lasting a few seconds. Thereafter, bleaching proceeds more slowly. A brief laser pulse can be used for examination or measurement of fluorescence without errors due to photobleaching.

Recovery may occur if the specimen is kept in the dark, preferably at a low temperature. At very low temperatures, recovery of fluorescence in crystalline material may be complete, that of biological material less so.

Photoactivation reactions occur because the excited state of a molecule tends to be much more reactive than its ground state, and behaves as a quite different molecule. While in general prolonged irradiation leads to fading of fluorescence, in some cases fluorescence is increased, or the excited form of the dye may undergo some reaction with the tissue, leading to firmer binding to the stained site. Excitation of fluorochromes used as a vital stain can produce greatly increased toxicity; an application of this phenomenon is 'zap axotomy', selective killing of all or part of single neurons achieved by filling the cell with a fluorochrome and irradiating the desired region with blue or ultraviolet light to excite the fluorophore.

Quenching

Quenching is the reduction of fluorescence by a competing deactivating process resulting from the presence of other molecules in the system. Four common types of quenching are recognized: temperature, oxygen, impurity and concentration quenching. The latter is due to non-linearity in the relationship betwen fluorophore concentration and fluorescence intensity (see Chapter 8). The term quenching is also used to describe the alteration of the fluorophore by a chemical process, e.g. the borohydride reduction of formaldehyde-induced fluorescence (see Volume II, Chapter 12), and previous treatment which abolishes fluorescence for whatever reason. The quenching of fluorescence in solutions has been studied in some detail; see Guilbault (1973), Udenfriend (1962, 1969) and Pringsheim (1949).

Temperature quenching is a decrease in fluorescence as temperature increases. The degree of temperature dependence depends on the compound; it is usually about 1% per degree Celsius but can be as high as 5% (e.g. tryptophan, Rhodamine B). The effect is believed to be due to increased molecular motion and increased frequency of collisions, resulting in increased probability of transition to the ground state before fluorescence can occur.

Impurities may decrease, increase, or have no effect on fluorescence. An excited molecule may transfer its surplus energy to an adjacent molecule, which itself becomes excited and may subsequently return to its ground state by any of the pathways which have been described above. If the absorbing molecule is normally fluorescent and the second molecular species non-fluorescent, the result is quenching of the fluorescence. Because transfer of energy from the donor to an acceptor molecule may be very efficient, and can apparently take place over distances as great as one micrometre, a very small concentration of an impurity can produce a high degree of quenching. On the other hand, if the donor molecule is non-fluorescent and the acceptor is fluorescent, the fluorescence of the acceptor may be greatly increased by the presence of the donor molecules if these are more efficient at absorption of photons. This latter process is known as *sensitized fluorescence*.

Oxygen quenching. One of the most notorious quenchers is molecular oxygen, which causes a reduction in fluorescence and completely abolishes phosphorescence. Oxygen in solution at a concentration of 1 mM typically reduces the fluorescence intensity by about 20% (Guilbault, 1973). In fluorescence microscopy, this probably accounts for some at least of the deterioration which usually occurs in fluorescent preparations on storage. Photobleaching, and fading during storage, are lessened by inclusion of a reducing agent in the mounting medium.

Concentration quenching. At low concentrations of the fluorophore, as the concentration increases so does the intensity of fluorescence, because more light is absorbed. However, at very high concentrations, the fluorescence may be reduced, and the phenomenon is known as concentration quenching. Quantitative aspects of concentration quenching are mentioned in Chapter 8.

Prior treatment may be directed towards abolishing or altering the fluorescence in part of the specimen. An example is pre-staining of nuclei with Haematoxylin before staining with Thioflavine T for amyloid; the Haematoxylin prevents fluorescent staining of nuclei by the Thioflavine T, which (being a basic dye) would otherwise be the most prominent staining. Similarly, Crystal Violet applied to isolated cells quenches surface but not internal immunofluorescence.

3

The optical system of the fluorescence microscope

This chapter and Chapters 4, 5 and 6 are concerned with the structure and function of fluorescence microscopes suitable for qualitative fluorescence microscopy. The present chapter is concerned with the optical and mechanical ('brass and glass') arrangements required for illumination and magnification of the object. The chapter is divided into four sections, dealing respectively with general principles, the illumination system (excluding the excitation filters, which are dealt with later in the chapter), the observing or magnifying system, and the filters. Other chapters will describe commercial instruments (Chapter 4), and instrumentation for quantification, microspectrofluorometry and video and scanning microscopy (Chapter 8); the latter are described in more detail in *Quantitative fluorescence microscopy* (Rost, 1991).

It is assumed that the reader is familiar with conventional transmitted-light microscopy: for further information see Bradbury (1984), Beyer (1977), James (1976) and Zieler (1972, 1973). Excellent booklets, of general interest but oriented towards particular manufacturers, include those of Abramowitz (1985; Olympus), Determann & Lepusch (1973; Leitz) and Zeiss (1967).

The usual structure of a fluorescence microscope is shown in Figs. 3.1, 3.2 and 3.3, which may be compared with the photographs of actual microscopes in Chapter 4 and with the block diagram of Fig. 1.1. As a starting point, Fig. 3.1 represents a conventional microscope for transmitted-light observation. The diagram is not of any one particular microscope, although it is based on the Leitz Diaplan (see Chapter 4). The diagram is meant to be representative of a typical medium grade or laboratory microscope. The microscope will be seen to consist mechanically of three main parts: the stand with stage and substage, the body, and the lamphouse. The major optical components are labelled in the diagram.

The simplest possible modifications for fluorescence microscopy are shown in Fig. 3.2, which shows the same microscope adapted for fluorescence microscopy using dia-illumination, i.e. with the excitation light from below using the conventional illumination system. The modifications for fluorescence are three in number: replacement of the tungsten-filament lamp by a *gaseous-discharge arc lamp* (e.g. mercury arc lamp), insertion of an *excitation filter* in the lamphouse filter slot, and insertion of a corresponding *barrier filter* in the nosepiece filter slot. For the sake of completeness it

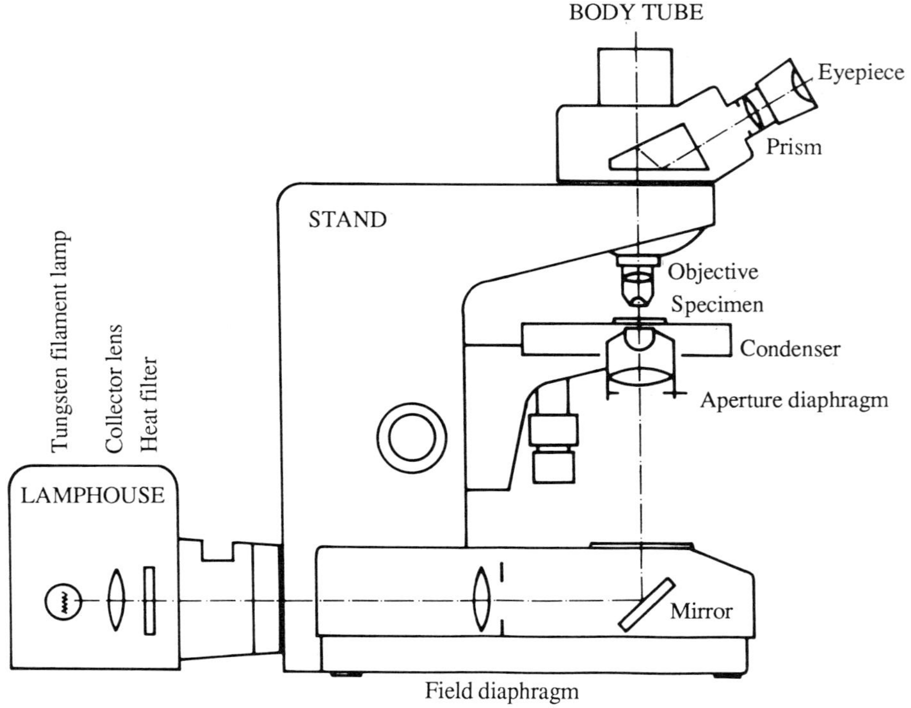

Fig. 3.1. Diagram of a conventional microscope for transmitted light, showing the main optical components.

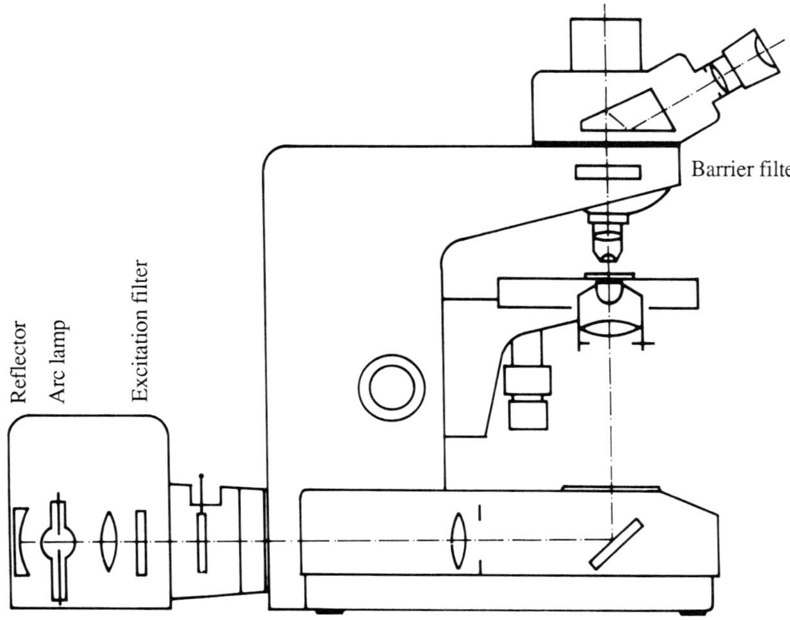

Fig. 3.2. The same microscope as in Fig. 3.1, modified for fluorescence microscopy (see text). The modifications are labelled.

should be added that it is usual also to exchange the condenser for a darkground condenser; and that in principle it is sometimes possible to use a tungsten-halogen lamp for fluorescence. The diagram also shows a reflector in the lamphouse, behind the arc lamp.

The third diagram (Fig. 3.3) shows the same microscope again, this time with the addition of an epi-illuminator unit, which directs the excitation light down through the objective, by means of a semi-transparent mirror. The epi-illuminator contains a series of beam-controlling lenses, as well as field and aperture diaphragms. The excitation and barrier filters, together with the semi-transparent mirror, are conveniently mounted in a single unit. Such units are interchangeable, each unit containing appropriately matched filters and mirror for a particular application. The semi-transparent mirror is normally *dichromatic*, meaning that it preferentially reflects short wavelengths (i.e. the excitation light) and transmits longer wavelengths (i.e. the fluorescence).

The tungsten-filament lamp has been replaced in the lower position, for simultaneous or alternate examination of the specimen by transmitted light, phase-contrast, interference-contrast or darkground illumination (not shown).

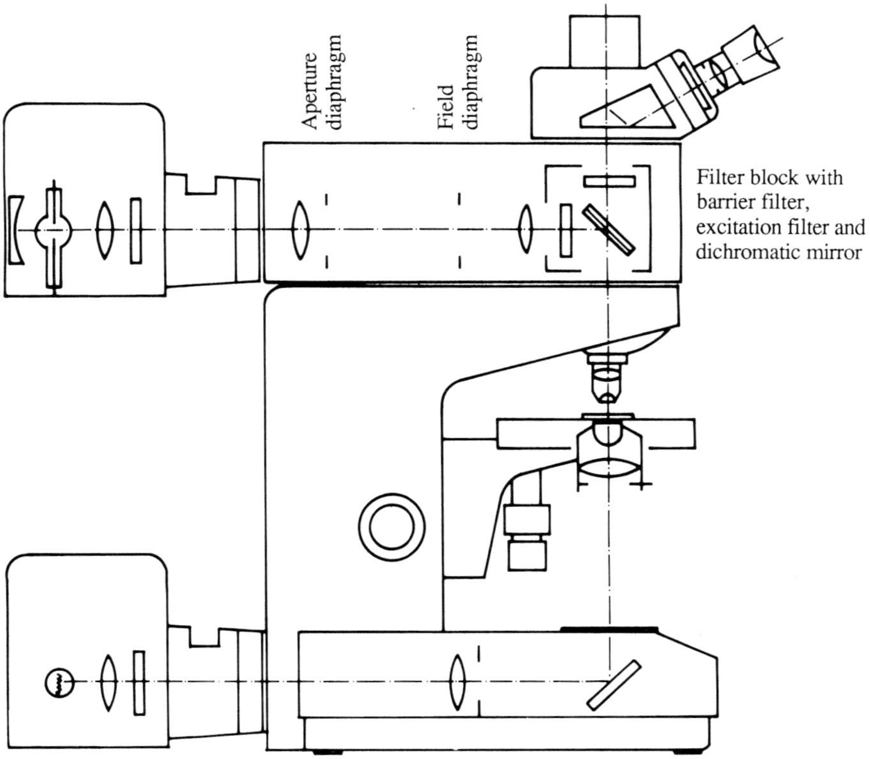

Fig. 3.3. The same microscope as in Figs. 3.1. and 3.2, modified for fluorescence microscopy, with an epi-illuminator unit added between the stand and the body tube. The diagram omits focussing and centration control knobs of the lamphouse for the arc lamp.

GENERAL PRINCIPLES

The optical system of a fluorescence microscope consists of two parts: the *illumination system*, from the lamp to the specimen, including the excitation filter; and the *observation system*, from the specimen to the eye, including the barrier filter.

In the illumination system, light is generated by a light source (lamp). Light from the lamp is collected by a collector lens in the lamphouse and focussed through a field diaphragm into the condenser, which focusses the light onto the specimen. Somewhere in the illumination system, the excitation filter selects the appropriate wavelengths for excitation. Often, there is a mirror behind the lamp to reflect light forwards into the collector lens, thereby approximately doubling the amount of light utilized for excitation. This mirror forms an image of the light source next to the light source itself, thereby appearing to double the width of the light source.

The observation system consists of the objective, ocular (eyepiece), and any intermediate lenses between the objective and the ocular. The objective produces a real inverted image of the object in the plane of the ocular's diaphragm; this intermediate image is further magnified by the ocular, producing a virtual image to be examined by the eye.

The illumination and observation systems are similar and symmetrical about the plane of the object. In principle, they are to some extent interchangeable – the objective can be used as a condenser, and vice versa.

The illumination system produces in the object plane a reduced (demagnified) image of the light source; its brilliance is dependent upon the square of the numerical aperture (NA) of the system (generally determined by the aperture iris diaphragm of the condenser) and upon the square of the demagnification of the image of the lamp (the smaller the image, the brighter, but only to a limited extent; the image can never be more intense than the source). Similarly, the observation system focusses an image of the specimen upon the eye; the brilliance of the image depends upon the square of the NA of the system (determined by the objective) and inversely upon the square of the image magnification (determined jointly by the objective and ocular, together with any intermediate magnifying device).

THE ILLUMINATION SYSTEM

It will be recalled that the object of the illumination system is primarily to illuminate the object with sufficiently intense light of an appropriate wavelength to excite fluorescence. In addition, the illumination system may be required to provide alternative forms of illumination, such as phase-contrast to enable unstained parts of the specimen to be identified, or excitation at a second wavelength to display a second fluorophore. The illumination system will be discussed systematically, commencing with the light source and then tracing the optical path to the specimen. Although the barrier filters form a part of the magnification system, all filters are discussed together in the latter part of the chapter, because excitation and barrier filters always form a matched pair and must be considered together.

The basic requirements of an illumination system for ordinary transmitted-light microscopy are as follows (Hartley, 1974):

1. It must collect a good deal of light, and direct it into the field of view to provide sufficient intensity for a bright final image; and
2. The light collected must be delivered in a condition to permit the objective to achieve its utmost resolution. This requires a lens system equivalent to the objective in aperture and correction.

In fluorescence microscopy, of these two requirements the former (1) is of greatest importance. For reasons to be explained in Chapters 6 and 7, the fluorescent image should almost always be made as bright as possible, and one way to achieve this is to provide the brightest possible illumination. On the other hand, the optical correction of the illumination system is of relatively minor importance, since it affects only the evenness of illumination over the field. Resolution is not affected by the aperture of the illumination system; moreover, excess of aperture over that of the magnification system does not usually lead to glare as it does in transmitted light microscopy. Uniformity over the entire observed field depends on corresponding uniformity of illumination; this uniformity of illumination must relate both to intensity and to wavelength distribution.

Subsequent sections of this chapter will discuss the components of the optical system of fluorescence microscopes systematically, commencing at the light source and following the light path to the ocular. Actual commercial realization of the principles involved will be described in the next chapter.

Light sources

The light source provides radiation for illumination, and consequent excitation. Its essential characteristic is a sufficiently intense output at the required wavelength. Further useful characteristics, such as stability, will be discussed below. In the early days of fluorescence microscopy, it was believed that only true ultraviolet (UV) could be used for excitation. This implied, at first, the use of carbon arcs – which were unstable, messy, and requiring frequent attention. These were superseded by high-pressure mercury arc and xenon arc lamps. With the realization that true UV is not always required, powerful tungsten filament lamps were found to be suitable for some applications. Most recently, lasers have been applied, particularly in quantitative instrumentation where pulses of a strictly monochromatic light are advantageous.

Lamps can be compared objectively in terms of their total light output, spectral distribution of emitted light, intensity of light at a particular wavelength, and the brilliance of the incandescent area. Examples are discussed in Chapter 4; relevant units of measurement were explained in Chapter 2.

Tungsten-filament lamps

Tungsten-filament lamps consist, as the name suggests, of a tungsten filament in a transparent bulb; the filament is heated to incandescence by an electric current. The envelope may be of glass or of silica (quartz); the latter is better able to withstand high temperatures. The bulb contains an inert gas (such as argon) which does not react with the filament even at high temperatures. Because of the high temperature of the filament, some of the tungsten evaporates and deposits on the inside of the envelope; this has two deleterious effects, namely a gradual blackening of the bulb (reducing the lamp output) and a corresponding diminution of the substance of the filament, which ultimately fuses. This loss of metal from the filament is reduced in lamps, commonly known as quartz-iodine or tungsten-halogen lamps, which have iodine vapour in the bulb and a quartz envelope which can run at a high temperature (over 260 °C); the vapour has the effect of causing the vaporized tungsten to be re-deposited on the filament. In consequence, the filament of these lamps can be run at a much higher temperature and emit a greater proportion of short-wavelength (blue) light.

Incandescent filament lamps of the tungsten-halogen (quartz-iodine, QI) type have been successfully used for fluorescence microscopy. The major advantages of such lamps are economy, convenience and portability. Unfortunately, these lamps have a much lower output at short wavelengths than any of the arc lamps, and in any case have a much lower brightness. Therefore, they are really only suitable for the demonstration of quite bright fluorescence (e.g. from a fluorochrome) using green or red light excitation. Tungsten-halogen lamps cannot be used for UV excitation because the lamp has no significant emission in that region. This is due to the spectral emission of the filament corresponding to that of a black body at about 3200 K. However, Cripps, Hawgood & Magnus (1966) found that they could examine the autofluorescence of porphyrins using a tungsten-halogen lamp, without the rapid photobleaching (fading) associated with mercury arc lamps. With present-day technology, this photobleaching would be minimized by choice of appropriate filters, including UV-suppressing filters and if necessary a neutral-density filter.

Gas-discharge lamps

Gas-discharge lamps consist of two electrodes, between which an arc forms, in a quartz (silica) envelope, containing an appropriate gas. On application of a high enough voltage to start an arc, a small number of electrons present in the gas are accelerated by the electrostatic field until their energies become large enough to ionize the gas by collisions with neutral particles. The ions act as new current carriers, and the ionization rate increases in a cascade effect. The electric current keeps the electrons, ions and neutral molecules in continuous motion. The gaseous molecules become excited and then radiate photons in returning to the ground state.

At low pressures, the gas radiates only at certain characteristic wavelengths, together with a continuum related to the temperature of the arc. At higher pressures, the resonance lines are supplemented by other spectral lines corresponding to transitions between higher energy levels. The spectral lines are not monochromatic except at very low pressures; at high pressures the emission bands broaden.

It is a characteristic feature of arc lamps that they require a very high voltage to strike the arc, which can thereafter be maintained by a low voltage. Modern power supply units provide the ignition voltage automatically. Mercury arcs may be operated from either alternating current (AC) or direct current (DC), provided in either case that an appropriate type of lamp is chosen. For qualitative fluorescence microscopy, AC is generally used; for microfluorometry, a highly stabilized DC current must be used. AC power supplies consist essentially of a device to generate a high voltage for ignition, and either a transformer or a choke to limit the current through the lamp and the power consumed. Details of lamps currently available are given in Chapter 4.

The only gas-discharge lamps in common use in fluorescence microscopy are the high-pressure mercury and xenon lamps. Theoretically, other vapours could be used; for example, cadmium has been suggested for immunofluorescence using FITC, because of its lines at 468, 480 and 509 nm (Bals & Velculescu, 1966). Zinc might also be useful. The combination of mercury and iodine has been used for lamps for projection, but these lamps do not appear to offer any particular advantages for microscopy.

Mercury arc lamps are widely used for fluorescence microscopy. Their suitability is based on their intense light emission, and the relative simplicity of the power supply required. The spectral distribution of light from high-pressure mercury arcs is shown in Fig. 3.4. It will be seen that the major part of the energy of the emitted light is concentrated in a few narrow bands, corresponding to mercury emission lines, with a much weaker broad spectrum radiation. The major bands are at wavelengths of about

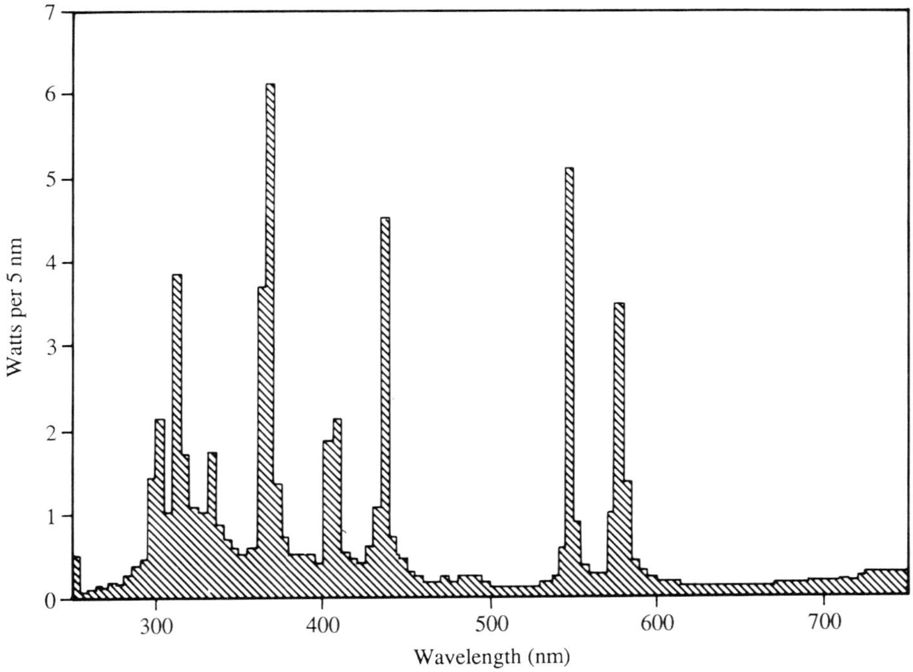

Fig. 3.4. The spectral distribution of light emitted from a high-pressure mercury arc lamp. The graph shows the absolute spectral power distribution of a Philips CS 200W-2 lamp. Redrawn after manufacturer's data, courtesy of Philips Lighting Industries.

Table 3.1. *Relative intensity of spectral lines in a mercury-vapour lamp*

Line wavelength (nm)	Relative intensity compared to 546 nm line
365·0	0·74
365·5	0·17
366·3	0·12
404·7	0·74
435·8	1·40
546·1	**1·00**
577·0	0·14
579·0	0·15

365, 405, 436 and 546 nm (Table 3.1). Each of these bands can be selected with a filter, and each one has specific applications for excitation in fluorescence microscopy.

Xenon arc lamps are similar in construction and operation to mercury arcs. However their spectral characteristics are quite different, see Fig. 3.5. There is a broad band of emission in the visible region, together with a substantial emission in the UV. Because of the absence of prominent spectral lines, xenon arcs give a substantially 'white' light, which also makes them suitable for colour photography and for all forms of microscopy. For fluorescence microscopy, this spectral distribution is both an advantage and a disadvantage. On the one hand, the even distribution of light enables the selection (by filter or monochromator) of any desired wavelength in the visible or UV regions. This applies particularly in the UV region, where mercury arcs are limited to the 365 nm line, which may or may not be suitable for a particular purpose. On the other hand, a sharp spectral line is more easily selected with a simple filter, whereas the xenon arcs tend to require more complex filter systems. The capital cost of xenon arcs is greater than that of mercury arcs. However, the life of individual lamps is proportionally greater, so that the actual running cost per hour may be less for xenon arcs.

Lasers

A laser is a device for producing pure monochromatic light in a coherent beam. Lasers are not at present appropriate for general observational microscopy, but are ideal for quantification: they will therefore be described in Chapter 8.

The lamphouse

The lamphouse is essentially a mechanical arrangement to hold the lamp and collector lens, often with a reflector, heat filter and sometimes with other filters. The lamphouse itself is always of metal, with provision for ventilation. All lamps used for fluorescence microscopy give off a substantial amount of heat. Arc lamps, particularly xenon arcs, also produce ozone by the action of UV on atmospheric oxygen. For this reason, an exhaust system should be provided for xenon and large mercury arc lamps, because the ozone is toxic.

The lamp is mounted in a socket, which should be supported by a system which will provide centring adjustments both horizontally (from side to side) and vertically. In a good lamphouse, these adjustments can be made with insulated knobs on the outside of the lamphouse. Good design of the lamphouse should obviate any necessity to wear asbestos gloves while making these adjustments!

A reflector may be provided behind the lamp, to collect rays which would otherwise be wasted and to focus them into an image near the original arc. In principle, the amount of light available can be almost doubled by this means. The reflector must be adjustable, so that its image may be placed near the arc but not coincident with it nor on the bulb (see above). The collector lens is placed in front of the lamp, and should be adjustable for focus. It is described in detail below.

A heat filter is usually mounted within the lamphouse, after the collector lens but before any other filters. Its mounting must be loose, to allow expansion as it becomes hot.

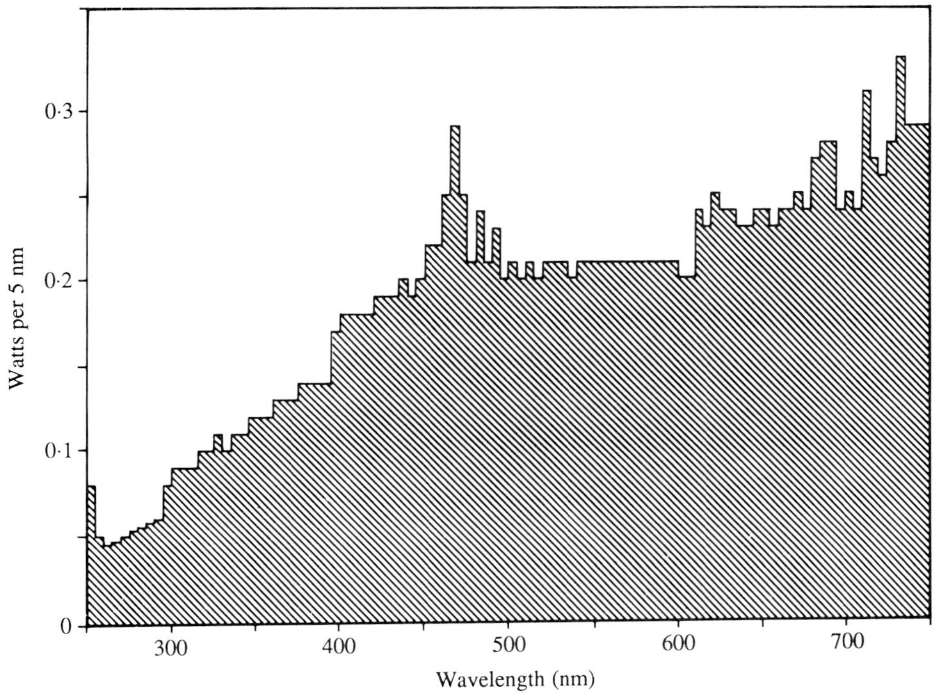

Fig. 3.5. The spectral distribution of light emitted from a xenon arc lamp. The graph shows the absolute spectral power distribution of a Philips CSX-150 (150 W) lamp. Redrawn after manufacturer's data, courtesy of Philips Lighting Industries.

Holders may be provided in the lamphouse for excitation filters. If so, the holders must be able to be swung in and out of the light path for selection of filters as required.

The collector lens functions to collect as much as possible of the light emitted by the lamp, and to concentrate it into a beam directed at the condenser. In most systems, the collector lens acts as the field lens for Köhler illumination (see below). It is essential that the focal length of the collector lens be appropriate for the particular lamp, so that an image of the lamp of appropriate size (i.e. completely filling the condenser aperture) is projected into the aperture of the condenser. For this reason, different collector lenses are required for different arc lamps, depending upon the dimensions of the arc (see Table 4.9). Some collector lenses have one surface modified (ground-glass or 'orange-peel') to act as a diffuser. This is not necessarily a good idea; while a separate diffusing glass can be added inexpensively if required and easily removed if not, changing the collector lens is much more difficult and expensive. For use in the UV range below about 360 nm, a quartz-glass collector lens is required, since optical glass does not transmit adequately. For the 365 nm mercury line, there is no need for quartz.

In the case of epi-illumination systems, where the objective acts as condenser, the above statements still apply: for 'condenser' read 'objective'. This form of illumination is described below.

Mirror-houses

A mirror-house is a unit attached to the microscope stand and supporting one or more lamphouses; the mirror-house contains movable mirrors, whose function is to provide a choice of alternative light paths, so that an appropriate light source can be selected and/or so that a single light source can be used alternatively for either dia-illumination or epi-illumination. For example, a mercury arc lamp and a tungsten lamp may be used, the former being required for both epi-illumination and dia-illumination, the latter only for dia-illumination. An example of this is the Leitz Mirror House 500, which has provision for up to four different lamps to be mounted on it simultaneously, of which two can be used for either dia-illumination or epi-illumination, while the other two are available one each for dia-illumination and epi-illumination, or may be shut off from the system. Such systems are particularly useful for quantitative studies.

Shutter

A shutter (usually an opaque slider) should be provided to block the light from the lamp when not required (for example, to avoid fading the specimen during intervals in examination, or while the specimen is being examined by phase-contrast). The shutter should ideally be placed in the light path between the lamp and the filters, so as to protect the filters from heat and excessive UV radiation. The shutter in that position tends to get hot, so it must have an insulated handle. An electrically operated shutter is desirable for quantitative work; ideally it should be rapid in operation and under computer control.

The field lens and field diaphragm

One of the functions of the illumination optical system is to concentrate the light from the lamp onto the specimen in such a way that the plane of the object is uniformly lit. In *source-focussed illumination* ('critical illumination'), an image of the light source is focussed on to the specimen. This gives a small, very brightly illuminated field, but unfortunately practical lamps do not have a uniformly incandescent surface, and moreover the source-focussed method usually illuminates only a small area at the centre of the field, suitable only for use at high magnifications. There are two methods for overcoming these difficulties, *Köhler illumination* and the use of a *diffuser*. The latter, in which a ground-glass or opal glass is placed in front of the lamp as a virtual source which is focussed on to the specimen, can give acceptably uniform illumination but at the expense of loss of light intensity and usually also, fluorescence apart, loss of resolution and contrast.

In Köhler illumination, a lens (the *field* or *Köhler lens*) is placed in front of the lamp such that the lamp is at the focus of the lens. The lens produces a uniform beam of parallel light. An iris diaphragm, the field diaphragm, is placed on the optical axis near the lens on the side away from the lamp; the diaphragm is thereby evenly illuminated. The diaphragm is used as a virtual source of light and is imaged by the condenser onto the specimen. The field diaphragm should be sufficiently far away from the lens for any dust on the lens not to be sharply focussed into the specimen plane.

A diffusion disc may be fitted, either permanently or on a swing-out holder, in the lamphouse, even if Köhler illumination is provided. The purpose of this is to provide a more even illumination. In general, such a disc is rarely required with a modern epi-illumination system, but may occasionally be useful with a darkground condenser. The diffuser produces a small reduction in light intensity, due to scattering at the ground surface and reflection at the smooth surface. In principle, the reflection is eliminated by putting the diffusion pattern on the surface of the collector lens (but see above) or (better) elsewhere, e.g. on the heat filter.

The condenser

There are essentially four ways in which the specimen may be illuminated:

1. From below (dia-illumination), with a conventional substage condenser, giving 'brightfield' illumination.
2. From below, with a darkground condenser.
3. From above (epi-illumination), with an epi-illumination device in which the light is reflected down through the objective, which acts as a condenser.
4. From above, with an epi-illumination device in which light is reflected, by an annular mirror surrounding the objective, down on to the specimen.

Conventional condensers

The use of a conventional condenser for fluorescence microscopy may be convenient and appropriate for scanning at low power, particularly if a suitable darkground condenser is not available. The conventional substage condenser gives bright illumination of a relatively large area. However, the UV or other light used for excitation is directed up through the objective, where it is not wanted. The method gives good illumination for low powers, and is relatively inexpensive in instrumentation, but suffers from low contrast (associated with the unwanted light entering the objective) and poor illumination for high powers, because it does not concentrate the light into the small area covered by high-magnification objectives, unless a 'zoom' condenser is used.

Highest fluorescence intensity is of course obtained with a condenser of high NA. If the nominal NA of the condenser is above 0·9, the full aperture of the condenser is only obtained if the slide is oiled to the condenser. This is always a very messy business. It may be less messy to substitute some other fluid for the immersion oil; even water will give a useful increase in aperture. For optimal results, the condenser must be adjusted correctly, as described in Appendix 3A.

Phase-fluorescence condensers

Prior to the development of epi-illumination, which allows the simultaneous use of fluorescence excitation and phase-contrast illumination, a special type of substage condenser was developed for this purpose. It consisted of a conventional phase-contrast condenser with a special phase plate. This latter incorporated an excitation filter, such that the cone of light through the phase ring was unfiltered ('white') while the remainder of the phase plate transmitted at the excitation wavelength only. The specimen was thereby illuminated both for phase-contrast (through the phase ring in the condenser) and for fluorescence (by the entire aperture of the condenser). The balance between the two forms of illumination could be varied by altering the voltage on the tungsten lamp used for phase-contrast; alternatively, if the phase rings incorporated polarizing material, the intensity of the phase-contrast system could be adjusted by rotating a polarizer in front of the lamp. This rather complex arrangement for combining phase-contrast and fluorescence had the disadvantage that separate phase plates in the condenser were required not only for different objective apertures but also for different excitation wavelengths. The method has been superseded by combining epi-illumination with phase-contrast or interference-contrast systems (see below).

Darkground condensers

A darkground condenser differs from a brightfield one in that the centre of the cone of illumination is blocked out, so that the illumination takes the form of a hollow cone of light which illuminates the specimen obliquely but does not enter the objective (see Fig. 3.6). The central stop of the condenser must be large enough to cover the aperture of the objective; hence, objectives of high NA require a darkground condenser with a large central stop. This in turn reduces the amount of light reaching the specimen. Darkground condensers are of two kinds. One is a conventional condenser with a central stop replacing the aperture diaphragm (Fig. 3.6a); the other is a special condenser with mirrors (Fig. 3.6b), which is free of chromatic aberration. The reflecting variety may be constructed to suit objectives of either high or low NA, but not both, so that the separate condensers are required for low and high powers. The low-NA variety gives a wider field and brighter illumination, but cannot be used with high-NA objectives.

A darkground condenser has the advantage that none of the light used for excitation is directed up into the objective (although of course some is deviated by the specimen and therefore reaches the objective to form the normal 'darkfield' image). In consequence, the requirements of the barrier filter are much reduced. On the other hand, the transmission of darkground condensers is much less than that of conventional

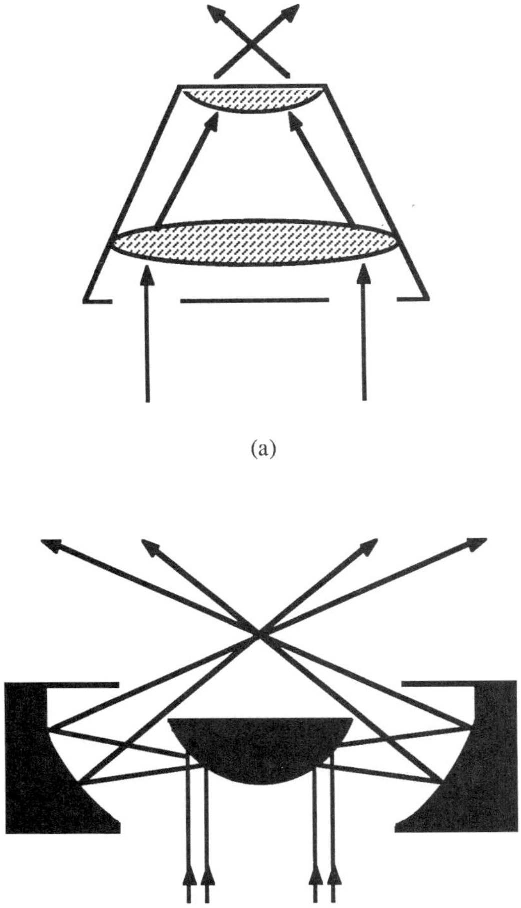

(a)

(b)

Fig. 3.6. Darkground condensers. (a) An otherwise conventional condenser with a dark stop; (b) a mirror condenser.

condensers, because of the large obstruction in the middle of the aperture. This problem is particularly evident in condensers designed for use with high-power objectives; and it is precisely with high-power objectives that the brightest illumination is required. Condensers intended for use with high-power objectives must generally be oiled to the slide (see above and Chapter 5), which is a nuisance.

Most darkground condensers suffer from a number of problems, particularly difficulty in obtaining even illumination of the field, small field size, critical focussing, and inability to accept the full light of the lamp because of the central stop. The problems of small field size and inability to accept the full light of the lamp are much reduced in the Tiyoda condenser (see Taylor, 1970).

Epi-illumination

It is necessary now to define two basic methods by which the specimen can be illuminated. In a conventional microscope, the specimen is illuminated from below (by a condenser) and examined from above (through an objective). Illumination of the specimen from the side opposite to the objective, implying light transmitted through the specimen, will be referred to as *dia-illumination*. Alternatively, light can be directed downwards onto the specimen from the same side as the objective, or even through the objective itself (Fig. 3.3); this will be referred to as *epi-illumination*. Of course, in an inverted microscope the up/down directions are reversed.

In principle, there are three methods by which epi-illumination can be achieved: by a concave circular mirror surrounding the objective, by a semi-transparent mirror placed above the objective, and by a prism placed above the objective and blocking part of its aperture. Each of these has advantages and disadvantages: the first was employed earliest, while the second method is now almost universally employed. Light not absorbed or reflected by the specimen is directed downwards, and should in principle never return upwards.

The principle of epi-illumination as usually employed for fluorescence microscopy is as follows. Light from the light source is directed onto a semi-transparent mirror situated in the body tube of the microscope over the objective lens. Light is reflected by the mirror down through the objective onto the specimen. The objective thereby serves as the condenser, focussing the light onto the exact area being examined. While the light used for excitation is absorbed by the specimen or passes through, light from the fluorescence is collected by the objective, passes through the semi-transparent mirror and reaches the ocular. Appropriate primary and barrier filters are of course present.

Epi-illumination was first considered in relation to the examination of vitally-stained tissues, and later considered for microfluorometry, for which it has certain theoretical advantages which will be considered in Chapter 8. The current popularity of the method is due largely to Professor J.S. Ploem, who in collaboration with Leitz developed practical systems (Ploem, 1967a).

The epi-illumination mirror is the key component of epi-illumination systems. Its function is to reflect the excitation light into the objective, while allowing light from the objective (the fluorescence) to pass to the eyepiece. The mirror must be semi-transparent, in order to transmit the fluorescence as well as to reflect the light for excitation. Because these are at different wavelengths, modern mirrors are designed to reflect preferentially at the excitation wavelength, and to transmit preferentially at the fluorescence wavelength.

These mirrors are variously known as mirrors or as beamsplitters; and those which reflect or transmit preferentially at different wavelengths are variously called 'dichromatic' or 'dichroic'. The latter term is unfortunate, as the word 'dichroic' normally refers to a quite different property, related to polarization, in which a substance absorbs light differently according to the plane of polarization of the incident light. Since the functions of the mirror or beamsplitter are primarily to reflect light down through the objective, and secondarily to refrain from posing an obstruction to light passing up from the objective, I prefer the term 'mirror', without having any real objection to the other term. Therefore, in the present work, the term 'dichromatic mirror' is used.

A dichromatic mirror gives a significant increase in efficiency, which can be explained as follows. First, consider a mirror which reflects 50% and transmits 50% at all wavelengths, with no absorption. Then the intensity of the excitation will be reduced by 50% at reflection, and that of the fluorescence will equally be reduced 50% on transmission. This results in a total reduction to 25%, compared to the ideal mirror reflecting 100% at the excitation wavelength and transmitting 100% at the emission wavelength. Therefore, in principle, a fourfold increase in efficiency would appear to be possible; in practice, a threefold increase can be obtained. It should be noted also that a glass or quartz plate, uncoated, reflecting say 10% and transmitting 90%, has a total efficiency of 9%.

The dichromatic mirrors used in epi-illuminators are constructed in the form of interference coatings on glass, which must be of excellent optical quality because it is placed in the magnification system between objective and ocular. Each mirror is characterized by a wavelength, above which it preferentially transmits and below which it preferentially reflects. The transition wavelength must lie between those of the excitation and barrier filters. Manufacturers usually designate mirrors according to this wavelength, possibly preceded by letters to indicate the type or make of filter (see Chapter 4).

Theoretically a similar system could be based on a dichromatic mirror with reversed characteristics, i.e. reflecting long-wavelength light and transmitting short wavelengths; however, the microscope magnifying system would have to be designed with a right-angle bend in it. The concept may prove useful for some purposes as in inverted microscopes and for microfluorometry.

Comparison of epi- and dia-illumination

Compared with transmitted light, epi-illumination has the following advantages:

1. Opaque specimens can be examined.
2. The transmitted-light pathway can be used independently for alternative or simultaneous examination in phase-contrast, darkfield or brightfield.

Epi-illumination by a semi-transparent mirror above the objective offers the following additional advantages:

3. Excitation filter, dichromatic mirror and barrier filter can conveniently be assembled into a single unit. This allows quick changes of filter combinations, and prevents operator error in matching excitation and barrier filters.
4. The illumination is concentrated on, and confined to, the field of view. Light is neither wasted, nor unnecessarily directed at other parts of the specimen (which may fade).
5. The diameter of the exit pupil of an objective is smaller in diameter than that of a substage condenser, and is more easily filled by an image of the arc.
6. Objective and condenser are always accurately aligned because they are one and the same.
7. Immersion fluid, if needed, is used only on one side of the specimen, between coverglass and objective: there is no messy oiling of the substage condenser.
8. For quantification, only epi-illumination gives reproducible and ideal conditions (see Chapter 8).

Disadvantages of epi-illumination

There are two disadvantages of epi-illumination:

1. An epi-illuminator is required, increasing the capital cost of the equipment.
2. The brightness of the image with objectives of low NA (low-power objectives) may be less than can be obtained with a substage condenser of high NA. Special low-power objectives of high NA are made for this reason, but they are expensive, have a short working distance and require an immersion fluid.

It is of interest to compare the relative brightness of the image as seen with various systems. Epi-illumination through the objective gives increasingly bright images as the magnification of the objective is increased, whereas the reverse usually holds with dia-illumination. In general, the intensity of illumination of the specimen plane depends upon the square of the NA of the condenser, and also on the square of the demagnification of the image of the light source (i.e. as the image of the field diaphragm is made smaller and smaller, it becomes brighter according to the square law). Similarly, the brilliance of the image of the specimen as it reaches the ocular is directly proportional to the square of the NA of the objective, and also inversely proportional to the square of the magnification. Hence, with dia-illumination, changing objectives (without changing the condenser) results in changes according to the NA and magnification of the objectives. With epi-illumination, the same considerations apply in respect of the image, but also, because the objective acts as its own condenser, the intensity of illumination and therefore of the final image depends directly upon the square of the NA and magnification of the objective. As the magnification of the objective increases, the image of the object is magnified more but simultaneously the image of the light source is demagnified by a corresponding amount; these effects cancel

Table 3.2. *Effect of changing objectives: epi-illumination and dia-illumination*

This table illustrates, for some typical objectives, the theoretical effects of changing objectives when using either epi-illumination or dia-illumination, based on the NA and magnification of the objectives. In epi-illumination, it is assumed that the illumination system is adjusted when changing objectives so that an image of the light source is focussed into the 'exit' pupil (rear focal plane) of each objective. Comparison is basically valid within each column, not between the dia-illumination and epi-illumination columns. However, the dia- and epi-illumination columns can be compared approximately, it being assumed that the substage condenser corresponds in magnification to the ×4 objective, and has a NA of 0·9. Numbers are given to two significant figures only; the intensity with the ×4 objective and epi-illumination is arbitrarily taken as 1.

Objective magnification	NA	Relative intensity of image	
		Epi	Dia
4	0·10	1	80
10	0·25	39	80
25	0·50	625	51
40	0·65	1800	34
100	1·30	29 000	22

out and therefore the brilliance of the final image is independent of the magnification and depends only on the fourth power of the magnification. Theoretical figures are given in Table 3.2. In practice, the effect of changing objectives is affected by differences in the diameter of the rear aperture of the objectives (exit pupil of the objective; see Chapter 4). However, it will be seen that, in general, epi-illumination gives increasingly bright images with higher-powered objectives, whereas with dia-illumination the reverse tends to be the case.

THE VIEWING SYSTEM

The viewing system is that part of the optical system which produces a magnified image of the object, filters out the light used for excitation, and possibly (by further filtration) limits the perceived range of fluorescence wavelengths to a specific region. The viewing system normally includes the objective, ocular, intermediate magnifying system if any, and barrier filter. Apparatus for photography and photometry is dealt with in other chapters.

The overall magnification is determined jointly by the objective, the ocular, and the intermediate magnifier (if any), and is equal to the product of the respective magnifications. Given that all these three components are changeable, it follows that any given overall magnification can often be achieved in more than one way, using different combinations of magnifications of objectives, intermediate lenses, and oculars. The brilliance of the final image is inversely proportional to the square of the overall magnification.

The objective

The prime purposes of the objective are to provide resolution of the object and initial magnification. The brightness of the final image is also determined in part by the objective, according to the square of the NA and inversely as the square of the magnification. In epi-illumination, the objective acts as a condenser; and again its NA and magnification affect the brilliance of the illumination and therefore of the image, although in this case the brilliance is directly proportional to the square of its magnification, since this produces a corresponding demagnification of the light source.

Objectives are available in great variety, differing in magnification, NA, chromatic correction, flat-field correction, other corrections, the type of immersion fluid (if any), and other characteristics. Some are made specifically for fluorescence microscopy; many that are not can still be used very successfully. It used to be said that the simplest possible objectives (achromatic) should be used to avoid autofluorescence of the fluorite glass used in more highly corrected systems, and because of light loss at the many air–glass interfaces. In my experience, however, these are rarely a problem; I have successfully used semi-apochromatic (fluorite), apochromatic, planachromatic, and planapochromatic objectives. It is true that light losses may be greater in the more complex objectives, but this may be more than compensated for by increase in NA.

The resolution of the system depends primarily on the NA of the objective. In fluorescence microscopy, resolution is rarely a problem. When maximal resolution is required, the practical considerations involved are essentially the same as for ordinary microscopy, except that the NA of the condenser is no longer relevant. The fluorescence microscopist will usually be using the highest NA objective available, to maximize the brightness of the image. Provided that the fluorescence is bright enough, visualization of fine details is aided by using a higher ocular magnification (say ×8 or ×10) than would otherwise be optimal.

Working distance of objectives is generally related to the NA, being least for objectives with highest NA. Some types of objectives are designed with a particularly long working distance (e.g. the new Olympus LB objectives), and are useful for dealing with thick specimens. Usually, difficulties arise only with immersion objectives.

Depth of field is the thickness of the specimen which is acceptably sharp at a given setting of the focus control; depth of focus is the range over which the focus control can be moved while maintaining acceptable sharpness of a specified object plane. The term *axial resolution* is also used. The depth of field is approximately inversely proportional to the NA of the system, and also is decreased somewhat by increasing the total magnification; see Zieler (1973).

The optical corrections of objectives have the same implications in fluorescence microscopy as in ordinary microscopy. Flat-field objectives are always preferable for photography. Otherwise, in principle the more highly corrected objectives (apochromats) have a higher NA than achromatic objectives and therefore provide a brighter image, particularly with epi-illumination. However, the simplest achromatic objectives usually have fewer air–glass interfaces and no fluorescent glass, and may thereby achieve a higher light transmission and darker background. In my experience, flat-field objectives seem to be particularly suitable for epi-illumination.

Ultraviolet-transmitting objectives are only required for epi-illumination with particularly short wavelengths (below 360 nm). They are therefore essential for some types of microspectrofluorometry (see Chapter 8), but are never required for routine purposes. Just as well, because they are very costly!

Built-in barrier filters were at one time provided on the front of some Leitz objectives. This prevented the excitation light from entering the objective, thereby reducing autofluorescence. Of course, only illumination by dia-illumination was possible, and the barrier filter, being fixed, could not be changed.

Objectives for uncovered specimens, corrected for use without a coverslip, are intended for use with uncovered smear preparations.

Immersion objectives are of two types: those intended for use with conventional coverslipped slides, and low-power water-immersion objectives intended for examining specimens under water. The former usually exploit the possibilities of increased NA as a result of increased refractive index of the medium between the objective and specimen. The immersion medium may be conventional non-fluorescent immersion oil, silicone oil, glycerol, or water. Water-immersion objectives can be used for studying objects under water in a vessel without a lid.

The diameter of the objective's exit pupil is of importance in epi-illumination, because this is also the entrance pupil for illumination, and must be filled by an image of the light source. If this image is smaller than the pupil, uneven illumination will result; if the image of the light source is greater than the pupil, light is wasted and the intensity of the fluorescence correspondingly reduced. The diameter of typical objectives varies from about 12 mm (low powers) to about 5 mm (high powers); see Chapter 4, particularly the tables of specifications of objectives.

Tubelength

Modern objectives are usually designed either for the standard 160 mm tubelength (see Fig. 3.7) or for infinite optical tubelength (intermediate image at infinity; see Fig. 3.8). The latter are designed in such a way that the light leaving them is parallel, so that accessories such as epi-illumination mirrors, filters and phase-rings can be placed above the objective without introducing complications such as spherical aberration. Infinity-optics therefore offer advantages for fluorescence microscopy. A separate 'tube lens' serves to produce an intermediate image in the normal position, and normal oculars are used. Ordinary objectives can be used for phase-contrast, as the phase rings are placed as accessories in the body tube. Infinity-corrected optics are currently made by Reichert and Zeiss (Oberkochen) for general use, including fluorescence (see Chapter 4), and by other manufacturers primarily for metallographic purposes.

Intermediate magnifying devices

Nosepieces sometimes have an auxiliary lens in them. Leitz in particular tend to use this device; it is a correcting lens to adjust the optical tubelength, i.e. compensating for changes in the actual or mechanical tubelength brought about by the design of the nosepiece. From the point of view of the fluorescence microscopist, these lenses may be a nuisance because they possess at least two air–glass interfaces at which light is lost.

Magnification changing devices can be added as accessories to several types of microscope. Usually the purpose of these devices is to enable the optimal magnification

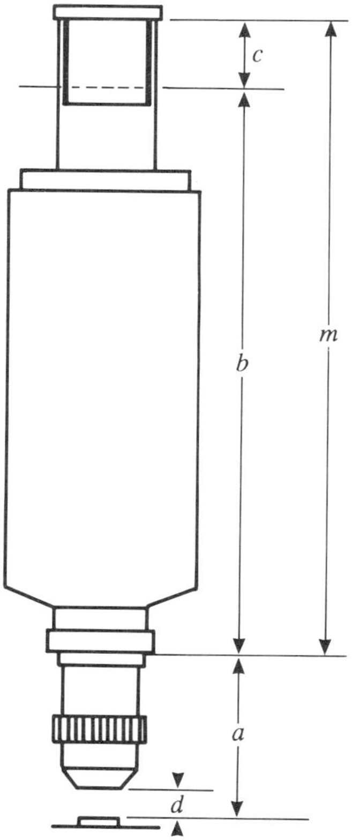

Fig. 3.7. Mechanical dimensions of a standard microscope body tube; DIN standards in parentheses. a = adjustment length of the objective (45 mm); b = image distance of the objective (150 mm); c = adjustment length of the eyepiece (10 mm); d = free working distance; m = mechanical tubelength (160 mm).

to be obtained for photography. They consist either of a rotating turret with several lenses which can be turned into the optical pathway, or alternatively of a zoom system. The fluorescence microscopist is strongly advised to have nothing to do with either, and particularly not with the zoom devices; they decrease the brightness of the image, both by increased magnification and by the presence of optical elements which cause losses, mainly by reflection at air–glass interfaces.

However, an intermediate reducing lens (i.e. giving a magnification less than unity) will tend thereby to increase image brightness by decreasing the overall magnification.

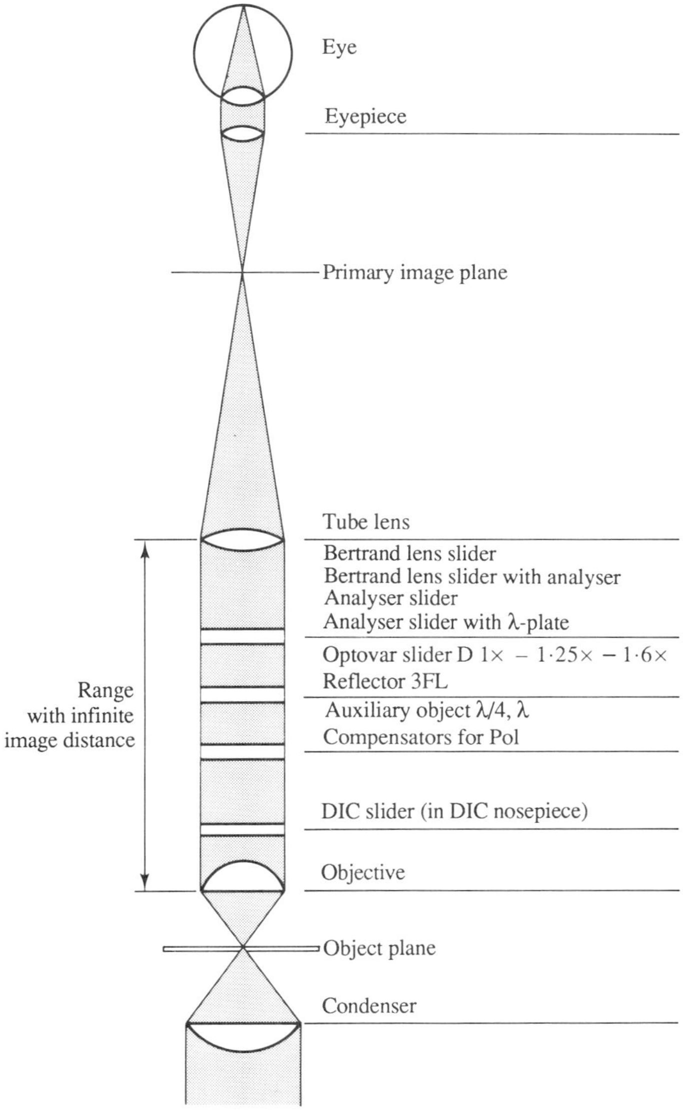

Fig. 3.8. Infinity-corrected optical system. Diagram courtesy of Carl Zeiss Pty Ltd, Sydney.

Table 3.3. *Relative brightness of oculars of different magnification*

Oculars are compared to those of the usual ×10 magnification.

Magnification	4	6·3	8	10	15
Relative brightness	6·3	2·5	1·6	1·0	0·4

For example, the Leitz ×0·8 lenses theoretically should increase the brightness by a factor of 1·56, slightly reduced by losses in the lenses.

Other intermediate optical components

Fluorescence microscopes almost invariably have a barrier filter in the body tube, somewhere between the objective and the ocular, and in most cases also a semi-transparent mirror associated with an epi-illumination system. Sometimes the filter is placed in the eyepiece, near its field diaphragm. The introduction of a thickness of glass or other material in the optical system may be expected on theoretical grounds to introduce spherical aberration unless the rays at that point are parallel rather than diverging. This requires the type of objective which is infinity-corrected rather than corrected for a tubelength of 160 mm. Alternatively, the optical system could be designed to include a slab of glass, representing either the barrier filter (of fixed thickness) or a clear glass plate.

Oculars

It has already been said that oculars should usually have low magnification, to maximize the intensity of the final image; however, a magnification of ×8 or ×10 is required to utilize the full resolution of the objectives. The special fluorescence objectives designed with low magnification and high NA are theoretically capable of being used with even higher ocular magnification to achieve their theoretical resolution, but of course at the expense of the wide field of view and bright image for which they are intended.

The optimal magnification for oculars for fluorescence microscopy is determined by the need to balance the need to maintain image brightness as high as possible (requiring as low a magnification as possible) against the value of making full use of the resolving power of the objective (requiring a magnification, usually, of about 8–10). The relative brightness of the image with oculars of commonly available magnifications is shown in Table 3.3.

The usual criteria for determining the ocular magnification, using normal transmitted light, are:

1. The minimal useful magnification: sufficiently high magnification to enlarge the smallest detail resolvable by the objective to at least the minimal size resolvable by the observer's eye.

2. Avoidance of 'empty magnification': no magnification higher than that required for the eye to resolve the smallest detail resolvable by the objective.
3. Magnification low enough to show the full field of view covered by the objective.
4. Exit pupil diameter not greater than the diameter of the pupil of the observer's eye, otherwise light is lost and the aperture of the system becomes limited by the eye.

The field of view is limited by the diameter of the intermediate image diaphragm in the ocular, the diameter of the latter being limited by the diameter of the body tube. Therefore, the full field of which any objective is capable can be visualized with an ocular with a magnification of about ×10 (normal tube) or ×6·3 (wide tube). Further reduction in magnification does not increase the field of view, and it is for this reason that lowest-power eyepieces are usually made only for fluorescence or projection.

The *field number* of an ocular is the diameter of its field stop: or, more precisely, the theoretical diameter of a field stop which would contain, without a field lens, the same field of view as does the actual stop with the field lens. The field number can be divided by the objective magnification to obtain the diameter of the field (in mm) in the specimen plane as seen through the ocular.

A cone of parallel bundles of rays, each one of which originates from a different object point, emerges from the ocular and passes through a circular area, small in diameter, which is called the *exit pupil* of the microscope. The exit pupil is an image of the aperture stop (back focal plane) of the objective. The pupil of the observer's eye must be placed around the exit pupil in order to accept all the image-forming rays. Under dim light conditions, the eye's pupil dilates to a maximum of about 7 mm, contracting in brighter light. The diameter of the exit pupil must therefore not exceed 7 mm, otherwise light will be wasted. The diameter of the exit pupil depends in part on the objective, and in part on the magnification of the ocular; being inversely proportional to the magnification of the ocular.

The diameter of the exit pupil in millimetres is given by the formula:

$$\varnothing = 500\ A/M$$

where $\varnothing$ = diameter of exit pupil (mm), A = NA of objective, M = total magnification of microscope, 500 = twice the distance of optimal viewing (250 mm). The diameter of the exit pupil, for a typical objective and ×10 ocular, is usually of the order of 1 mm; with ×4 ocular, this becomes 2·5 mm. However, with the special fluorescence objectives of high NA, and a ×4 ocular, the exit pupil may exceed 5 mm (see Table 3.4). Since the pupil of the eye, under bright light conditions, may contract to as little as 1 mm diameter, part of the process of adaptation of the observer's eyes is to allow the pupil to expand sufficiently to encompass the entire exit pupil of the microscope. It is also possible to calculate the *lowest power ocular* which can be used without its exit pupil exceeding the 7 mm maximal diameter of the pupil of the eye.

Binocular heads

There used to be a body of opinion that binocular heads should never be used for fluorescence microscopy, because of light losses. On theoretical grounds, since the light

Table 3.4. *Diameter of exit pupils*

The table lists the diameter (⌀) in mm, of the exit pupil for some Leitz objectives and two magnifications of oculars.

			Exit pupil ⌀ (mm) Ocular	
Objective	Magnification	NA	×4	×10
Pl Fluotar	6·3	0·20	4·0	1·6
	10	0·30	3·8	1·5
	16	0·45	3·5	1·4
	25	0·60	3·0	1·2
	40	0·70	2·2	0·9
	100	1·32	1·7	0·7
NPL Fluotar Fluoreszenz	10	0·45	5·6	2·3
	25	0·75	3·8	1·5
	40	1·30	4·1	1·6

from the objective is split into two, the intensity of each must be halved, with further reduction due to light losses in the splitting system. However, with modern optical systems, the losses at the splitting system appear to be acceptable under most conditions. Otherwise, the total available light (to both eyes) is the same, and probably this is the most important factor. In my experience it appears easier to accommodate if both eyes are used, and I have always preferred to use a binocular head, in order to be able to observe with both eyes.

Trinocular heads belong properly to the chapter on photography, but it should be said here that only those trinocular heads which offer the choices of 100% light through the observation tube and 100% up to the camera are suitable for observation and photography of fluorescence.

FILTERS

It has already been indicated that a fluorescence microscope requires a system of two complementary filters, of which the first selects the appropriate wavelength for excitation, and the second blocks that wavelength while permitting the longer wavelengths of fluorescence emission to pass to the eye or camera. This filter system is central to the operation of the microscope; both the excitation and barrier filters must be appropriate for the fluorophore being examined as well as matching each other.

Spectral characteristics of filters

The shapes of the spectral transmission curve of a filter may be classified as indicated below.

Neutral density filters give a nearly constant transmission over a wide range of wavelengths. In fluorescence microscopy they are used occasionally to decrease the intensity of the excitation light, e.g. to reduce fading.

Shortpass (SP) filters preferentially transmit at short wavelengths, below a particular transition point, and transmit minimally at longer wavelengths (see Fig. 3.9). Because they transmit at high frequencies they can also be termed highpass filters. This type is required for use as excitation filters.

Longpass (LP) filters preferentially transmit at long wavelengths, above a certain transition point, with minimal transmission at short wavelengths. Because they transmit at low frequencies, they are also known as *lowpass* filters. This type is used for barrier filters.

Bandpass (BP) filters have a bell-shaped transmission curve, transmitting maximally at a particular frequency (or wavelength) and less at shorter and longer frequencies (wavelengths). A bandpass filter can be formed by combining a longpass filter with a shortpass filter (Fig. 3.9). Bandpass filters are characterized by the following (see Fig. 3.10):

λ_{max} = position within the spectrum of the peak transmission, i.e. the wavelength at which transmission is highest.

T_{max} = maximum transmission value, i.e. the transmission at λ_{max}.

Bandwidth = the spectral range over which the transmission is equal to or greater than half the maximum transmission (this is often called the half-bandwidth, HB).

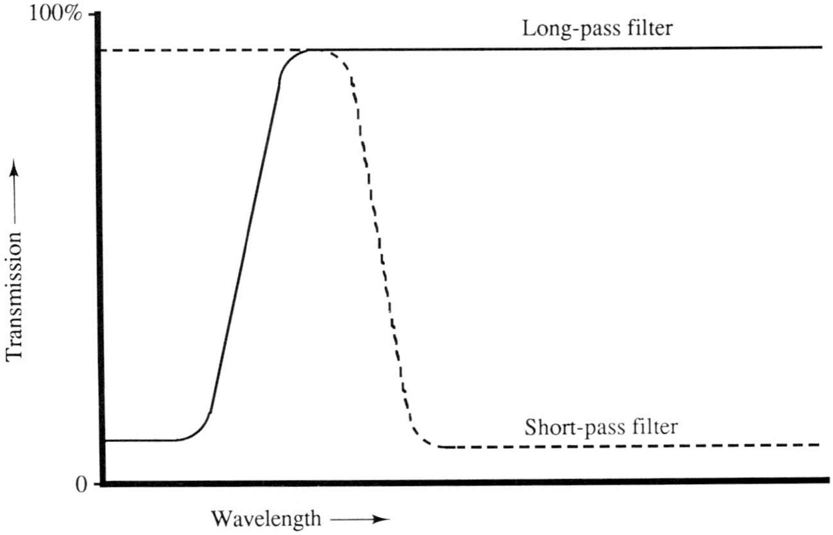

Fig. 3.9. Illustrating shortpass and longpass filters, and the combination of one of each to form a bandpass filter (broken line).

Background transmission = transmission at wavelengths removed from the peak, and at which minimal or near zero transmission is required.

Bandcut filters have characteristics inverse to those of bandpass filters: they transmit at all wavelengths except over a narrow range.

Types of filter

Several optical devices are available for monochromation (i.e. selective transmission of a specified wavelength band). These are: coloured glass, gelatine or liquid filters, interference filters and mirrors, and prism and grating monochromators.

Glass filters are of two basic classes. In the first, the colour is due to the presence of ions in true solution in the glass. In the second, known as temperature-coloured glasses, the colour is due to submicroscopic crystals, the optical effect of which depends on their composition and size. The final characteristics of the glass are obtained by heat treatment of an originally clear glass.

The transmission characteristics of the two types are quite different. Glasses with ionic coloration have an approximately bell-shaped transmission curve; that is, they

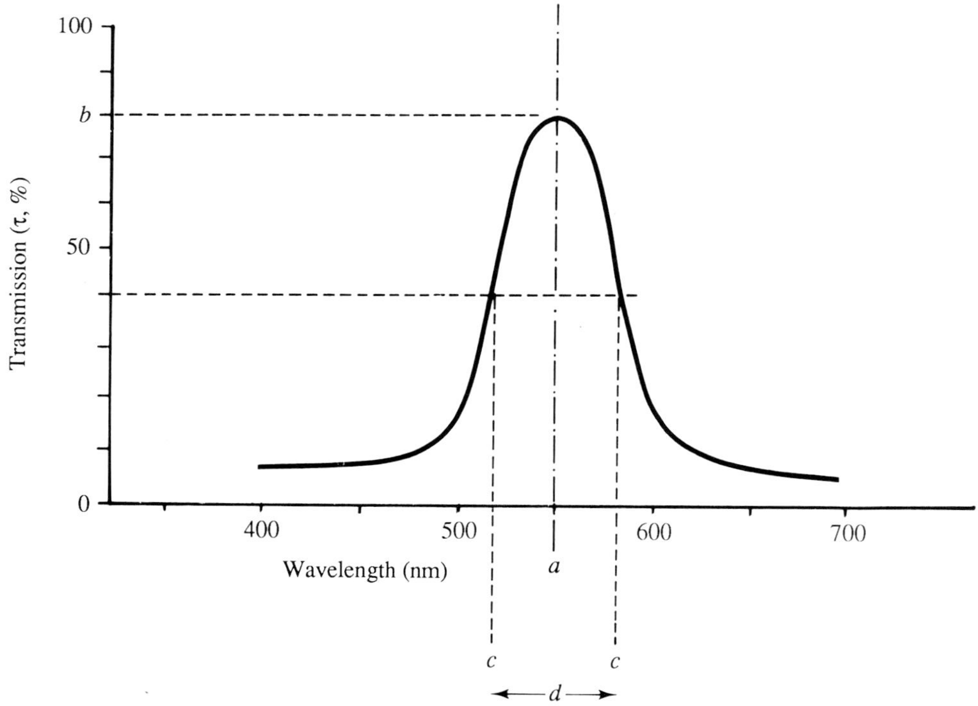

Fig. 3.10. Characteristics of an hypothetical bandpass filter. a = wavelength of peak transmission (550 nm); b = peak transmission (80%); c = wavelengths at which half of peak transmission occurs (520 and 580 nm); d = half-bandwidth = 580 − 520 nm = 60 nm (very broad).

transmit preferentially in a particular, relatively narrow region. This renders them suitable for use as excitation filters. The obtainable range of such curves is limited. On the other hand, temperature-coloured glasses are characterized by a long-wave region of high transmission and a short-wave region of lower transmission, which happily happens to be the characteristic required of a barrier filter. The cut-off wavelength, between the regions of high and low transmission, can be varied by changing the heat treatment; hence, a great variety of optical characteristics can be obtained.

Fluorescent glasses contain inorganic fluorescent substances. These glasses can be used as standards for fluorometry. Some of these glasses have transmission characteristics which, apart from the fluorescence, would make them suitable for barrier filters, and unfortunately some have actually been used for this purpose; their fluorescence does not assist in maintaining good contrast in the microscope image!

Gelatine filters are essentially similar to dye-glass filters, except that the dye is dissolved in a sheet of gelatine rather than glass. The most widely used gelatine filters are the Wratten filters supplied by Kodak. Similar filters are available from Ilford, and some from other manufacturers. For protection, gelatine filters should preferably be cemented between two layers of optical glass. Gelatine layers are incorporated into some commercial barrier filters.

Interference filters consist of two semi-transparent mirror surfaces separated by a thin layer of transparent material (the dielectric). In practice, they normally consist of a sheet of glass, coated on one side first with one thin semi-transparent coating of metal (e.g. silver), then the intermediate layer, followed by the the second semi-transparent metallic coating. To protect the coatings, a second glass is cemented on. Interference between successive reflections between the reflective coatings results in transmission of light of the desired wavelength, while other wavelengths are reflected, except that light is also transmitted at wavelengths which are integral multiples of the primary (first-order) wavelength. On casual inspection, an interference filter therefore appears as a mirror. The thickness of the dielectric layer, and its refractive index, critically determine the wavelengths of the transmission bands, and the amount of background transmission. The bandwidth and the maximal transmission value (T_{max}) depend largely on the optical characteristics of the reflective coatings. The more highly reflective the coatings, the narrower is the range of wavelengths transmitted with appreciable intensity. The wavelength of maximum transmission by a given interference filter varies somewhat according to the angle of illumination; λ_{max} is displaced in the short-wavelength direction with increasing angle of incidence (further from normal incidence) and at the same time the transmission is slightly reduced. With light beams strongly convergent or divergent, the effect is to broaden the bandwidth. When light is polarized, the displacement depends on whether the polarization is parallel or normal to the plane of incidence.

Practical interference filters always require a supplementary filter to remove higher-order transmission bands and to decrease the background. The supplementary filter is usually a coloured-glass filter cemented to one side; such combination filters show one mirror surface and one coloured surface.

Interference filters available at the time of writing (1989) have not, in general, as high a maximal transmission as glass filters, and neither do they have as high an opacity in regions remote from their transmission maxima; also, most interference filters have small pin-holes in the metallic coatings which transmit unwanted light. The great advantages of interference filters are that they can be made for any wavelength (whereas the range of wavelengths for which glass filters are available is small), and that they can be made with a very narrow bandwidth.

Interference filters can be used as excitation filters, provided that they are adequately protected from overheating. The mirror surface must always face the source of illumination, as this reflects away much of the light which would otherwise need to be absorbed, generating heat. They should also be preceded by a heat filter, and if a set of filters is used simultaneously the interference filter should always come last. They find occasional use as barrier filters for examining specific fluorophores such as FITC.

Interference mirrors are used mainly as dichromatic mirrors in epi-illumination devices, but can be used for fluorescence excitation (Ploem, 1967b).

Liquid filters were used in the early days, at least until the 1950s; they are now mainly of historical importance. However, it may be worth noting that distilled water can act as a heat filter, and otherwise transmits down to about 194 nm. Aqueous copper sulphate, in distilled water or alkaline solution, can be used as a red-stopping filter; for this purpose it has been supplanted by BG38 glass. Aqueous potassium nitrate could be used as a UV-stopping filter, absorbing from 400 nm downward.

Monochromators, both prism and grating, are not really suitable for general qualitative microscopy; they are expensive, and tend to have a relatively low light transmission, and a small aperture. They are appropriate for quantitative microscopy.

Filter blocks

In the majority of modern epi-illumination systems, the entire filter system (excitation filter, dichromatic mirror, and barrier filter) is conveniently accommodated in a single interchangeable unit, following the design of Ploem.

Excitation filters

Filter systems for excitation are of two basic types: broad-band and narrow-band. For broad-band excitation, the filter system usually consists of dyed-glass filters, transmitting light over a substantial bandwidth. A common example is the use of BG12 glass for blue-light excitation. This type of excitation may be used when the excitation and emission maxima of the fluorophore are widely separated. The main advantages are cheapness and high intensity of the light. Broad-band excitation is used also in double-staining techniques, if two fluorophores of differing excitation maxima are to be excited simultaneously (e.g. FITC and TRIC, requiring blue and green excitation).

Narrow-band excitation is used whenever the excitation and emission peaks are closely spaced, and also when it is desired to obtain optical specificity by preferentially exciting one specific fluorophore while avoiding excitation of other fluorophores. The corresponding disadvantages are greater cost and a reduced amount of light available for excitation.

In practice, the excitation filter usually consists of a series of individual filters, some of which are glass and one or more may be of the interference type. The bandwidth of the excitation is best controlled by using two separate filters, one of the SP type to determine the long-wavelength limit, and the other of LP type to determine the short-wavelength limit. The two filters together form a bandpass filter. In epi-illumination systems, the mirror (if dichromatic) also forms part of the excitation filter, acting as an SP excitation filter.

Broad-band excitation filters

These consist typically of three components: a coloured-glass filter with peak transmission in the required area, a heat filter, and a red-stopping filter. To these may be added a UV-stopping filter (similar optically to a barrier filter). Some currently available glasses are described in Chapter 4.

Barrier filters

Barrier filters are also known as secondary or suppression filters. They should be substantially opaque at the wavelength used for excitation, while transmitting as much as possible of the fluorescence emission (Plate 2 a, b). Practical filters are usually made of one or more layers of 'temperature-coloured' glass. In dichromatic epi-illumination systems, the dichromatic mirror acts as a barrier filter (in addition to its other functions).

For the examination of specific fluorophores, particularly in immunofluorescence, sometimes the barrier filter is chosen with a narrow-band transmission only around the emission peak of the specific fluorophore. An interference filter, supplemented with glass, is used for this purpose. As compared to a conventional barrier filter, this may increase the optical specificity of the system (see Chapter 7); on the other hand, a misleading result may be obtained because emission spectra tend to be broad, and the use of a narrow-band barrier filter gives everything a monochrome appearance irrespective of its true colour. In general, optical specificity is better obtained by a narrow-band excitation filter than by a narrow-band barrier filter.

Barrier filters made from 'temperature-coloured' glasses are liable to some inconsistency in their manufacture. Another potential cause of difficulty is that some of these glasses are themselves fluorescent (see above), spoiling an otherwise dark background in the image. This can be checked by holding them in a beam of short-wavelength light, e.g. by putting the filter on the microscope stage.

4

Commercial fluorescence microscopes

The purpose of this chapter is to give a brief overview of currently available standard equipment for fluorescence microscopy, and to offer some guidance to the prospective purchaser as to how to set about choosing suitable equipment for a particular purpose. Some earlier equipment is referred to in Volume II, Chapter 16. Published information on commercial fluorescence microscopes is to be found mainly in manufacturers' catalogues, brochures and house journals such as the Leitz *Scientific and Technical Information* (*Mitteilung für Wissenschaft und Technik*) and *Zeiss Information*.

In this chapter I shall attempt to describe briefly some of the current commercial fluorescence microscopes. Naturally, the list is not exhaustive. Because of my work in microspectrofluorometry with a Leitz instrument, I am more familiar with that brand than with any other; also, the majority of the microscopes in my present department are Olympus. However, I value my friends in other firms too!

At the time of writing, the commercial microscope scene is dominated by four firms: the Leica group, Nikon, Olympus and Zeiss (Oberkochen). These firms all make equipment specifically for fluorescence microscopy, as do several other firms.

CHOOSING A MICROSCOPE

The first step is to define exactly what needs the new microscope must satisfy. One should make a list of precisely what is to be done with it, and of all relevant technical and other considerations. As regards technical aspects, the following keywords and phrases may assist in this process: upright/inverted, multipurpose/dedicated, multiuser, compatibility with existing equipment, image quality, dim/brilliant fluorescence, alternation with conventional transmitted light or phase-contrast or darkfield, photomicrography, quantification, automation. In my experience, non-technical considerations may include the amount of money available or likely to be obtainable, the quality of service to be anticipated from manufacturers' agents, and the possible preference of grant-giving bodies for a locally made product. Having decided what needs are to be satisfied, one should shop around. Appendix 2 gives some suggestions for a checklist for comparison of individual makes and models.

LEITZ

Leitz was one of the pioneers in commercial fluorescence microscopes, and has taken a particular interest in this field, particularly in respect of quantitative instrumentation. Current (1990) microscopes suitable for fluorescence microscopy are the Aristoplan, Diaplan, Laborlux, Biomed, and Fluovert.

The Aristoplan is the current major Leitz microscope, released in 1986. While strongly resembling its predecessor, the Orthoplan, the Aristoplan features a more modular design, and a field diaphragm situated in the base of the microscope with Köhler illumination for objectives down to ×1·6 without re-focussing the condenser.

There is an integrated epi-illumination system. Filter modules are simply inserted into the light path in the body tube. The viewing height is unchanged and there is no necessity to move objectives from one nosepiece to another. Lamphouses are interchangeable with other current Leitz microscopes. The Aristoplan epi-illuminators accept a separate series of the standard Leitz filter blocks (see below, especially Table 4.1). Mercury or xenon arc lamps of up to 100 W, as well as tungsten-halogen lamps, are accommodated in an interchangeable lamphouse, type 103Z, specially designed for fluorescence. This lamphouse is equipped with a reflector, centrable lamp base, condenser (collector) lens, and a flip-in centring screen at the side. The lens has an apochromatic correction, obviating the need to re-focus the lens at each change of excitation wavelength. This lamphouse is used on all current Leitz fluorescence microscopes (except the Orthoplan, now used only for fluorometry).

The Diaplan (Fig. 4.1) is a general-purpose microscope whose performance places it somewhere between routine laboratory microscopes and large research microscopes. Introduced in 1985, it conforms to the modular system of current Leitz microscopes. The Diaplan is, as far as I know, the first microscope to have both field and aperture diaphragms in the base of the microscope.

For fluorescence, two epi-illuminators are available: a '3-wavelength PloemOpak' with three interchangeable filter blocks (see Fig. 4.2), and a 'single-wavelength PloemOpak' which accepts a single filter block which can be exchanged through a trapdoor in the side. Both epi-illuminators rest on the flat top of the microscope stand, the binocular or trinocular head being placed on top of the illuminator. Filter blocks are standard, and lamphouses are interchangeable with the Aristoplan.

The Laborlux S model is intended for routine laboratory use. Model D is shown in Figure 4.2. A PloemOpak epi-illuminator is fitted above the top of the stand, the binocular tube being shifted to the top of the illuminator unit. There are two models of illuminator. The single-wavelength version ('l-lambda PloemOpak', illustrated) accepts one filter block, which can be exchanged through an opening in the side of the housing. The filter block is exchanged for a transmitted-light element for other illumination methods, e.g. brightfield, if necessary. The A block (for UV excitation) has an almost colourless barrier filter and mirror, and for visual observation can be left in place for transmitted-light observation; in photomicrography with transmitted light, it may give a faint yellowish colour cast. Simultaneous bright-field, phase-contrast or differential interference-contrast illumination is possible.

The Biomed model is the simplest and least expensive fluorescence microscope in the Leitz range, and is intended for routine laboratory or educational use. It accepts the 1-wavelength and 3-wavelength PloemOpak epi-illuminators, and can be fitted for photomicrography and phase-contrast.

The Orthoplan microscope, introduced in 1964, is still current for quantitative instrumentation and those in use are likely to be around for some years. The Orthoplan was the major Leitz general-purpose microscope for two decades (1964–1986). It was particularly important for microfluorometry, for which purpose it is still (1991) supplied. This microscope stand pioneered the rectangular shape with a flat top and

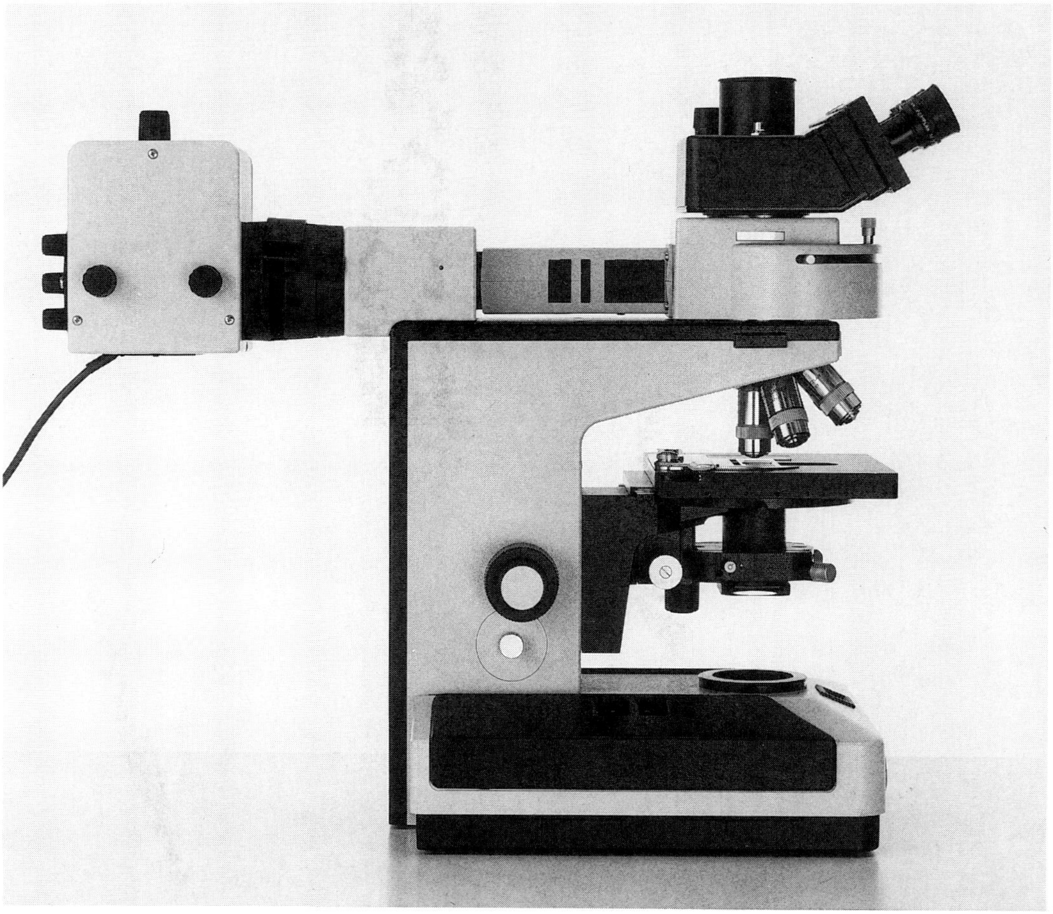

Fig. 4.1. The Leitz Diaplan microscope, with three-wavelength epi-illuminator. The lamphouse (LH103Z) is equipped with a 50 W mercury lamp. In this case a trinocular body tube is fitted, to accept a camera or photometer. The substage condenser has been replaced by a light trap. Photograph courtesy of Ernst Leitz Wetzlar GmbH.

rear to facilitate fitting of accessories. This pattern lends itself particularly to fluorescence microscopy, because of the ease with which epi-illumination can be achieved. In the Orthoplan, the flat rear of the stand is provided with two bayonets to accept interchangeable lamphouses; the lower one for transmitted light and the upper one for epi-illumination. The epi-illuminator, one of the PloemOpak series, slides into the nosepiece slot and has an integral nosepiece.

The illumination system is based on modules of lamphouses and mirror-houses. There are four basic lamphouses, designated 100, 100Z, 250, and 500. The Lamphouse 100 is meant for tungsten-halogen lamps only, the other three for gas-discharge lamps. The 50 and 100 W mercury arc lamps and the XBO75 are accommodated in the Lamphouse 100Z; an appropriate collector lens is required for each lamp type. A heat

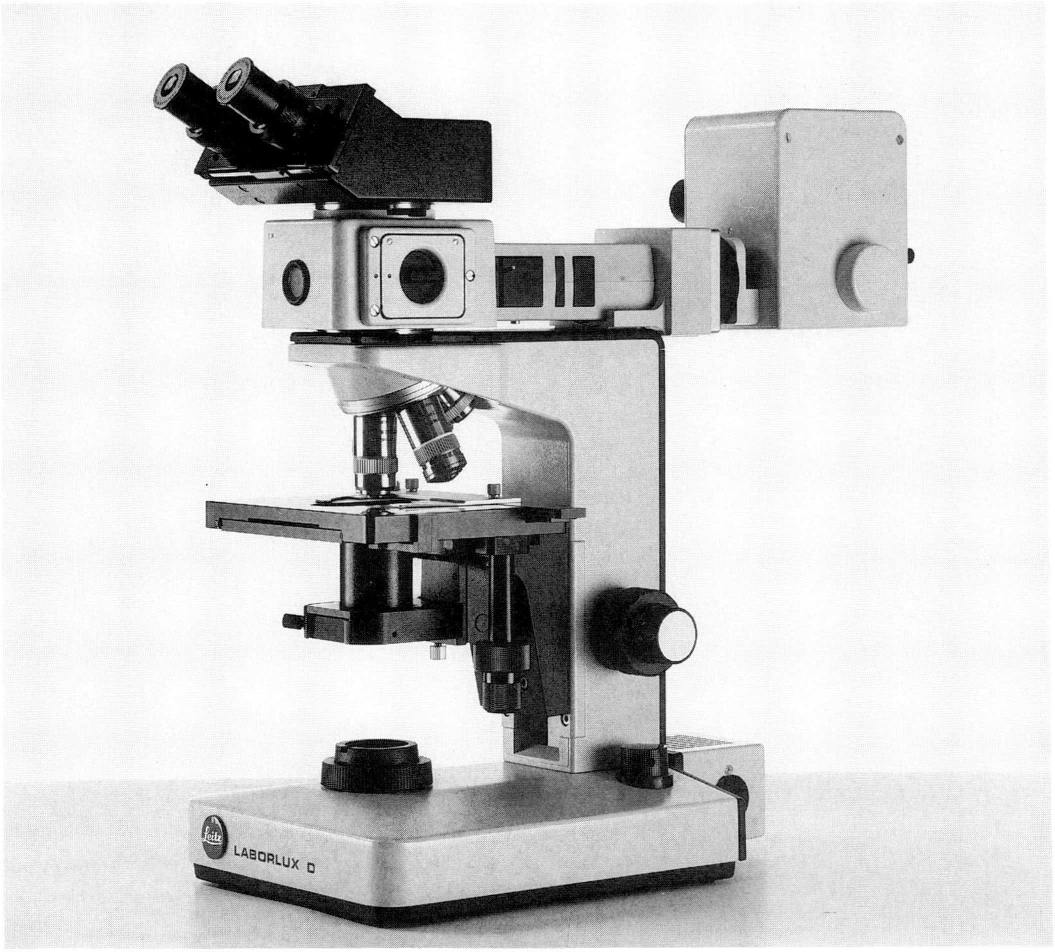

Fig. 4.2. Leitz Laborlux D microscope with single-wavelength epi-illuminator, lamphouse LH102Z and substage light-trap. The circular object at the front of the epi-illuminator is a centring screen to facilitate centring the lamp. Photograph courtesy of Ernst Leitz Wetzlar GmbH.

Table 4.1. *Leitz filter blocks for PloemOpak epi-illuminators*

The table below lists the technical data for the current range of Leitz filter blocks. Combinations can also be assembled by the user by purchasing the block containing only the dichromatic mirror, and separate excitation and barrier filters. In the table, all figures refer to the wavelength in nm. Bandpass filters are indicated by the letters BP and are characterized either by the centre wavelength and the half-bandwidth (shown as BP 525/25, for example) or by the short and long wavelength half-transmission points (e.g. BP 450–490). For the longpass filters (LP), the figures refer to the edge wavelength.

Filter block	Excitation range	Excitation filter	Dichromatic mirror	Barrier filter
A	UV	BP340–380	RKP400	LP430
A2	UV	BP270–380	RKP380	BP410–580
B2	UV + violet	BP340–410	RKP455	LP470
D	UV + violet	BP355–425	RKP455	LP460
E3	Blue	BP436/7	RKP475	LP490
G	UV + violet + blue	BP350–460	RKP510	LP515
H3	Violet + blue	BP420–490	RKP510	LP515
I3	Blue	BP450–490	RKP510	LP515
K3	Blue	BP470–490	RKP510	LP515
L3	Blue	BP450–490	RKP510	BP525/20
M2	Green	BP546/14	RKP580	LP580
N2	Green	BP530–560	RKP580	LP580
N2.1	Green	BP515–560	RKP580	LP580

Notes:

A2: bandpass barrier filter to suppress red fluorescence.

I3: standard set for FITC.

K3: narrow-band blue excitation, more specific for FITC than I3, slightly weaker excitation.

L3: designed for specific examination and microfluorometry of FITC, but may give false positives (see Chapter 7).

M2: very specific 546 nm excitation, suitable for microfluorometry.

There is no standard block for specific violet excitation with the Hg 406 line. Presumably the required result can be achieved using blocks B2 or D, together with a BP410 or equivalent (e.g. a Schott glass) added to the excitation filter to remove the unwanted UV. There is a separate block for transmitted light, to be inserted instead of the filter blocks. Blocks containing only the mirror (TK 455, TK 510, or TK 580) are available for use with separate or the user's own excitation and barrier filters. A separate range of filters in a rotating changer is available for transmitted-light fluorescence.

filter is installed permanently, and a slot is provided for excitation filters. The Lamphouse 250 accommodates the 200 W mercury and 150 W xenon arcs, and has provision for four sets of interchangeable excitation filters in addition to a fixed heat filter. Filters are interchanged by a system of levers, so that the operator does not need to touch the hot filters. The Lamphouse 500, which is always combined with the Mirrorhouse 500, is very large and attaches to both rear bayonets of the Orthoplan.

Mirror-houses enable two or more lamphouses to be attached to the stand simultaneously, and/or to enable the same lamp to be used alternately for epi- and dia-illumination. The Mirrorhouse 500 accommodates up to three or four lamphouses (according to the model), and, by a system of movable mirrors controlled by three levers, either of two lamps can be used for epi-illumination while the other can be used simultaneously for dia-illumination. By combining the Mirrorhouse 500 with one or more of the other mirror-houses, an almost unlimited number of lamphouses can be attached simultaneously.

The PloemOpak 2.1 is the fluorescence epi-illuminator for the Orthoplan. This replaces the normal nosepiece; it incorporates a nosepiece, a revolving turret to accept filter systems, and field and aperture diaphragms and associated optics. The filters are accommodated in interchangeable blocks, each of which contains the appropriate excitation filter, dichromatic mirror, and barrier filter for one type of illumination. The characteristics of current blocks are listed in Table 4.1. Four blocks can be accommodated in the turret simultaneously. One of the blocks can be the blank block for use with normal transmitted light without fluorescence. In operation, blocks are interchanged by turning a knob; this changes the excitation and barrier filters and the dichromatic mirror simultaneously, so that quick changes can be made from one excitation wavelength to another. Separate excitation filters can be inserted into the lamphouse if required, and an additional barrier filter in a slide inserted into the stand above the nosepiece.

The presence of the PloemOpak does not disturb the capability of the microscope stand to accept other accessories, such as a camera (see Chapter 6), photometer (see Chapter 8), and heating/cooling stage (useful for kinetic studies of enzymes, see Volume II, Chapter 15). Simultaneous phase-contrast microscopy requires only the use of appropriate objectives and condenser. For simultaneous differential interference-contrast (DIC), the requirements are as follows. The polarizer and first Wollaston prism (see Chapter 5) are held in a DIC condenser unit. The second Wollaston prisms, one for each objective, are attached between each objective and the nosepiece. Objectives for which DIC is not required are held on extension tubes to maintain parfocality. The analyser is attached to the top of the PloemOpak unit.

The Fluovert inverted microscope (Figs. 4.3, 4.4) is specially designed for fluorescence with epi-illumination. At first sight, the instrument resembles an ordinary, upright microscope except that the nosepiece is underneath the stage. The body tube is on top of the stand, exactly as in non-inverted Leitz microscopes. This is made possible by the Reinheimer principle (see Beck, 1984), whereby the illuminating and image-forming rays cross at least once, thus allowing both dia-illumination and epi-illumination on an inverted microscope with an upright stand form. Standard Leitz interchangeable body tubes can be fitted, and accept normal attachments such as a photomicrographic camera or photometer.

The fluorescence epi-illuminator of the Fluovert is built into the base in the form of a drawer which can be opened to exchange filter blocks (Fig. 4.4). Filter blocks held in the illuminator are selected by means of a knob. Simultaneous bright-field, phase-contrast or DIC illumination is possible.

Optical components

Leitz filters for fluorescence microscopy are normally fitted into standard interchangeable blocks, each containing a matched set of excitation filter, dichromatic mirror and barrier filter. The user can assemble special blocks, by purchasing a block containing only the dichromatic mirror, and then adding matched excitation and barrier filters appropriate to the special application. The standard blocks are listed in Table 4.1 (above).

In recent years, Leitz has been changing its optical system, so that the objectives and oculars conform to the internationally accepted DIN standards for tubelength and adjustment lengths of objectives and eyepieces (see Fig. 3.7). The actual changes which took place recently were two: reduction of the tubelength to 160 mm, and reduction of the adjustment length of the eyepieces (i.e. distance of the intermediate image from the

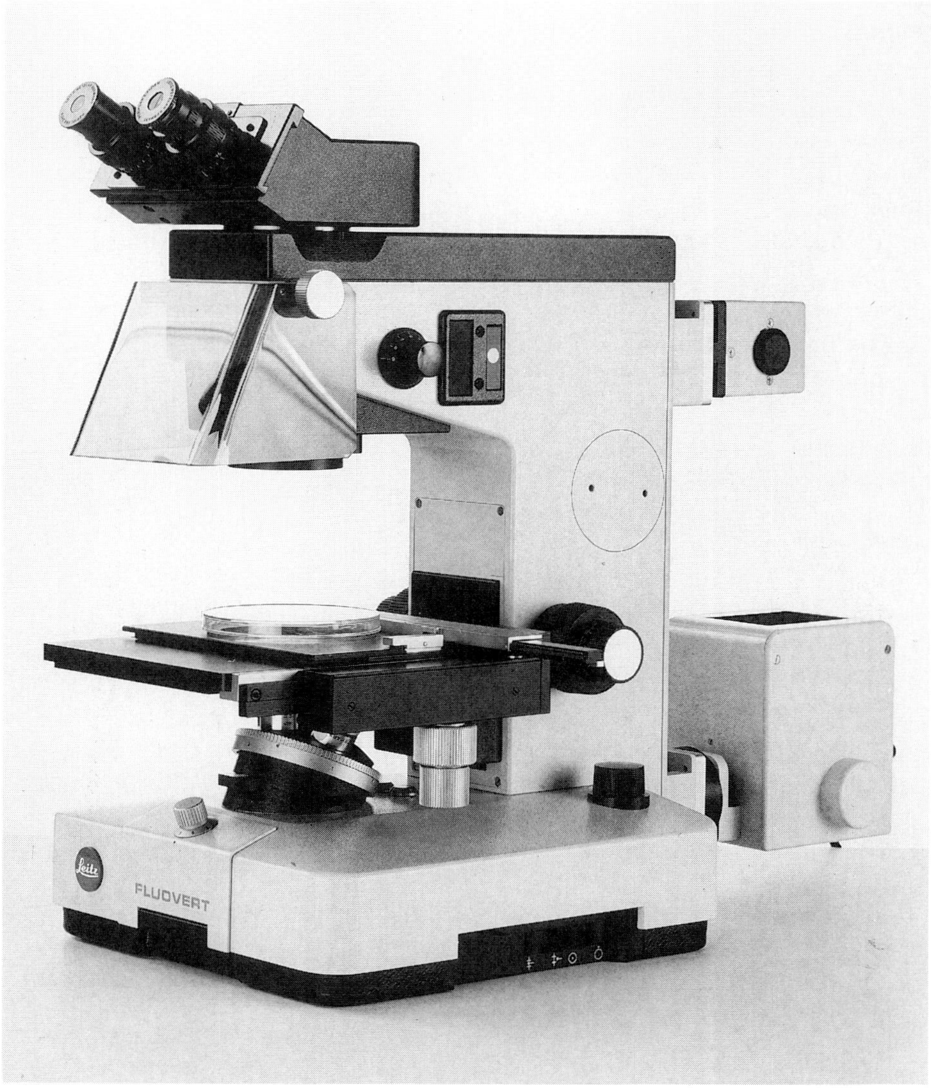

Fig. 4.3. The Leitz Fluovert inverted fluorescence microscope. Photograph courtesy of Ernst Leitz Wetzlar GmbH.

top of the body tube, see Fig. 3.7) from 18 mm to 10 mm. This has necessitated new microscope stands with 160 mm tubelength, and also a new series of objectives and oculars. At the time of writing (1990) the changeover is complete except for the more complex photometric equipment which is still based on the Orthoplan stand; this equipment will soon be available for the new microscopes.

A wide range of objectives and eyepieces is available both for the standard 160 mm tubelength and for the 170 mm tubelength of the Orthoplan. The change to standard tubelength enables objectives and oculars of other makes to be used where appropriate, thereby widening the range of available optics. The possibilities of interchanging Leitz equipment of 160 mm and 170 mm tubelength are summarized in Table 4.2. To use the old-style eyepieces on the new microscopes, without loss of parfocality, an 8 mm collar should be inserted below the eyepiece flange to raise the eyepiece so that the intermediate image falls at the correct level (this is the difference between the old 18 mm

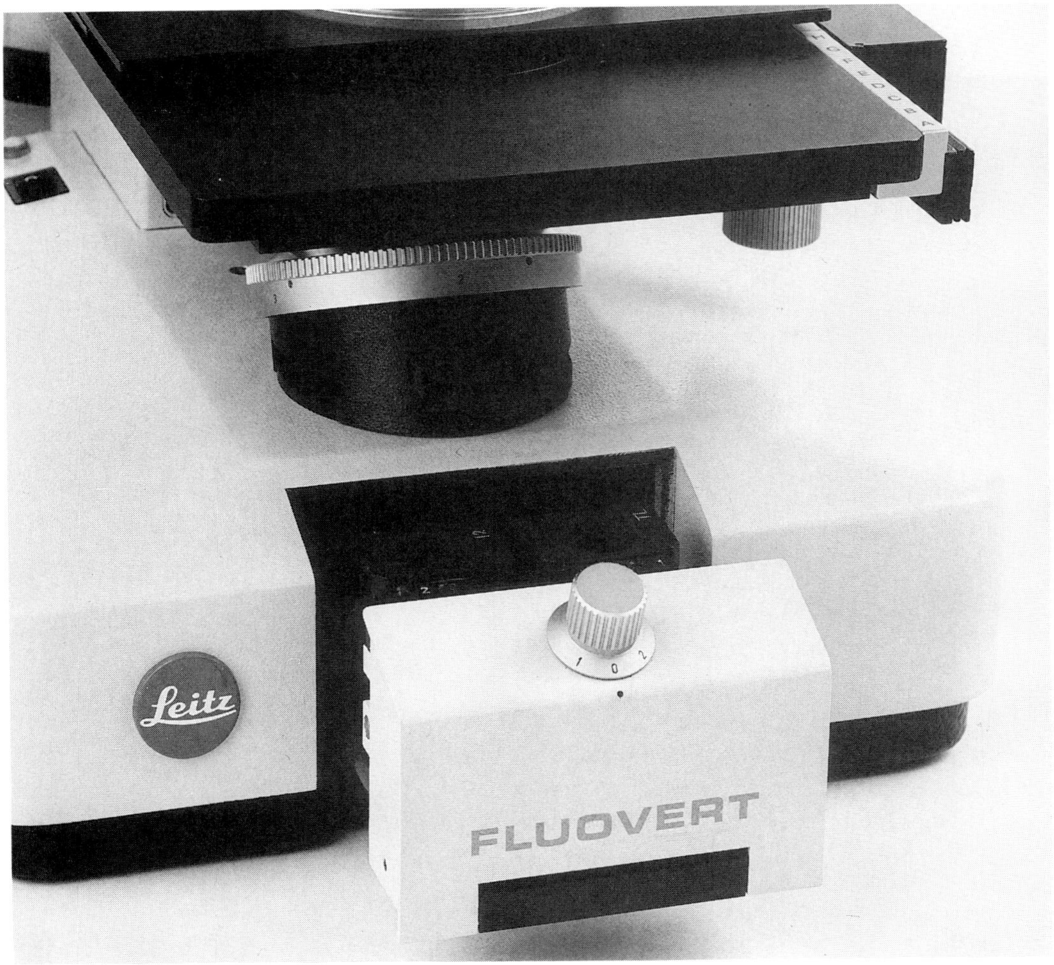

Fig. 4.4. The epi-illuminator drawer of the Leitz Fluovert microscope. Photograph courtesy of Ernst Leitz Wetzlar GmbH.

Table 4.2. *Interchangeability of Leitz oculars and objectives of 160 mm and 170 mm designed tubelength, on microscopes with tubelengths of 160 or 170 mm*

Designed tubelength	On microscope with 160 mm tubelength	On microscope with 170 mm tubelength
Eyepieces		
160 mm	Designed use	Not recommended: decreased working distance, loss of parfocality
170 mm	Use with 8 mm collar	Designed use
Objectives		
160 mm, ×16 or more	Designed use	Slight loss in parfocality, no other problem
160 mm less than ×16	Designed use	Not recommended: decreased working distance, loss of parfocality[a]
170 mm ×16 or more	Slight error in parfocality, no other problem	Designed use
170 mm less than ×16	Usable, parfocality lost	Designed use

Notes:
[a] When changing from low-power objectives to high-power or other objectives with small working distance, the latter objective or the specimen may be damaged.

adjustment length and the new 10 mm length). Leitz agents should be able to supply an 8 mm collar (code no. 519 653); otherwise, one can be made easily enough, even of cardboard or 'Dymo' tape. As regards objectives, the combined effect of the changes from old to new standards resulted in changing the intermediate image distance from 152 mm to 150 mm. As this is only a matter of 2 mm, the only significant effect of using the 'wrong' objective type is a loss of parfocality of low-power objectives; objectives with a magnification of 16 or more are not significantly affected. The use of the new eyepieces and low-power objectives on the old microscope stands is not recommended because the objectives have to be focussed closer to the object; in the case of low-power objectives the re-focussing required could, when changing back to higher power, result in damage to the objectives or specimen.

Objectives available include special high-aperture oil-immersion and water-immersion objectives with a high numerical aperture and low autofluorescence, specially designed for fluorescence microscopy (Table 4.3). Characteristics of some Leitz objectives are listed in Table 3.4. Eyepieces include low-power ×6·3 and ×4 (the latter for the Orthoplan only) to maximize fluorescence intensity.

The Orthomat E photomicrographic camera is the obvious camera for photomicrography of fluorescence with a Leitz microscope, although some other camera systems are available to fit, including the Wild system described below. The Orthomat E generally sits on top of an FSA trinocular tube, which has a built-in reflecting system which

Table 4.3. *Leitz objectives particularly recommended for fluorescence microscopy*

Type	Magnification/NA	Type
Oil immersion		
NPL Fluotar	10/0·45 Oil	Fluorescence
NPL Fluotar	25/0·75 Oil	Fluorescence
NPL Fluotar	40/1·30 Oil	Fluorescence
	63/1·30 Oil	Fluorescence
Water immersion		
	25/0·60 W	Fluorescence
	50/1·00 W	Fluorescence
	100/1·20 W	Fluorescence
Dry		
EF	40/0·75	Fluorescence
Phase-contrast		
NPL Fluotar	25/0·75 Oil	Phaco 2 fluorescence
NPL Fluotar	40/1·30 Oil	Phaco 3 fluorescence
	40/0·75	Phaco 2 fluorescence
	63/1·30 Oil	Phaco 3 fluorescence

makes possible simultaneous observation of images of the specimen and the photographic field graticule. A low-power projection eyepiece is required, e.g. the Periplan GW 6·3× 'red dot' type. On the Aristoplan the Orthomat E operates as an integrated module. The exposure can be determined either manually (range 0·01 s to 99 min) or automatically. The photometer system measures from either the full frame or from a spot whose location can be moved within the field. This facility allows the photometric spot to be set to cover the significant part of the object after the picture has been composed. Both the location of the measuring spot and the frame graticule are bright rather than dark, and are superimposed on the microscope image in both eyepieces. This is achieved by the reflection of the graticule and movable spot by a prism in the binocular phototube. Rotation of the Orthomat E for optimal framing, easily achieved when mounted on a trinocular tube, produces automatic corresponding rotation of the graticule image. Similarly, if a zoom eyepiece (×5 to ×12·5) is used, the graticule varies in size according to the magnification setting. The markings, being bright rather than dark, can be clearly seen against a dark field. The image of the graticule can be switched off when not required, avoiding distraction when the microscope is used for visual observation. Automatic compensation for reciprocity failure is achieved with six settings according to film type. Unique, as far as I know, to the Orthomat is provision for compensation of the exposure reading for the proportion of bright field, which can be set in the range 20% to 100% in 20% steps for light and dark background (see Chapter 6 for explanation of this feature). Data backs for 35 mm and Polaroid formats are available.

The Wild MPS46/52 automatic camera system can be attached to any Leitz microscope. It has provision for spot or centre-weighted metering, using a beamsplitter which sends 90% of the light to the sensor and 10% to the focussing telescope; when the exposure key is touched, the beamsplitter is swung out of the way, allowing 100% of the light to the camera during exposure. Keys on the control unit are illuminated.

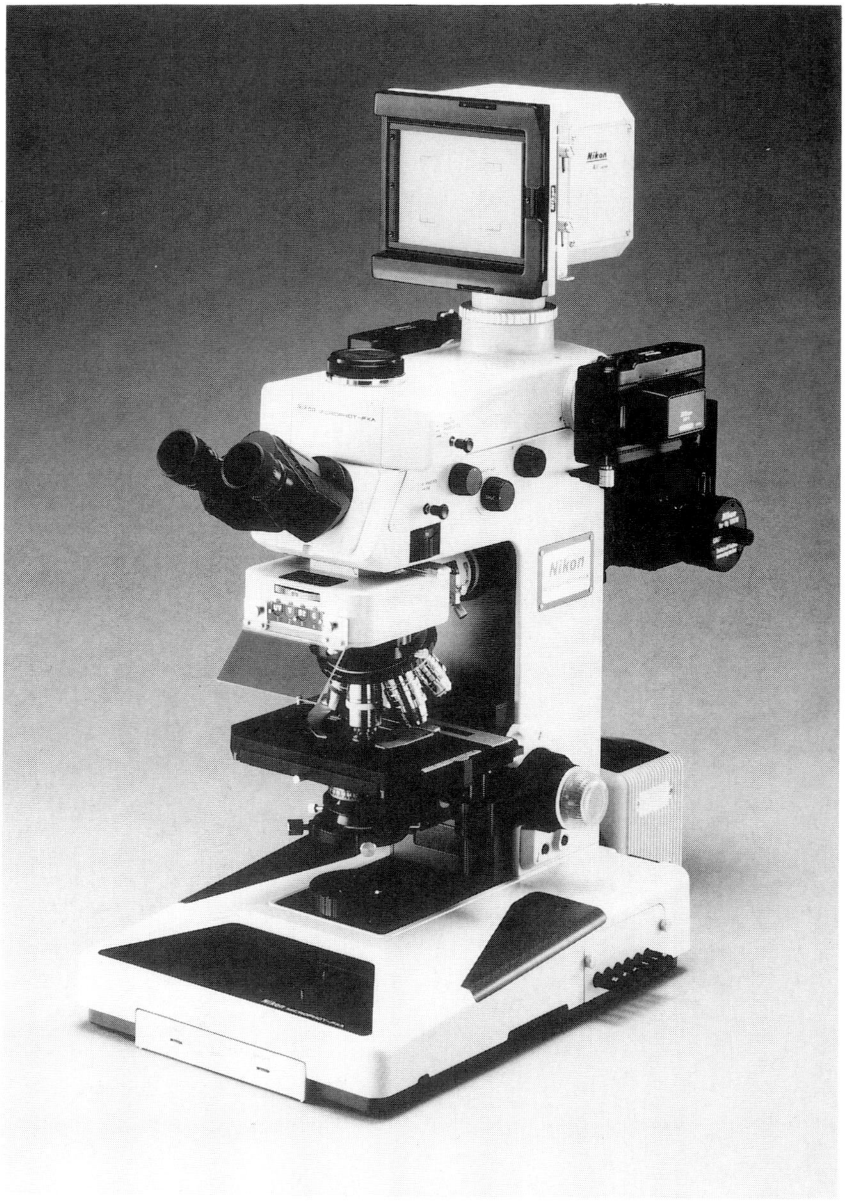

Fig. 4.5. Nikon Microphot-FX fluorescence microscope. Photograph courtesy of FSE Pty Ltd, Sydney.

NIKON

Nikon (pronounced *Nick-on*, not *Nye-kon*) microscopes and cameras are made in Japan by Nippon Kogaku K.K. Nikon fluorescence microscopes have standard equipment for both epi- and dia-illumination, phase and interference contrast, photomicrography, quantification and television microscopy. Both conventional and inverted microscopes are available.

The Microphot-FXA and the Microphot (Figs. 4.5, 4.6) are the top models, the former having advanced equipment for photomicrography. The fluorescence epi-illuminator uses a 100 W mercury lamp, and carries up to four interchangeable filter blocks in a rotating turret. This attachment features improved transmission in the ultraviolet (UV) (340 nm). Phase-contrast or differential interference contrast (DIC) can be used in conjunction with the epi-illuminator. For dia-illumination, a 200 W mercury lamp is used; a mirror-house accepts also the 100 W tungsten-halogen lamp for general brightfield microscopy.

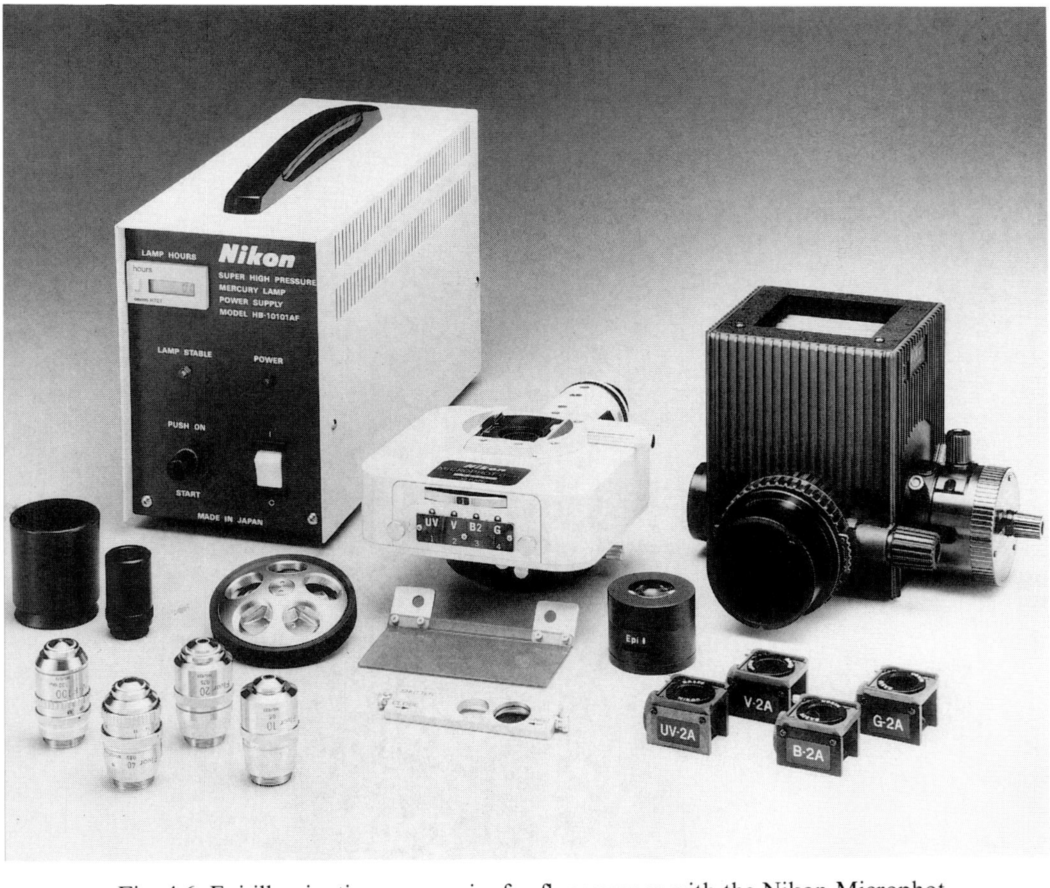

Fig. 4.6. Epi-illumination accessories for fluorescence with the Nikon Microphot fluorescence microscope. Photograph courtesy of FSE Pty Ltd, Sydney.

For photomicrography, the Microphot-FXA functions as a single-lens reflex camera with an instant return mirror. It accepts three camera bodies, normally two 35 mm cameras and one large-format. A third 35 mm camera can be substituted for the large-format camera. The two main 35 mm cameras are DX-compatible, i.e. they automatically sense the ISO speed rating when loaded with appropriate cassettes. A data back is attachable to each of the 35 mm cameras, to record information on the film. This information can include the measured intensity of the fluorescence of the specimen. Exposure is measured either from the centre 30% of the frame or from a 1% spot which is movable. During exposure, 100% of the light of the image reaches the film plane. The eyepiece graticules are illuminated. A separate port with standard ENG mount is provided for a video camera or photometer unit.

The Microphot-FXA has a considerable potential for automation. An autofocus attachment provides automatic focussing with low-power objectives (×1 to ×20). A motorized nosepiece and motorized condenser provide quick changes of magnification. A remote switching unit provides centralized control of automatic options, and there is provision for linking to a computer for fully automatic control.

The Optiphot and the Labophot (Figs. 4.7, 4.8) are general-purpose laboratory microscopes. An epi-illuminator (Fig. 4.9), attached above the stand, holds two filter blocks, and uses a 100 W mercury lamp. Simultaneous phase-contrast is possible.

The Diaphot TMD inverted microscope (Figs. 4.10, 4.11) has an epi-illumination attachment for fluorescence. The epi-illuminator has slide-in filter blocks, and a lamphouse for a 100 W mercury lamp. Each slide-in filter block contains excitation and barrier filters and dichromatic mirror. As in the Microphot series, the illuminator features improved transmission in the UV (340 nm). The oculars project forwards and upwards from the front of the microscope, somewhat above stage level. The front of the microscope also has a standard Nikon camera bayonet for photomicrography, e.g. with an F3 camera body.

Accessories

Epi-illumination filter blocks, similar to those of the Leitz PloemOpak, each contain an excitation filter, a dichromatic mirror, and a barrier filter. Each of the three elements can be interchanged. Currently (1990) 11 types of excitation filter, 5 dichromatic filters, and 13 barrier filters are available. Standard combinations are listed in Table 4.4.

Objectives of several kinds are available. *CF-Fluor* and *CF UV-F* objectives, specially developed for fluorescence microscopy, are designed for high UV transmission, high numerical aperture (NA), and low autofluorescence. *CF Fluor-DL* objectives are phase-contrast objectives with high NA for use in fluorescence epi-illumination systems, giving high-resolution phase-contrast as well as bright fluorescence. *CF Plan Apochromat* objectives have high NA, and are suitable for V, BV, B and G excitation filter sets. *CF Plan Achromat* objectives have lower NA but give adequate performance with B and G excitation systems.

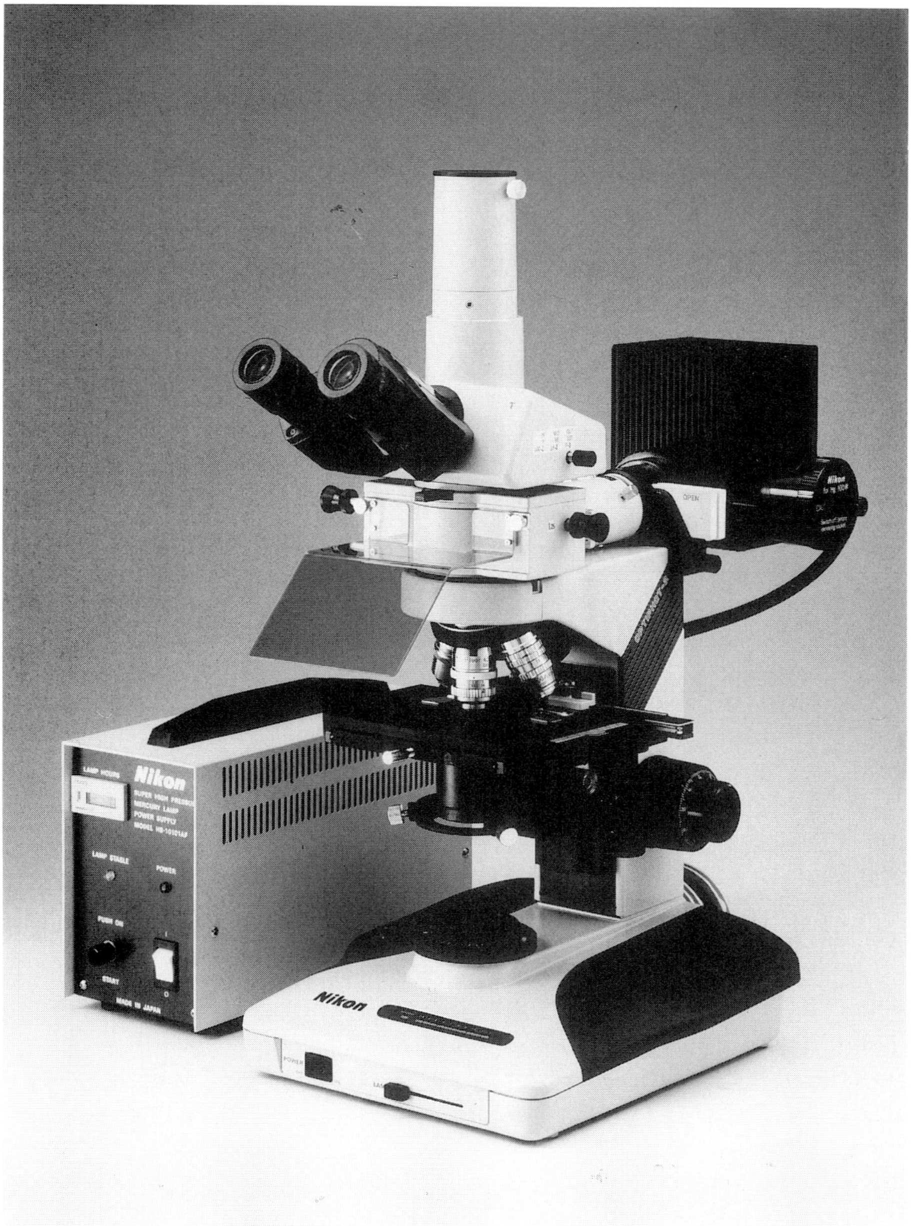

Fig. 4.7. Nikon Optiphot microscope with fluorescence epi-illuminator. Photograph courtesy of FSE Pty Ltd, Sydney.

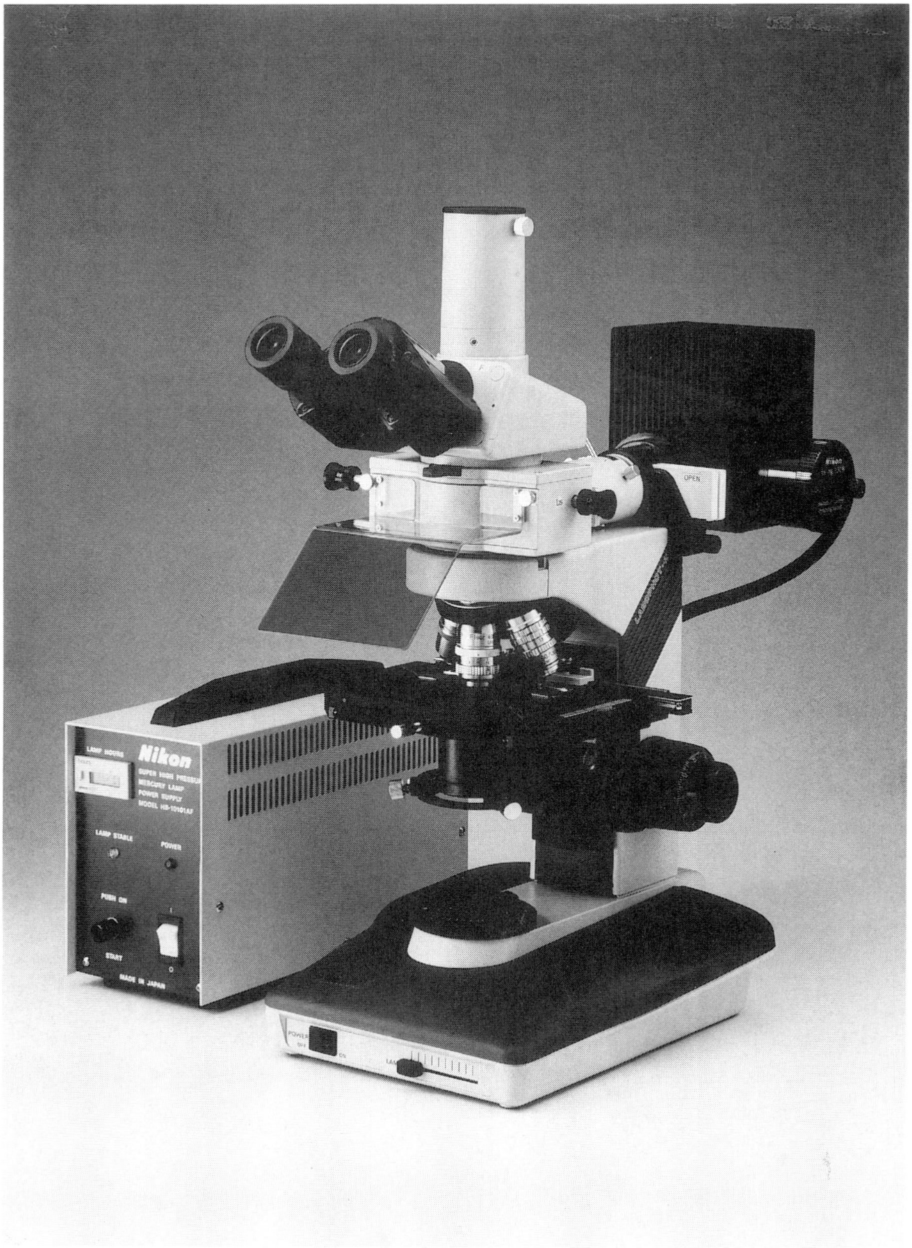

Fig. 4.8. Nikon Labophot microscope with fluorescence epi-illuminator. Photograph courtesy of FSE Pty Ltd, Sydney.

OLYMPUS

Olympus microscopes are made in Japan by the Olympus Optical Company Ltd, Tokyo. There are two main series of upright laboratory microscopes in their current range: the new *Vanox AH2* (comprising AHBS and AHBT) and the *BH-2* series (comprising BHS, BHT and BHTU models). There was an older Vanox microscope, very different from the new model. In addition their inverted microscope IMT-2 can be used for fluorescence with epi-illumination.

The BH-2 is illustrated in Figs. 4.12 and 4.13. An epi-illumination system for fluorescence mounts between the arm and the body tube, and is additionally supported by a clamp on the arm. At the free end of the illuminator, the lamphouse attaches; this normally holds a 100 W mercury lamp, but an alternative housing for a 50 W tungsten-halogen lamp can also be used. The arc lamp is mounted directly on the inside of the lamphouse door, in a simple and elegant fashion between metal strips; provision is made for centring. There is a focussable collector lens but no reflector. The horizontal arm of the illuminator is equipped with aperture and field diaphragms, a shutter to block off the light, and holders for excitation filters. The vertical, body tube part of the

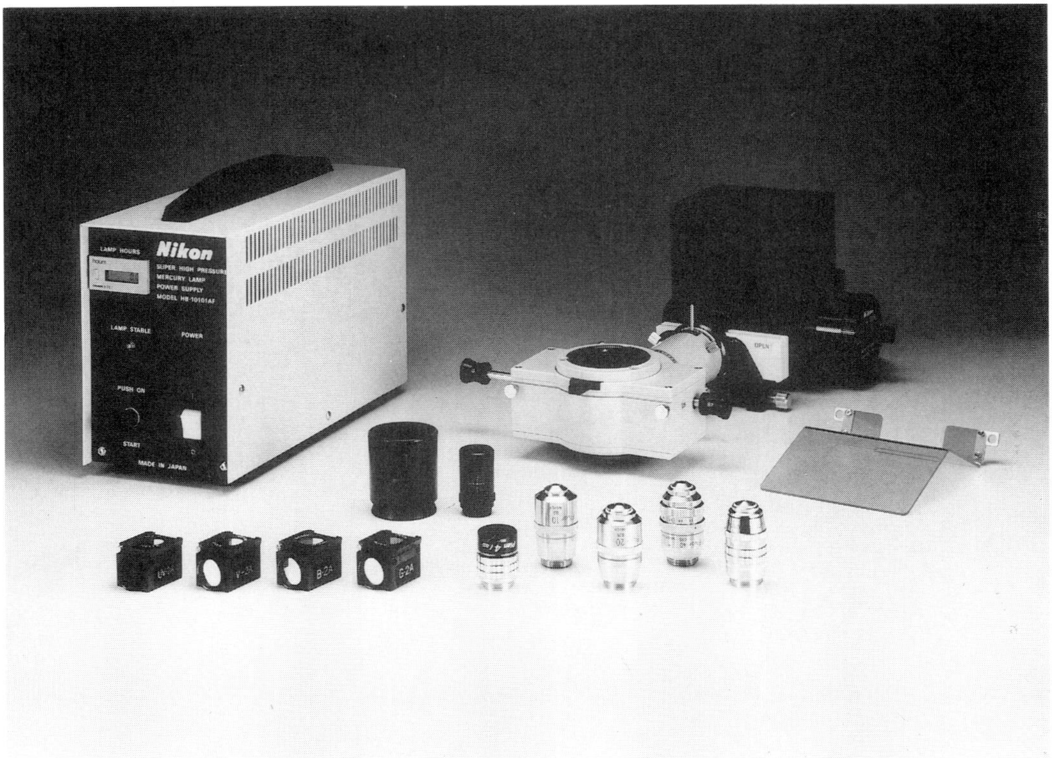

Fig. 4.9. Nikon fluorescence epi-illuminator attachment and accessories for Optiphot and Labophot microscopes. Photograph courtesy of FSE Pty Ltd, Sydney.

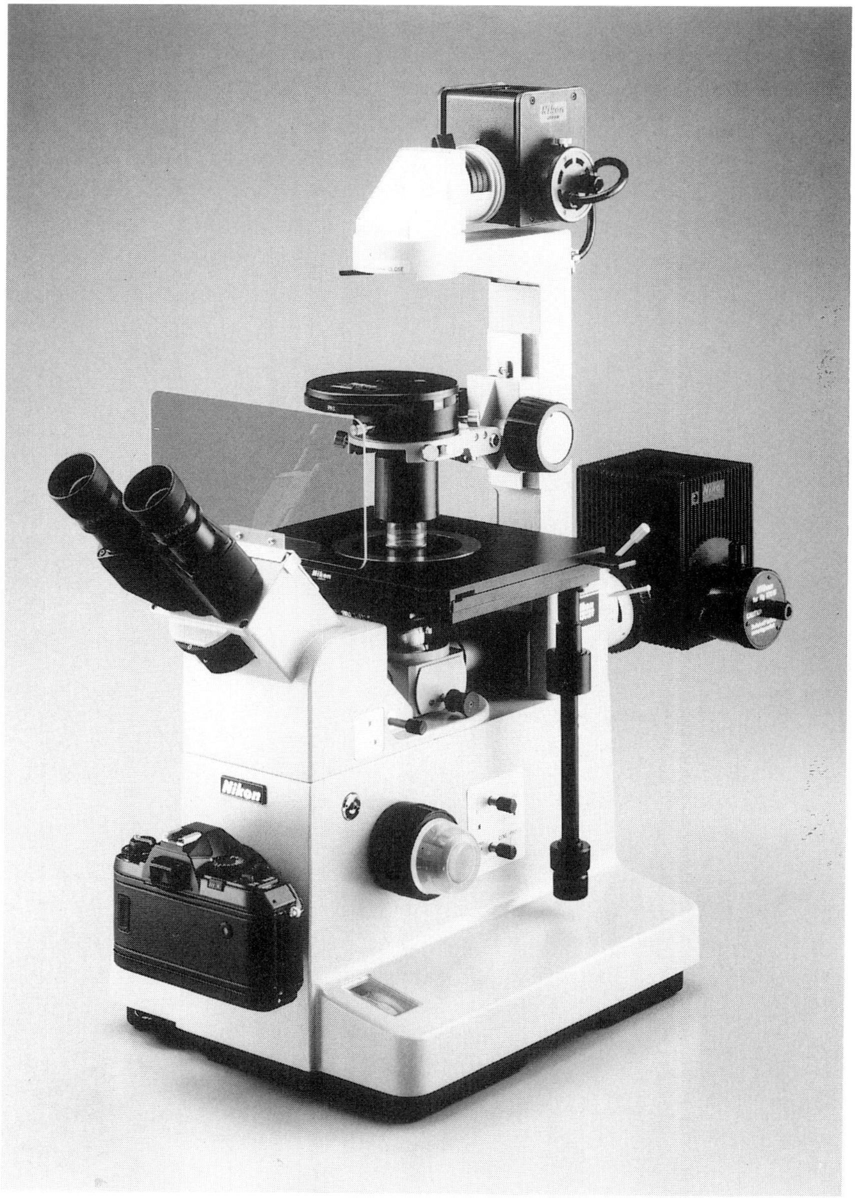

Fig. 4.10. Nikon Diaphot TMD inverted microscope equipped for fluorescence. Photograph courtesy of FSE Pty Ltd, Sydney.

illuminator carries holders for dichromatic mirrors, a slot for the barrier filter, and an attachment for a UV-protective shade.

Provision is made for excitation in the UV, violet (V), blue (B), and green (G). Excitation filters and dichromatic mirrors are each grouped in pairs (UV–V, V–B, B–G). Changing between different wavelengths within standard pairs (e.g. UV–V) is not too difficult, involving only sliding the excitation filter and dichromatic mirror slides from one position to another and exchanging the barrier filter; changes beyond that involve unscrewing and re-attaching the retaining thumb screws for the excitation filter slides to change slides, and changing the dichromatic mirror housings. There is also an individual filter set for FITC excitation only. A range of eight separate barrier filters is provided. Their code numbers (e.g. L-410, Y-455, O-515, R-610) indicate the visual appearance (L = colourless; Y, O, R = yellow, orange, red, respectively) and the wavelength at which the filter has 50% transmission.

A feature of the Olympus system is a set of special objectives, with non-fluorescent glass and high apertures (see Table 4.5). The eyepiece magnifications available are $\times 5$, $\times 8$, $\times 10$, $\times 15$ and $\times 20$. The $\times 5$ eyepiece (NK5$\times$) has a field only slightly bigger than that of the widefield $\times 8$ and $\times 10$ eyepieces, but gives a much brighter image.

Equipment is also available for fluorescence microscopy with dia-illumination, using a darkground condenser, a separate set of excitation filters, and an attachment for the

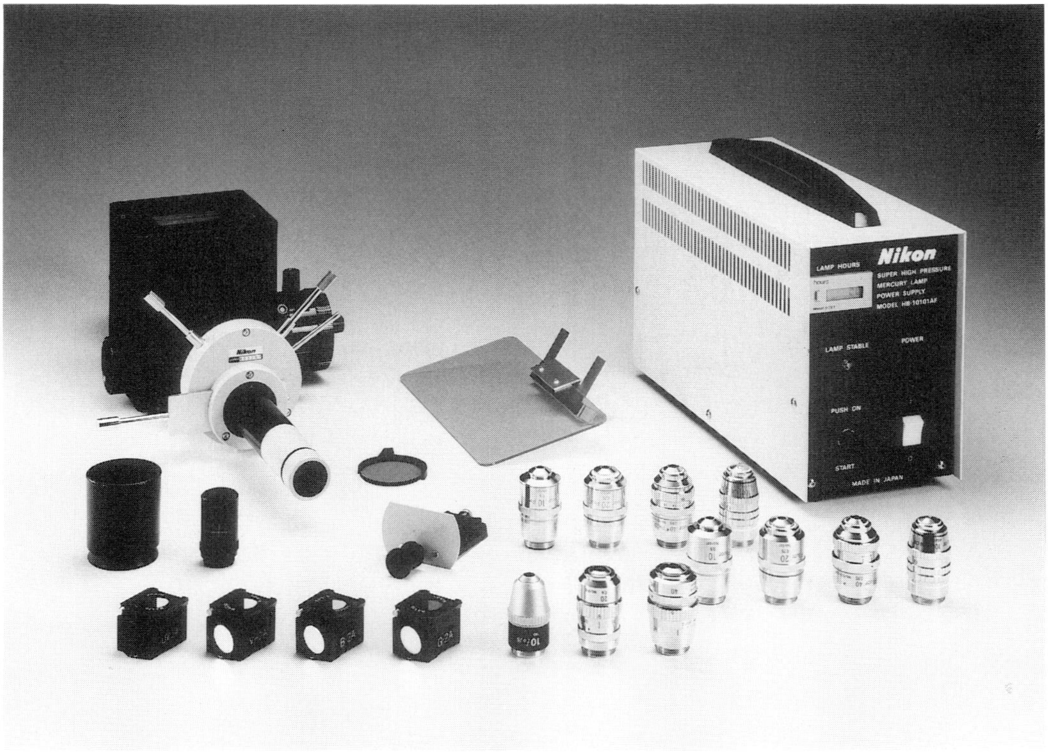

Fig. 4.11. Epi-illumination accessories for fluorescence with Nikon Diaphot TMD inverted microscope. Photograph courtesy of FSE Pty Ltd, Sydney.

Table 4.4. *Nikon filter sets*

Excitation	Name	Excitation filter	Dichromatic mirror	Barrier filter
UV	UV-1A	365/10	400	400
	UV-2A	330–380	400	420
	UV-2B	330–380	400	435
Violet	V-1A	405/10	430	435
	V-2A	380–425	430	450
	V-2B	380–425	430	460
Blue-Violet	BV-1A	435/10	455	460
	BV-2A	400–440	455	470
	BV-2B	400–440	455	480
Blue	B-1A	470–490	510	520
	B-1E	470–490	510	520–560
	B-1H	470–490	510	515IF
	B-2A	450–490	510	520
	B-2E	450–490	510	520–560
	B-2H	450–490	510	515IF
	B-3A	420–490	510	520
Green	G-1A	546/10	580	580
	G-1B	546/10	580	590
	G-2A	510–560	580	590
	G-2B	510–560	580	610

Notes:
The filter sets supplied as standard are UV-2A, V-2A, BV-2A, B-2A, B-3A and G-2A.

barrier filter in a kind of intermediate body tube which fits between the microscope arm and the body tube proper.

The epi-illumination system (Fig. 4.12) with the special objectives gives bright and reasonably uniform illumination of the field, and a clear fluorescent image with only minimal instrumental background. Changing filters can be tedious, but for routine work at one or two excitation wavelengths the equipment is very satisfactory.

The trinocular phototube has the ideal provision for diverting light to the eyepieces only, to the camera only, or to both partly. Only the Olympus photomicrographic cameras can be fitted, unless a special adapter can be made.

The Vanox microscope (Fig. 4.14) has a larger stand, with built-in provision for epi-illumination as well as transmitted light. For fluorescence using epi-illumination, the arrangements are similar to those for the BH-2 described above. For dia-illumination, a mirror-house is attached at the rear of the stand; this supports the lamphouse,

Table 4.5. *Olympus objectives for fluorescence microscopy*

Tubelength: 160 mm. Coverglass thickness: 0·17 mm. Manufacturer's data.

Objective type	Phase	Magnification	NA	Immersion	Exit pupil ⌀, mm
D-Planapo 10× UV	a	10	0.40	dry	12·5
D-Planapo 20× UV	a	20	0.70	dry	10·9
D-Planapo 40× UV/R		40	0.85	dry	6·9
D-Planapo 20× UV/O		20	0.80	oil	12·4
D-Apo 40× UV/R10	a	40	1.30	oil	11·3
D-Apo 100× UV/R10	a	100	1.30	oil	4·9

Notes:
a Objectives indicated are also available with phase annuli for simultaneous epi-fluorescence and dia-illumination phase-contrast.

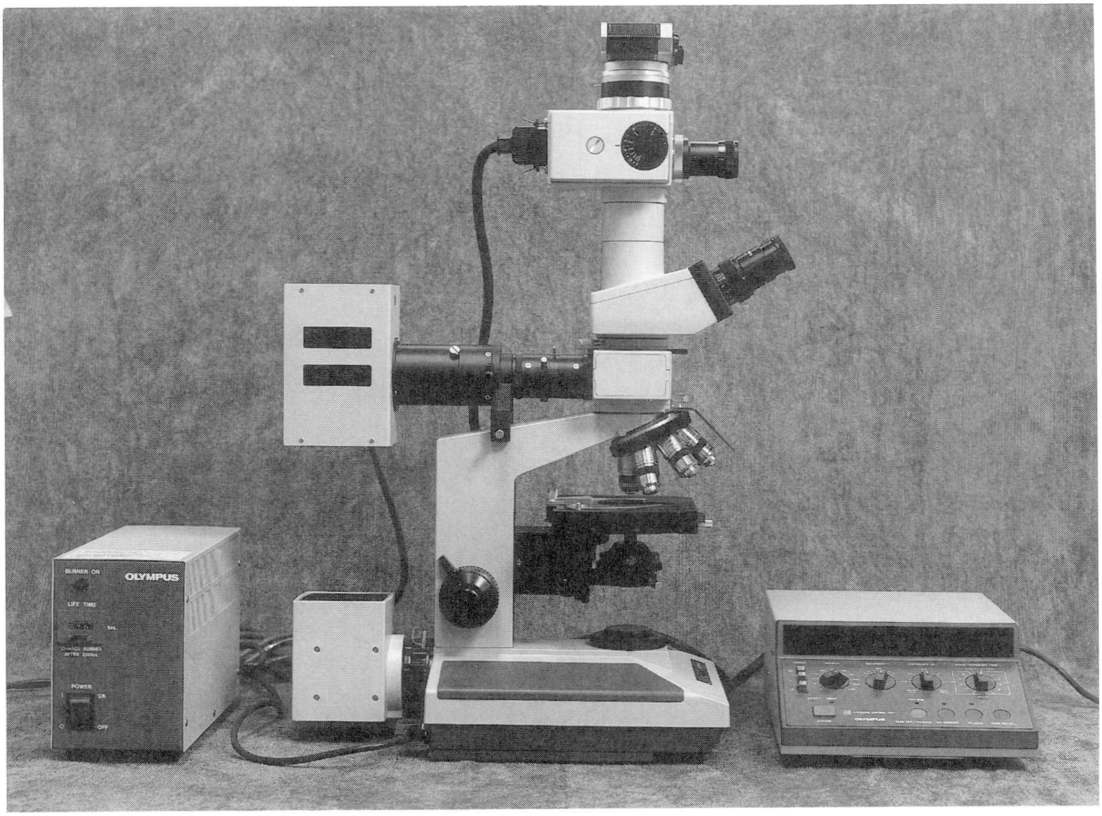

Fig. 4.12. Olympus BHS microscope with BH2-RFC fluorescence epi-illuminator and PM10-ADS photomicrographic camera. Left to right: power supply for arc lamp, microscope, control unit for camera.

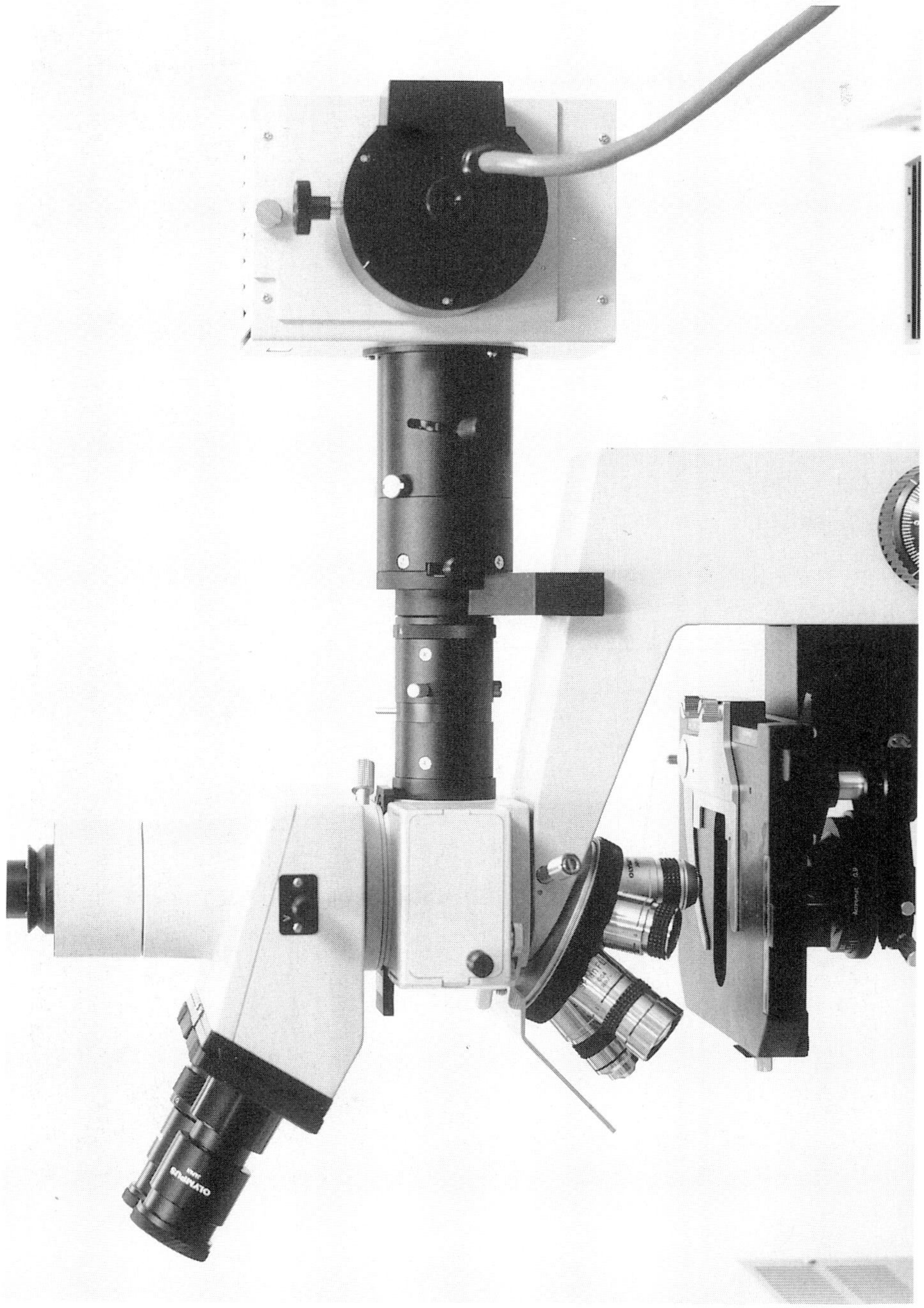

Fig. 4.13. Olympus model BH2-RFC fluorescence epi-illumination attachment for BH series microscopes.

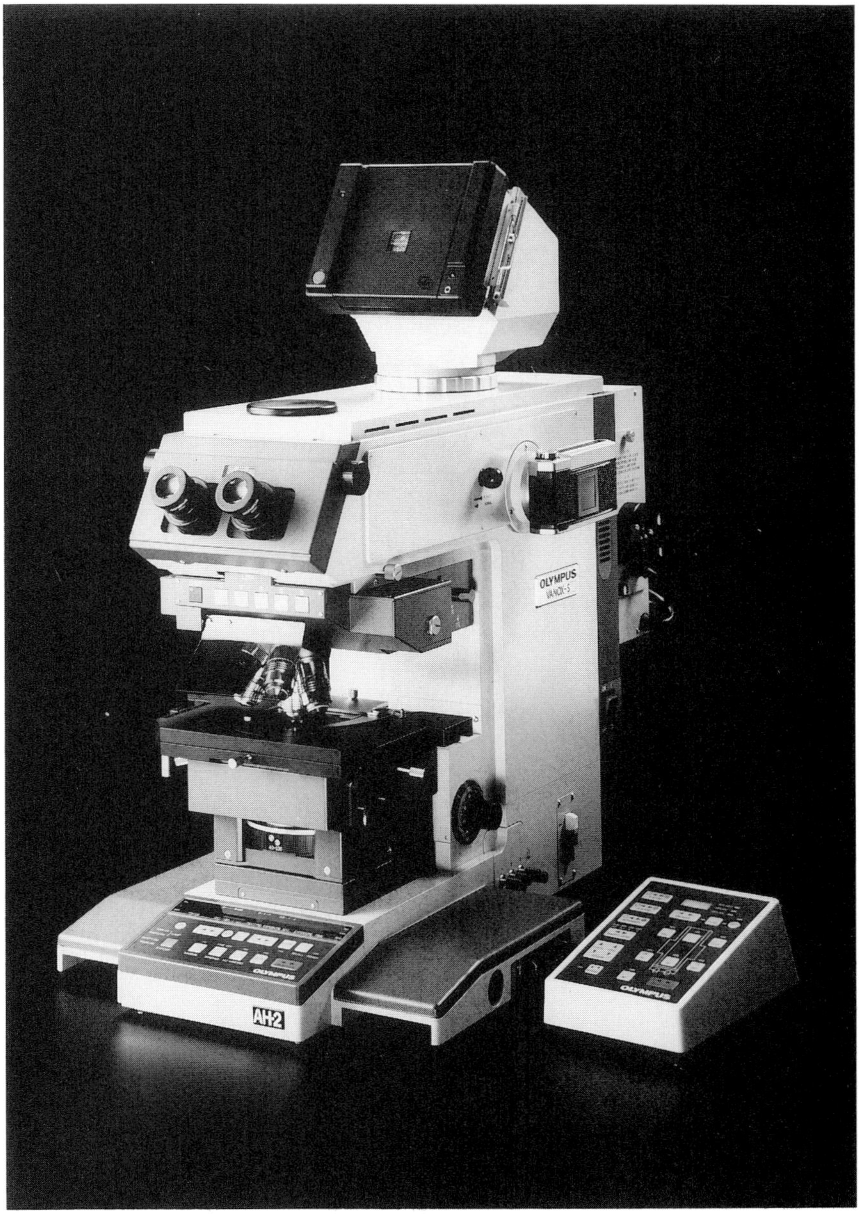

Fig. 4.14. Olympus Vanox AHBS fluorescence microscope. This model has automatic focussing of the specimen for low magnifications in brightfield transmitted light, and provision for three cameras (two 35 mm and one large-format). For routine dia-illumination, there is fully automatic adjustment of the substage light source when the objective is changed. Photograph courtesy of Selby-Anax Pty Ltd, Sydney.

provides for either epi- or dia-illumination, and also houses the excitation filters for dia-illumination. Barrier filters are held in a turret interposed between the body tube and the stand, or in a slide. The AHBS microscope features automatic focussing of the specimen in brightfield transmitted light.

Photomicrographic systems

The Olympus PM10 camera system (see Fig. 4.12) is designed to attach directly to a trinocular tube of the Olympus BH2 and older Vanox microscopes, but can be attached to almost any microscope with appropriate adapters. A projection ocular is required in the body tube. These oculars are available with magnifications of ×2·5, ×3·3, ×5, and ×6·7; I normally use only the lowest power (×2·5) for fluorescence, but have occasionally found the ×3·3 useful for obtaining a small increase in magnification. Note that the magnifications of the projection eyepieces refer to the magnification on to the film plane in 35 mm Olympus photomicrographic attachments; with the large-format camera the magnification is three times higher, and for visual use (not recommended, but quite possible) their magnification is approximately double that stated. A Polaroid adapter is available: the additional magnification onto the larger format is ×3.

The built-in exposure meter is very sensitive. The PM10–AD measures from 60% of the field; the PM10–ADS has spot metering. Six automatic settings for correction reciprocity failure are provided. Exposure can be made either manually (range 1 s to 40 min) or automatically. Compensation factors (e.g. for bright or dark field) can be set in the range ±2 stops (×0·25 to ×4). Film speeds can be set from 1·5 to 4000 ISO (35 mm) or 6 to 25 600 ISO (large format). All settings are displayed in large red fluorescent numerals, clearly visible in the dark and not spoiling the accommodation of one's eyes.

Focussing and composing the picture may be accomplished through a separate telescope. There are three beamsplitter positions for the focussing telescope: 100% light into the telescope, 100% to the camera, or part each way. The exposure meter obtains light from a beamsplitter, which cannot be swung aside. In the most recent model, a focussing telescope with illuminated graticule is available as an optional accessory.

The stage on the BH2 series microscopes can be rotated to compose the picture. Alternatively, the entire PM-10 camera assembly, with focussing telescope, can be rotated, but this procedure is unofficial and requires unclamping the camera mount.

The Vanox has an integrated photomicrographic system, with two 35 mm and one large-format camera ports and a separate video port. The finder graticule is visible in the normal eyepieces. The graticule is composed of dark lines, and the image brightness outside the picture frame is reduced; these features make the frame difficult to see against a dark field. Four projection eyepieces are mounted on a turret for easy selection: the lowest power, as in the PM10, is ×2·5. Exposure control is automatic or manual. The cameras are non-rotatable, but the stage can be rotated easily.

REICHERT

Reichert-Jung AG (Vienna), now part of the Leica group, makes one major microscope, the *Polyvar*, and collaborates with Cambridge Instruments in several smaller and simpler microscopes, including the *Diastar*.

The Polyvar (Fig. 4.15) is of the modern rectangular building-block style, with a basic stand to which special-purpose modules are attached. The outstanding feature of this microscope is that the objectives are of the infinity-corrected type (see Chapter 3). In other words, the light between the objective and the telelens forms a parallel beam. This greatly facilitates the insertion of a fluorescence epi-illuminator and other accessory optical systems, which can be inserted into the parallel beam in the body tube.

For fluorescence, an additional lamphouse for the arc lamp is attached at the rear. Both 200 W mercury and 150 W xenon lamps are available. The normal tungsten-halogen lamp remains attached. No provision exists for more than the two lamps simultaneously. However, a mirror-housing enables the use of either lamp for epi- and dia-illumination, and fluorescence excitation by epi-illumination can easily be combined with dia-illumination by brightfield or Nomarski interference contrast; phase-contrast requires a special set of objectives.

For epi-illumination, sets of filters (primary, dichromatic mirror and barrier filter) are held in blocks which are placed in a huge slot in the body tube. Each block takes the form of a rectangular box with a short handle. The slot is large enough to accommodate two blocks, fed one from each side (this requires that the handle be fitted on different sides). Currently, filter blocks are available for narrow-band excitation in the UV, violet, violet-blue, blue, and green; there are additional modules for several other excitation systems (Table 4.6).

For transmitted-light excitation, primary filters are held in rotating holders in the body of the microscope; two such rotating holders are provided, one from each side of the microscope. Several condensers are available.

Fluorescence excitation by epi-illumination can easily be alternated with examination by interference contrast. Simultaneous interference contrast is possible but the fluorescence image is much reduced in intensity.

Standard oculars are wide-field, with a field number of up to 30 mm. The lowest magnification is ×6·3 (with a field number of 25 mm; or ×10 with a field number of 30 mm), but these can be combined with ×0·8 in the magnification changer. Only infinity-corrected optics can be used.

The Reichert photomicrographic system has a number of excellent features. The graticule for the photographic format is projected into the binocular eyepieces as bright lines, which are very clearly seen against the dark background in fluorescence microscopy. The brightness of the graticule is variable over a wide range. The beamsplitter only has two positions, with either 100% to the oculars or 20% to the oculars and 80% to the film; there is no provision for transmitting 100% to the film. The graticule is only visible with the beamsplitter in the 80/20 position. The camera has a sensitive photometer for exposures from 1/125 s to 1 h, automatic or manual. The

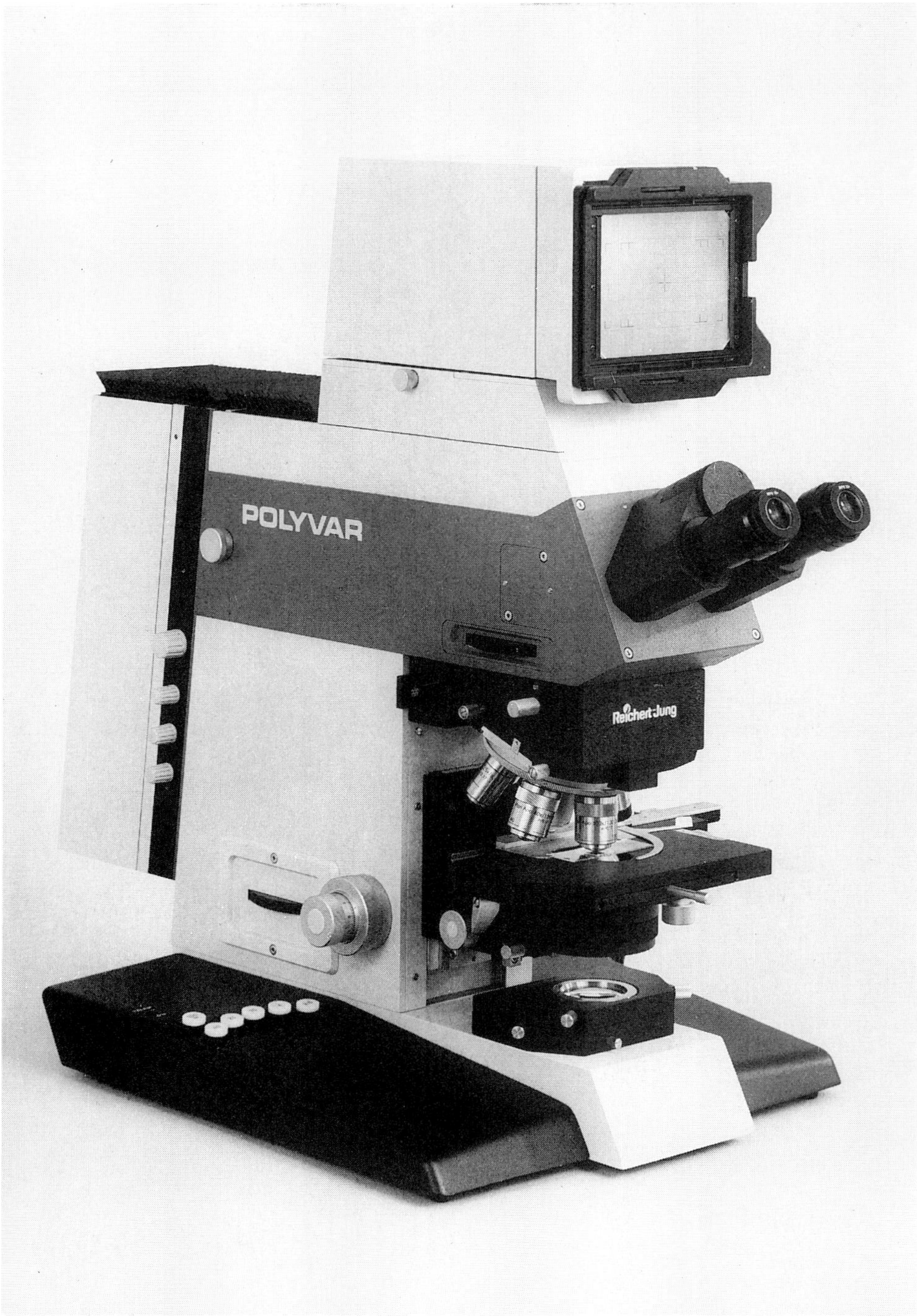

Fig. 4.15. Reichert Polyvar microscope. Photograph courtesy of HD Scientific Supplies Pty Ltd, Sydney.

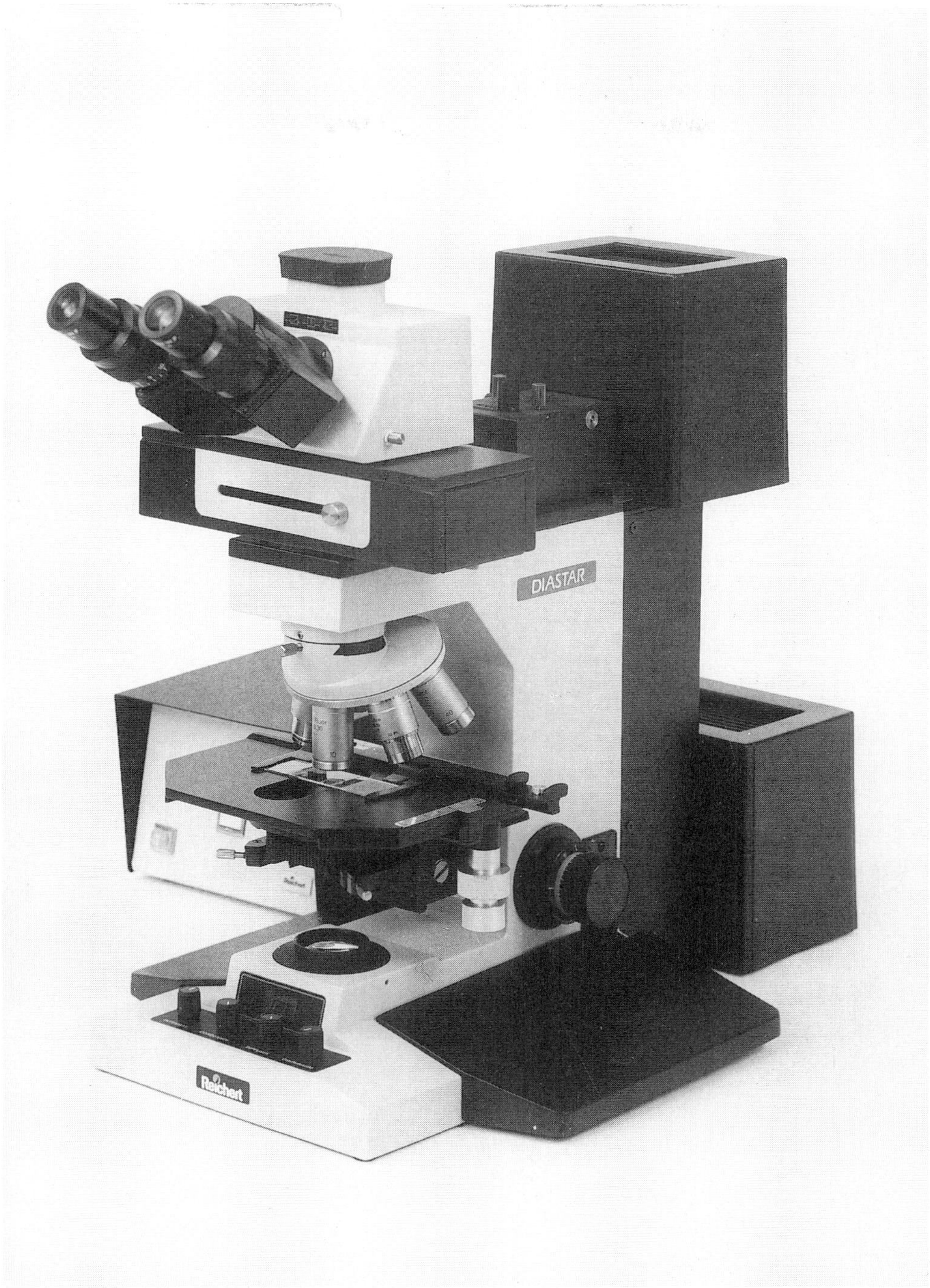

Fig. 4.16. Reichert-Jung Diastar microscope. Photograph courtesy of HD Scientific Supplies Pty Ltd, Sydney.

Table 4.6. *Reichert filter sets for epi-illumination*

Information courtesy of Reichert-Jung AG, Vienna and H.D. Scientific Supplies Pty Ltd, Sydney.

Excitation	Filter code	Excitation filter	Dichromatic mirror	Barrier filter
UV	IU1	BP 330–380	DS 420	LP 418
Violet	IV1	BP 405/8	DS 420	LP 418
Violet	IV3	BP 430–446	DS 460	LP 470
Violet + blue	IV2	BP 390–450	DS 460	LP 475
Blue	IB1	BP 455–490	DS 500	LP 515
Blue	IB2	BP 455–490	DS 500	BP 515–560
Blue	IB3	BP 475–495	DS 510	LP 520
Blue	IB4	BP 475–495	DS 510	BP 520–560
Green	IG1	BP 546/10	DS 580	LP 590
Green	IG2	BP 510–560	DS 580	LP 590
Green	IG3	BP 530–585	DS 610	LP 615

exposure meter's photodiode normally measures from the entire field; a model with spot metering and a photomultiplier is also available.

The Diastar (Fig. 4.16) is a simpler model, manufactured in the USA. Like the Polyvar, it has infinity-corrected objectives. Coarse and fine focussing are carried out with the nosepiece; the stage and the body tube are both fixed to the microscope stand to maximize stability and to facilitate manipulation of the specimen. Photomicrography is accomplished through a trinocular body tube, of which there are two versions: both give either full light to the camera or full light to the viewing tubes, and one also offers 80% to the camera and 20% to the viewing tubes. Fluorescence can be examined by either epi-illumination or dia-illumination. For epi-illumination, matched sets of excitation and barrier filters and a dichromatic mirror are assembled into units ('Clusters'), two of which can be accommodated in the illumination unit at any one time.

ZEISS (OBERKOCHEN)

Carl Zeiss (Oberkochen) has a long record of making excellent microscopes, including fluorescence microscopes. In 1986 Zeiss started the conversion of research microscopes to incorporate infinity-corrected optics and a fully corrected intermediate image. To date (1990), many microscopes have been redesigned to incorporate the new optics. This conversion made it possible to dispense with transfer and compensation optics, i.e. eyepieces without compensation are now being used exclusively. Compatibility is therefore assured in the new microscope range, i.e. the position of the intermediate image and the parfocality of objectives remain unchanged without any necessity for special intermediate rings. The field of view is 25 mm in diameter in the Axiophot and Axioplan microscopes, and 20 mm in the Axioskop and the inverted microscopes.

The Axiophot (Fig. 4.17) is the latest major microscope of Carl Zeiss (Oberkochen). This microscope, introduced about 1986, represents something of a break with tradition. First, the base is narrow, the operator's hand resting on the bench while operating the focussing controls. The SI (system integration) architecture ensures a constant ocular height of 450 mm and therefore ergonomic working over long periods of time. Less obviously but more importantly, the objectives and oculars are corrected for an infinite tubelength (see Fig. 3.8), i.e. the light between the objective and ocular forms a parallel beam. This greatly facilitates the insertion of a fluorescence epi-illuminator and phase-contrast systems. Therefore, the Plan-Neofluar objectives can be used for all contrast methods (including phase-contrast, DIC, etc.). The fully corrected primary image, produced by the new, infinity-corrected objectives, can also be used directly for video microscopy without any need for compensating optics.

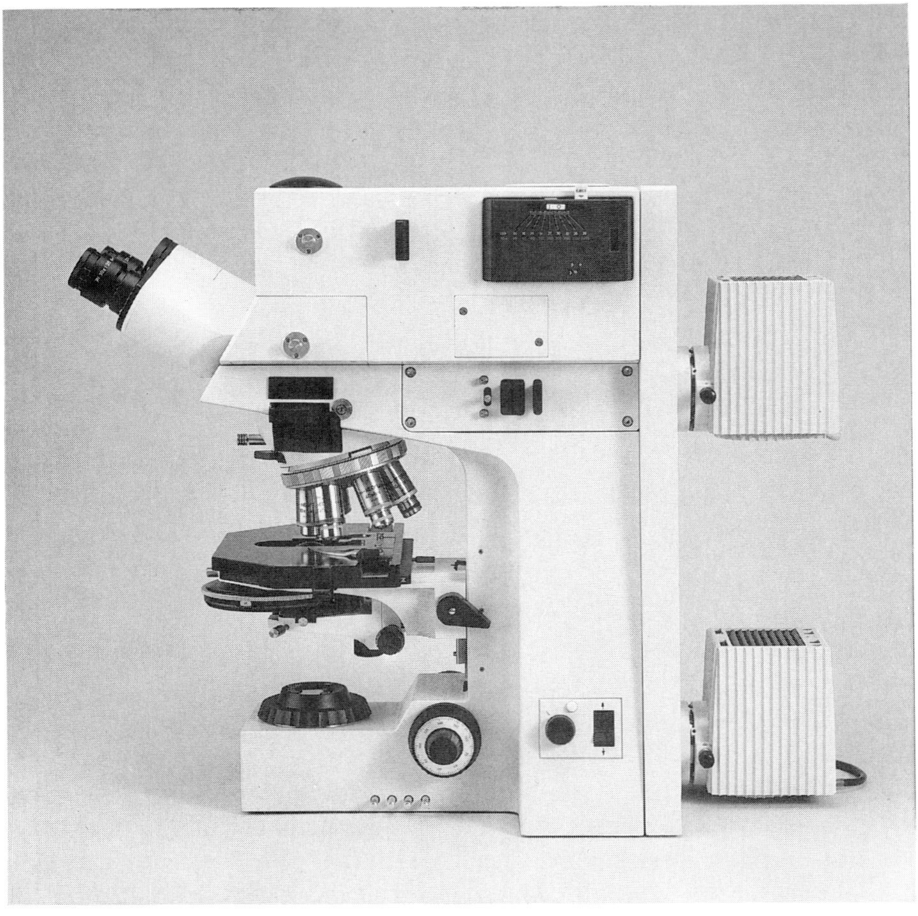

Fig. 4.17. Zeiss Axiophot microscope, with integrated photomicrographic module, equipped for fluorescence. Photograph courtesy of Carl Zeiss Pty Ltd, Sydney.

For fluorescence, the Axiophot is normally fitted with an attachable lamphouse for an HBO50 mercury lamp. The lamphouse has a three-element collector lens, to minimize need for adjustment when changing excitation wavelength. Moreover, Zeiss offers fluorescence illumination with 100 W mercury and 75 W xenon lamps; these two lamps are connected to the same power supply unit. For all usual applications, the 50 W mercury lamp is fully sufficient due to the new architecture of the microscopes. Filter and epi-illuminator mirror sets are held in a three-position slide which fits in the body in a wide slot above the nosepiece. The slide has provision for up to three sets of filters, or two filter sets and a blank position for examination in ordinary light. Additional excitation filters and a blanking plate can be fitted in slots in the top of the stand. Optical components for the epi-illumination system are available which, in conjunction with Plan-Neofluar objectives, enable epi-illumination at wavelengths down to 340 nm.

The camera module fits on the top of the stand. It has two interchangeable 35 mm camera bodies and one large-format camera. In the 35 mm format, because of the smaller camera magnification factor of ×2·5 in the Axiophot compared with ×3·2 in the older Photomicroscope III, the exposure times are shorter (by a nominal factor of $[3{\cdot}2/2{\cdot}5]^2$, i.e. 1·6) and a larger field is included on the film. In addition, improved optics leads to a total reduction in exposure times by a factor of 2 to 3 as compared to the Photomicroscope III (Gundlach & Wehland, 1988). A separate automatic control unit offers considerable versatility, including corrections for reciprocity failure. For focussing and composition, an illuminated graticule of variable brightness is reflected into the oculars. There is an additional port above the body tube to accommodate a video camera or photometer. For composing the picture with the rectangular stage, it may be helpful to remove the stage and replace it back-to-front, which increases its possible rotation to 140°. A data-recording back is available for the 35 mm camera.

Objectives of the new infinity and colour-corrected type suitable for fluorescence microscopy are listed in Table 4.7. Older Zeiss objectives can be used with a correction lens. The current lowest-power oculars are ×10.

The Axioplan microscope (Fig. 4.18) is similar to the Axiophot, the stand having a roughly pyramidal shape without the traditional wide base. Infinity-corrected optics are used. The fluorescence system is substantially identical to that of the Axiophot. The difference between the Axioplan and the Axiophot lies in the photomicrographic system, which is more extensive in the latter. It is possible to convert an Axioplan to an Axiophot by adding the appropriate modules.

The Axioskop (Fig. 4.19), introduced in 1987, has been designed for routine work in physicians' laboratories and clinical laboratories up to and including applications which formerly required a research microscope. These extended applications have been made possible by the systematic continuation of the philosophy of system integration and infinity-corrected optics. The Axioskop gives the same image quality as that obtained with a research microscope, since the same optical components are available for both instrument types. Although the Axioskop is fully compatible with the Axiophot and Axioplan microscopes – including filters and illumination systems for fluorescence – it has nevertheless been designed especially for routine work, and meets

Table 4.7. *Zeiss (Oberkochen) objectives for fluorescence microscopy*

Type: Plan-Neofluar or Plan Apochromat. Optical tubelength: infinite. Field-of-view number: 25. Parfocal length: 45 mm. Phase-contrast versions available, except ×16/0·50. Data courtesy of Carl Zeiss Pty Ltd, Sydney.

Magnification (×)	NA	Immersion	Type	Working distance (mm)	Exit pupil ⌀ (mm)	Object field (mm)
10	0·30		PN	5·6	9·9	2·5
16	0·50	Imm	PN	0·22	10·6	1·56
20	0·50		PN	1·3	8·2	1·25
20	0·60		PA	0·45	9·9	1·25
25	0·80	Imm, korr	PN	0·3	10·5	1·0
40	0·75		PN	0·33	6·1	0·63
40	0·9	Imm, korr	PN	0·13	7·4	0·63
40	1·30	Oil	PN	0·12	10·7	0·63
40	1·30 (iris)	Oil	PA	0·31	0·82 (max)	0·63
63	1·25 (iris or AP)	Oil	PN	0·1	6·5	0·39
63	1·40	Oil	PA	0·09	7·3	0·39
100	1·30	Oil	PN	0·06	4·2	0·25
100	1·30	Oil	PA	0·08	4·3	0·25

Notes:
⌀ = diameter, AP = antireflex phase, imm. = multi-immersion (oil, glycerol or water), iris = variable NA, korr = correction ring for thickness of coverslip and refractive index of immersion medium, PA = Plan Apochromat, PN = Plan-Neofluar.
For excitation in normal UV (365 nm), Plan-Neofluar objectives are recommended.

all the requirements made on an instrument in this field of application. The Axioskop has been designed for a 20 mm field of view and is equipped with a 50 W or 20 W halogen transilluminator. For fluorescence, the basic equipment allows the insertion of filter sets in up to three positions of the filter slider. A low-cost version with a suitable filter set and a clear aperture for transmitted-light phase-contrast has been especially created for FITC-labelled specimens, e.g. in medical routine diagnosis.

Inverted microscopes

The Axiovert (Fig. 4.20), developed in 1988, is the latest of the new microscopes from Zeiss. The Axiovert inverted microscopes are available in several different versions. The Axiovert 10 is the simplest type, allowing not only routine work but also complex experimental set-ups, such as micromanipulation and microinjection workplaces. Photomicrography is possible with a microscope camera attached via a binocular photo tube. A port for the attachment of a single-lens reflex (SLR) camera is a standard feature of the Axiovert 35. The Axiovert 35 MOT has a motorized nosepiece. The Axiovert 405 MOT is an inverted photomicroscope. With its integrated 4 in. × 5 in. and 35 mm cameras, it is the inverted counterpart of the Axiophot photomicroscope. The

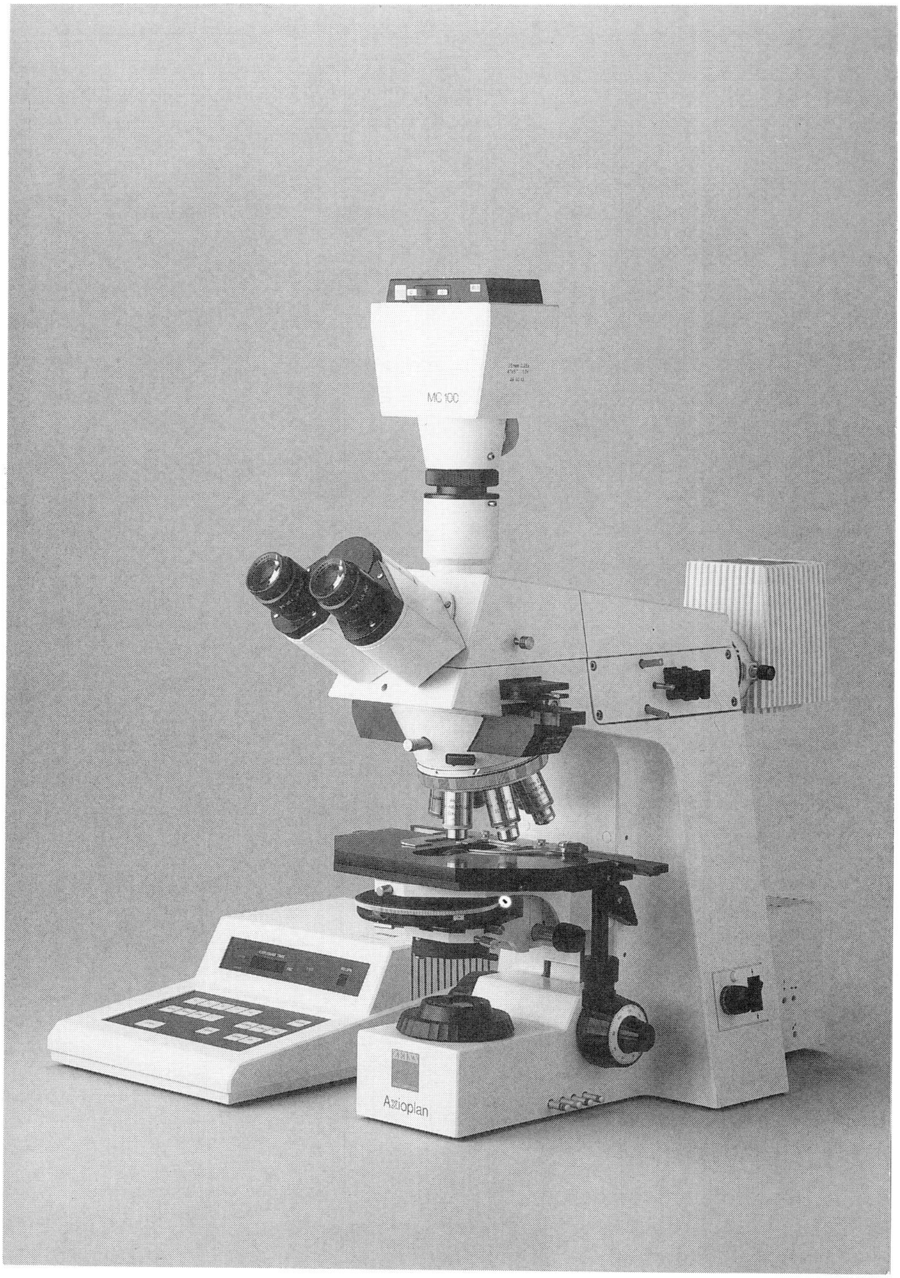

Fig. 4.18. Zeiss Axioplan microscope equipped with fluorescence epi-illuminator, HBO 50 lamp, and MC100 photomicrographic camera. Photograph courtesy of Carl Zeiss Pty Ltd, Sydney.

range of inverted microscopes has been equipped with the new infinity-corrected optics, their field of view is 20 mm and the standard transilluminator is a 100 W halogen lamp. This makes it possible to use almost all accessories of the new pyramid microscopes also with the inverted microscopes. Two filter positions can be equipped for fluorescence. The sliders are interchangeable. An interface is available to connect a laser to the Axiovert for excitation.

Other Zeiss microscopes

Fluorescence equipment is available for several other Zeiss microscopes, using the traditional 160 mm tubelength. The *Axiomat* is based on a modular system, designed for versatility and extreme rigidity. The *Universal* and the *Photomicroscope III* have a similar large stand of traditional pattern; the photomicroscope has a built-in automatic 35 mm camera, which complicates the fitting of an epi-illuminator. The *Standard* is designed as a routine laboratory microscope. Microscopes *ACM*, *UEM* and the *Axiotron* series have a very large stage (150 mm square) for large objects, and are designed for industrial applications and for microdensitometry and microfluorometry.

The Standard 20 is the special microscope for routine work. The Standard 20 superseded, in 1990, the Standard 16 (Fig. 4.21) and uses the same fluorescence

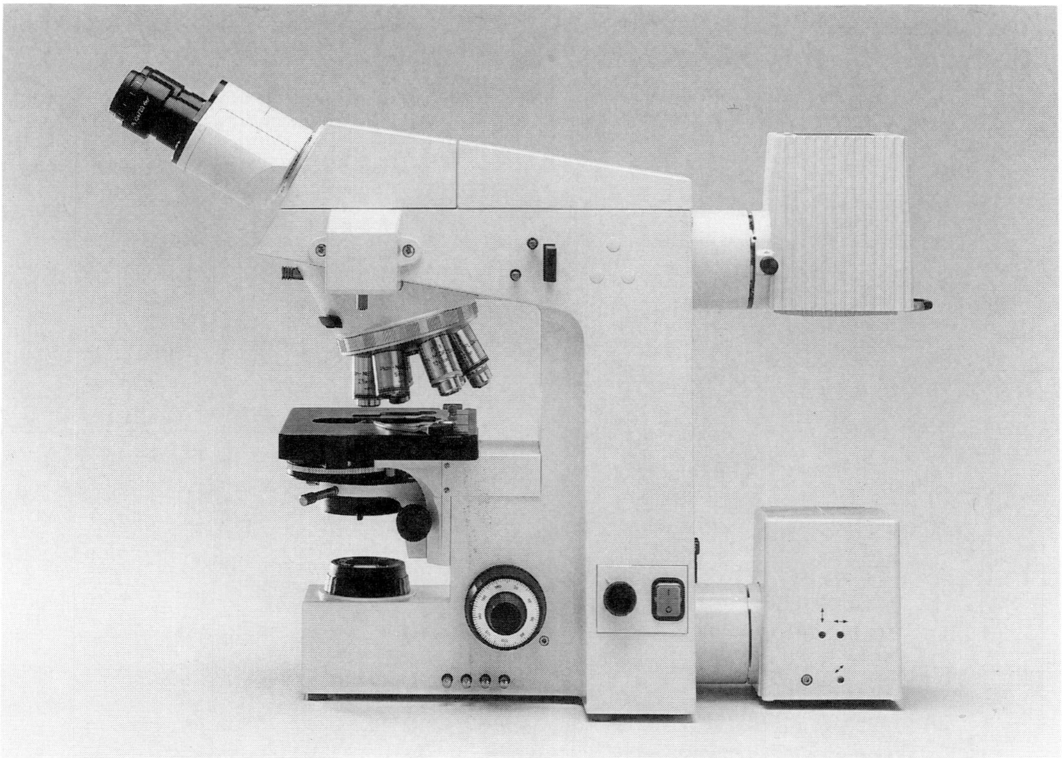

Fig. 4.19. Zeiss Axioskop microscope equipped for fluorescence. Photograph courtesy of Carl Zeiss Pty Ltd, Sydney.

equipment. It is used in physicians' and clinical laboratories and for the training of students. Since traditional objectives with a tubelength of 160 mm are used with this microscope, a fluorescence intermediate tube is required. The slider can be equipped with two filter combinations which are compatible with those used with the Universal, Photomicroscope, IM, IM 35 and ICM 405 microscope series.

Filter sets

Filter combinations for Zeiss microscopes are listed in Table 4.8. A wide variety of filter sets, consisting of an exciter filter, a dichromatic beamsplitter and a barrier filter, is available for all standard examination methods. For further special applications, the user can select the optimum combination from the range of filters available.

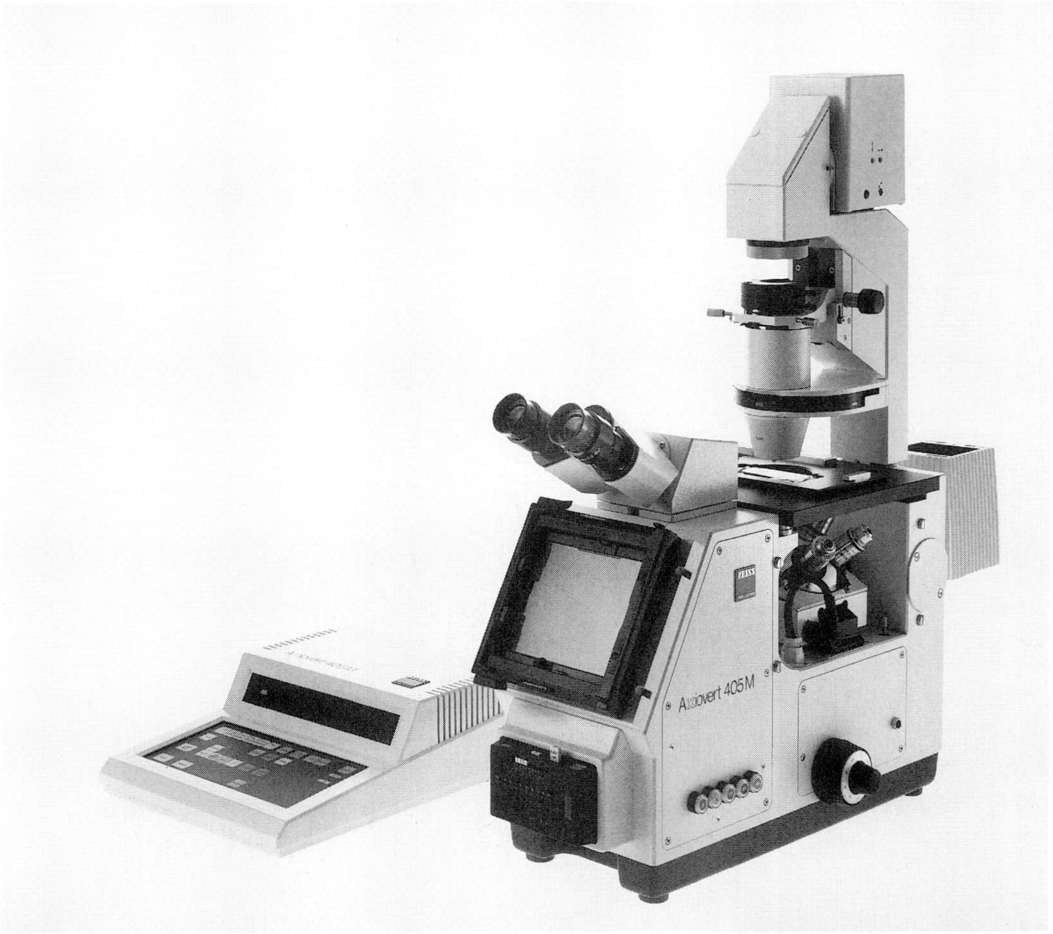

Fig. 4.20. Zeiss Axiovert 405 inverted microscope equipped with fluorescence epi-illuminator, 35 mm camera and large-format camera. Photograph courtesy of Carl Zeiss Pty Ltd, Sydney.

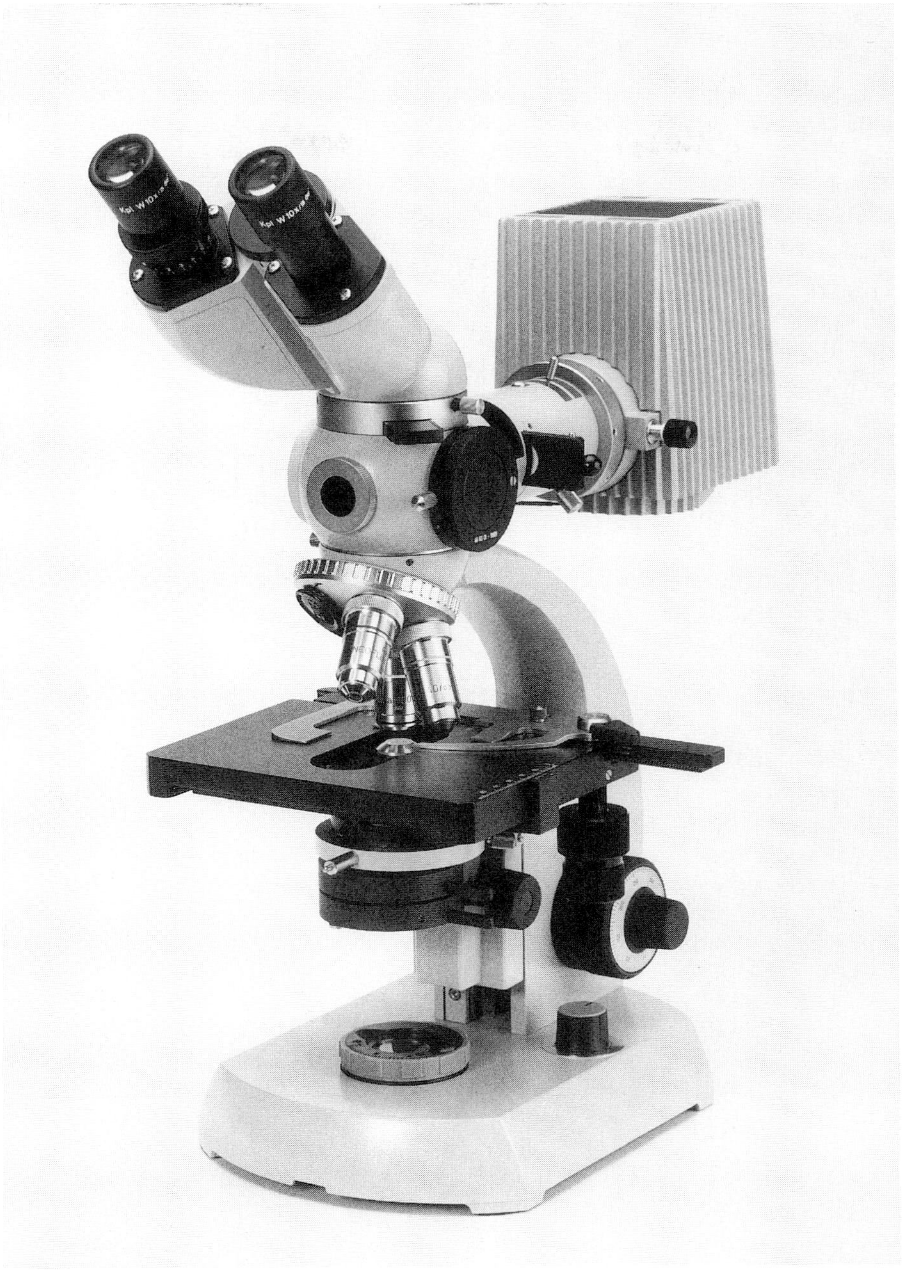

Fig. 4.21. Zeiss Standard 16 microscope, with IVFL epi-illuminator and HBO 50 lamp. Photograph courtesy of Carl Zeiss Pty Ltd, Sydney.

Table 4.8. *Filter combinations for Zeiss microscopes*

Filter sets for fluorescence with epi-illumination. The filter sets are always named according to the type of excitation. Data courtesy of Carl Zeiss Pty Ltd, Sydney.

(a) Filter sets for the Axioplan, Axiophot, Axioskop and Axiovert

Excitation mode	Set name	No.	Excitation filter	Beamsplitter	Barrier filter
1. Basic filter sets with conventional filters, mostly coloured glass. Excitation filters with broad pass-band, transmission curve almost bell-shaped; barrier filter pass-band starting at an unspecified distance from the main excitation wavelength. Code letter G.					
UV	G 365	02	G 365	FT 395	LP 420
Blue–violet	G 436	07	G 436	FT 510	LP 520
2. Selective excitation. Filter sets with high-performance narrow-band filters (interference filters). Excitation filters with transmission curves steep on both sides; barrier filters with pass-band starting close to main range of excitation filter wavelengths. Code letter H.					
(Double excitation for Quin 2 and Indo-1)		21	BP 340/10 BP 380/10	FT 395	BP 500–530
UV	H 365	01	BP 365/12	FT 395	LP 397
Blue–violet	H 436	06	BP 436/10	FT 460	LP 470
Blue	H 485	16	BP 485/20	FT 510	LP 520
Blue SB	H 485	17	BP 485/20	FT 510	BP 515–565
Green	H 546	15	BP 546/12	FT 580	LP 590
Green (for Phycoerythrin)	H 546	20	BP 546/12	FT 560	BP 575–640
3. The most intensive excitation. Filter sets with high-performance wide-band filters (interference filters). Excitation filters with transmission curves steep on at least the long-wave side; barrier filters with pass-band starting close to main range of excitation filter wavelengths. No code letter.					
Green	530–580	00	BP 530–580	FT 600	LP 615
UV–violet	390–420	18	BP 390–420	FT 425	LP 450
Blue–violet	395–440	05	BP 395–440	FT 460	LP 470
Blue	450–490	09	BP 450–490	FT 510	LP 520
Blue SB	450–490	10	BP 450–490	FT 510	BP 515–565
Green	510–560	14	BP 510–560	FT 580	LP 590

(b) Filter sets for the Standard, WL, and Universal microscopes and the Photomicroscope

Excitation mode	Set name	No.	Excitation filter	Beamsplitter	Barrier filter
1. Basic filter sets with conventional filters, mostly coloured glass. Excitation filters with broad pass-band, transmission curve almost bell-shaped; barrier filter pass-band starting at an unspecified distance from the main excitation wavelength. Code letter G.					
UV	G 365	02	G 365	FT 395	LP 420
Blue–violet	G 436	07	G 436	FT 510	LP 520

Table 4.8. (*cont.*)

Excitation mode	Set name	No.	Excitation filter	Beamsplitter	Barrier filter
2. Selective excitation. Filter sets with high-performance narrow-band filters (interference filters). Excitation filters with transmission curves steep on both sides; barrier filters with pass-band starting close to main range of excitation filter wavelengths. Code letter H.					
H 365	01	BP 365/12	FT 395	LP 397	
Blue–violet	H 436	06	BP 436/10	FT 460	LP 470
Blue	H 485	16	BP 485/20	FT 510	LP 520
Blue SB[a]	H 485	17	BP 485/20	FT 520	BP 515–565
Blue IFB[b]	H 485	19	BP 485/20	FT 510	LP 515
3. The most intensive excitation. Filter sets with high-performance wide-band filters (interference filters). Excitation filters with transmision curves steep on at least the long-wave side; barrier filters with pass-band starting close to main range of excitation filter wavelengths. No code letter.					
Green	530–580	00	BP 530–580	FT 600	LP 615
UV–violet	390–420	18	BP 390–420	FT 450	LP 450
Blue–violet	395–440	05	BP 395–440	FT 460	LP 470
Blue	450–490	09	BP 450–490	FT 510	LP 520
Blue	SB450–490	10	BP 450–490	FT 510	BP 515–565
Blue IFB	450–490	11	BP 450–490	FT 510	LP 515
Green	510–560	14	BP 510–560	FT 580	LP 590

Notes:

[a] Blue excitation with selective viewing (barrier filter transmits only green).

[b] With interference barrier filter.

Photomicrographic cameras

The MC 100 microscope camera, introduced in 1987, is designed for all applications and can be attached to all Zeiss microscopes. The total image magnification on to the film is equal to the product of the magnifications of the objective, the photoeyepiece (usually ×10), and a camera factor of 0·25 for the 35 mm format or 1·0 for the large format (4 in. × 5 in.). The exposure time can either be set manually or measured automatically in the range 1/100 s to 8000 s. Image brightness is continuously monitored during exposure, using a beamsplitter which takes 10% of the light. The camera features compensation of reciprocity failure (nine different settings), exposure time storage, the possibility of double and multiple exposures, automatic control of exposure series and automatic adjustment of exposure times when changing the format or film. The camera system is a multiformat, multifilm and multiuser system which allows the components to be interchanged as required, without having to make any entries on the control panel. The camera can be freely rotated on the binocular photo tube or fixed in preset positions for horizontal or vertical pictures. The exposure time can be compensated in whole-stop intervals from −2 to +2 stops; for compensation in smaller steps, the film speed setting should be varied (take care to reset it correctly afterwards). A data-recording back is available for the motorized 35 mm camera (the same as that used on the Axiophot).

Table 4.9. *Technical data of lamps suitable for fluorescence microscopy*

Power (W)	Voltage (V)	Current (A)	Light flux (lm)	Mean luminous density (sb)	Dimensions of the luminous area (l × b) (mm × mm)	Colour temperature (K)	Life (approx.) (h)
Tungsten-halogen lamps, 12 V							
100	12	8·0 (max)	2800			3100 at 7·4 A	50[a]
Xenon high-pressure arcs							
75	14	5·4	1000	40 000	0·5 × 0·25	6000	400
150	20	7·5	3000	15 000	2·2 × 0·5	6000	1200
450	18	25	13 000	35 000	2·7 × 0·9	6000	2000
Mercury high-pressure arcs							
50	L1: 39–45 L2: 34–39	1·3 1·45	2000	30 000	1·0 × 0·3	—[b]	100
100	20	5·0	2200	170 000	0·25 × 0·25		200
200	L1: 6143·4–3·9 L2: 6343·9–4·4		9500	33 000	2·2 × 0·5		200

Notes:
Data from Koch (1972) and manufacturers' published data. See also Table 3.2.
[a] Depends on operating current.
[b] No valid figure can be given for the colour temperature of mercury arcs.

ACCESSORIES

Accessories such as arc lamps, and to a lesser extent filters, are common to all brands of fluorescence microscope, and can therefore be considered collectively. Photometric equipment is discussed in Chapter 8.

Arc lamps

Mercury arc lamps are available from several manufacturers in various powers from 50 W to 1000 W: those used for fluorescence microscopy are 50, 100 and 200 W. Manufacturers' type numbers generally consist of some letters followed by the power in watts – e.g. Philips CS200 and Osram HBO200. The main characteristics of each type are tabulated in Table 4.9.

Lamps of 50 W and 200 W are supplied in two types, designated L1 and L2, differing slightly in their electrical characteristics (see Table 4.9). This designation is marked on each lamp, on the upper electrode terminal. The power supply unit must be appropriately adjusted; failure to do this may cause explosion of the lamp.

50 W mercury arc lamps are the most widely used for fluorescence microscopy at the present time, having superseded the 200 W lamps which are more expensive both in themselves and in the lamphouse and starter requirements. Their small physical dimensions and low cost render them suitable for routine laboratory use with fluorochromes such as FITC. However, the illumination obtained is not as bright as that obtainable with the more powerful lamps, and the lamps cannot be adequately stabilized for quantitative work.

100 W mercury arc lamps have a tiny and extremely bright arc. They are only suitable for use with epi-illumination systems. They are unsuitable for use with substage condensers, particularly dark-ground ones, because of the difficulty of obtaining an image of the arc sufficiently big to fill the aperture of the condenser. For the same reason, they are unsuitable for large-aperture systems. However, the 100 W mercury arc is suitable for microfluorometry: however, mercury arcs are notoriously less stable than xenon arcs and are in that respect less suitable. A highly stabilized power supply and special care are required to minimize instability.

200 W mercury arc lamps were the first to be widely applied to fluorescence microscopy. They are particularly suitable for use with dark-ground condensers, having a sufficiently large arc for its image to fill the aperture of the condenser, and a high light output. This type of lamp is rarely used for microfluorometry; however I have had some success, using a stabilized alternating current (AC) power supply.

Xenon arc lamps are currently available in 75 W, 150 W and 450 W sizes. My experience is that for FITC immunofluorescence the 75 W lamps are adequate but do not give as bright illumination as the 200 W mercury lamps. The 150 W type is particularly useful for microfluorometry.

Objectives

Magnifications of modern objectives are usually some decimal multiple of a number in the standard DIN geometric series 1, 1·6, 2·5, 4, 6·3, . . . 10, because, in combination with eyepiece magnifications from the same series, the total magnification can be easily calculated as another number in the same series (e.g. $6{\cdot}3 \times 16 = 100$). Less often, the intermediate values 1·25, 2, 3·2, 5 and 8 are used.

Manufacturers occasionally change the design of their objectives and oculars; in recent years, there have been several such changes, mainly associated with a change to the international (DIN) standards for tubelength and parfocal distance (parfocal distance or adjustment length is the standard distance from the object plane to the objective flange; the DIN standard, accepted by the majority of manufacturers, is 45 mm). Old and new objectives are often not compatible, and neither are old objectives and new oculars and vice versa. However, sometimes there is a degree of compatibility. Leitz used to provide a special adapter tube (code-named PLEZY) to lengthen an old-model objective to make it compatible with longer-barrel newer objectives. PLEZY contained a lens, to compensate for the increased tubelength. I had some experience with PLEZY and the old but excellent ×25 apochromatic objective; however, for microfluorometry I came to prefer a simple screwed tube, made in our own workshop, to avoid the use of a lens with its air–glass interfaces and consequent light losses and risk of fluorescent dirt; the correct length to maintain parfocality was calculated.

Filters

The nomenclature of filters is usually based on a combination of one or more letters and a number, e.g. BG12 and K490. Filters consisting of a single kind of coloured glass are designated by the type of glass, named arbitrarily by the manufacturer, e.g. the BG12 and UG1 glasses of Schott. However, Schott glasses of the temperature-treated type are designated according to the approximate wavelength at which a 3 mm thickness of the glass has 50% internal transmission. Interference filters are numbered according to the wavelength of peak transmission.

Filters supplied as excitation and barrier filters by Leitz and Zeiss (Oberkochen) are designated BP (bandpass), LP (longpass), and either SP (shortpass) or KP (Kurtzpass) according to function, together with the relevant wavelength or wavelengths. Some Zeiss (Oberkochen) bandpass filters are designated according to both the wavelength and the transmission bandwidth: e.g. BP 405/8 is a bandpass filter with peak transmission at 405 nm, and with a bandwidth of 8 nm. Barrier filters are designated according to the wavelength in nm at which they give 50% transmission. Thus the LP510 filter has 50% transmission at 510 nm, with barrier effect at lower wavelengths. Older Zeiss barrier filters were designated with a two-digit number representing in tens of nm the wavelength at which the transmission was 10%.

UG1 is probably the most technically satisfactory of the coloured glasses for excitation. Its transmission covers mainly the 365 nm mercury line; unfortunately it, like all the other glasses used for excitation filters, has a substantial transmission in the red which necessitates the addition of a red-stopping filter, BG38 as well as a heat filter (KG1 or KG3). In my experience, the combination of UG1 and BG38 is just as effective as a commercial 365 nm interference filter, and has the advantages of lower cost and less risk of damage by overheating.

BG12 is widely used for broad-band excitation in the blue region. It must be supplemented by red-stopping filter (BG38) and a heat filter. This combination can be used with any light source – mercury arc, xenon arc or tungsten-halogen. For the last of these, however, the transmission is rather low compared to that obtainable with a specifically designed interference filter. Using a mercury arc, if excitation with UV is not required, a UV-stopping filter should be added. BG12 is also usefully employed to supplement an interference filter for the 436 nm mercury line.

BG3 has been used for broad-band excitation in the violet, transmitting the 406 and 439 nm mercury lines and also some of the 365 nm line. The usual KG1 heat filter and BG38 red-stopping filter must be added, as should (if possible) a UV-stopping filter. BG3 is also used to supplement interference filters for the 406 and 436 nm lines.

Oculars

These days manufacturers tend to provide only a rather limited range of eyepieces. For fluorescence, the lowest available power should be available as well as the 'normal' ×8 or ×10 ocular. The latter would be expected to give the full resolution of the objective; the former gives a brighter image. The lowest-power oculars known to me at the present time are listed in Table 4.10.

Table 4.10. *Lowest-power oculars commercially available*

Magnification	Field number (mm)	Optical tubelength (mm)	Manufacturer
4	28	170[a]	Leitz
5	18·6	160	Nikon
5	21	160	Olympus
5	20	160	Zeiss (Oberkochen)
6·3	28 (18)	160	Leitz
6·3	24	Inf.	Reichert
6·3	18	160	Zeiss (Oberkochen)
8	18·6	160	Nikon
8	18	160	Zeiss (Oberkochen)
10	30	Inf.	Reichert
10	25 (20)	Inf.	Zeiss (Oberkochen)

Notes:
Inf. = infinity. Numbers in parentheses are alternative field numbers depending upon the model.
[a] Can be used in GW heads (i.e. 30 mm ⌀) on 160 mm tubelength microscopes with 8 mm spacer.

5

Setting up and using the fluorescence microscope

This chapter deals with the practicalities of employing a fluorescence microscope for a specific purpose. The process of setting up and using a fluorescence microscope generally involves the following steps:

1. Initial adjustment of the microscope using conventional transmitted-light microscopy, with a conventional slide as a test specimen.
2. Setting up the microscope for fluorescence microscopy, with a fluorescent test slide.
3. Selecting the correct filter combination for the fluorophore (or fluorophores) to be examined.
4. Viewing the desired specimen.
5. Adjusting the microscope for optimal visualization of the fluorescence.
6. Interpreting the results (see Chapter 7).
7. Recording the results by photomicrography (see Chapter 6).

SETTING UP THE MICROSCOPE FOR CONVENTIONAL MICROSCOPY

It must be emphasized that time spent in preliminary adjustment of the microscope with conventional brightfield lighting is always worth while, since it is easier to check for cleanliness, focus the condenser, and so forth using conventional brightfield illumination. Moreover, conventional transmitted-light illumination is often used alternately with fluorescence to examine the specimens.

It is assumed that the reader is already familiar with the process of setting up the microscope for Köhler illumination in transmitted-light microscopy. Those who are hazy in this area should consult a work such as that by James (1976), and also the manufacturer's literature in respect of the particular microscope. For convenience in quick reference, a summary of the procedure is given in Appendix 3A. It may be necessary to remove the barrier filters and the dichromatic mirror or to set the filter combination to 'UV' to avoid coloration of the field, which may be confusing during preliminary adjustments.

Table 5.1. *Immersion fluids for objectives*

Listed in order of refractive index (n_D) for the sodium D line (589 nm). See also mounting media in Volume II, Chapter 15.

Type	Refractive index, n_D
Water	1·333 at 20 °C
Olympus silicone	1·404 at 25 °C
Glycerol	1·47
Cargille[a] Type FF	1·4790 at 23 °C
Mersol[b]	1·511
Cargille[a] Type DF	1·5150 at 23 °C
Cargille[a] Type A (low viscosity)	1·5150 at 23 °C
Cargille[a] Type B (high viscosity)	1·5150 at 23 °C
Cargille[a] Type NVH (very high viscosity)	1·5150 at 23 °C
Cargille[a] Type VH (very, very high viscosity)	1·5150 at 23 °C
Cargille[a] Type 37 (for use at 37 °C)	1·5150 at 37 °C
Zeiss (Oberkochen)	1·515 at 23 °C (n_E = 1·518)
Olympus	1·5155 at 25 °C
Lenzol[b] (cedarwood oil)	1·516
H&W tropical grade[a]	1·517
Microil[b]	1·518
Uvinert[b]	1·519

Notes:

[a] Cargille recommend Type FF for fluorescence, having virtually no autofluorescence; Type DF recommended for highest resolution, meeting exact DIN specifications for refractive index and dispersion, but with slight ('very very low') greenish autofluorescence.

[b] Hopkin & Williams (H&W).

The microscope must of course have optical equipment suitable for fluorescence microscopy, as described in Chapter 3. At this point, the reader will be prepared to check that the the microscope is equipped with the following:

1. A suitable light source (mercury arc, usually).
2. The epi-illumination device (usually) and appropriate filter set(s).
3. Suitable objectives (large aperture; phase-contrast?).
4. Low-power oculars.
5. Non-fluorescent immersion fluid (for immersion objectives and condenser; see Table 5.1).
6. Assuming that photomicrography is anticipated, a trinocular head giving 100% transmission alternatively to either the oculars or to the camera.

SETTING UP FOR FLUORESCENCE

Depending upon the microscope, the operator may have to make certain decisions as to how to use the equipment. These usually relate primarily to the form of illumination –

should it be by epi-illumination, dia-illumination with a darkground condenser, or dia-illumination with a conventional condenser? Also, what filter combination to use?

Choice of illumination mode

Some microscopes have provision for excitation of fluorescence in more than one way. From what has been said above (Chapter 3), it will be evident that epi-illumination is usually the method of first choice, particularly if high magnification is required. Only for scanning at low magnifications is dia-illumination preferable. Dia-illumination with a darkground condenser is usually the second most popular choice; however, a brightfield condenser usually gives more even and brighter illumination, and can be used irrespective of the objective's numerical aperture (NA). The more advanced fluorescence microscopes are equipped with mirror-houses so that the microscope can be set up to enable epi- or dia-illumination to be used alternately or together as required. It may be useful to employ simultaneous or alternate phase-contrast or interference-contrast, rather than normal brightfield transmitted-light microscopy. Combined forms of illumination are described below.

Setting up a fluorescence microscope with dia-illumination

The procedures required are set out below step by step.

Dia-illumination, conventional brightfield condenser

This method is now rarely used; it is useful for scanning at low magnification, particularly if the field exceeds the diameter which can be illuminated by a darkground condenser. The following steps are required:

1. If a gas-discharge lamp (mercury or xenon arc) is to be used, see that it is installed on a mirror-house with a separate tungsten-filament lamp available for conventional microscopy (a tungsten-halogen lamp can sometimes be used for both). Procedures for setting up arc lamps are set forth separately, below. Set the mirror-house to select the tungsten lamp.
2. Set up the microscope for conventional transmitted-light microscopy, using a conventional slide preparation. If possible use a specimen which can be used also for fluorescence (see below and Table 5.3 for suggestions).
3. Insert the correct barrier and excitation filters.
4. Insert a fluorescent test slide and focus on the specimen.
5. Adjust the mirror-house to select the arc lamp; or, for tungsten-halogen lamps, turn up the power supply to the lamp (usually to its full rated current; but be careful to avoid overrunning the lamp).

If the numerical aperture (NA) of the condenser exceeds 0·9, and if the full NA of the condenser is to be achieved, it is necessary to place immersion oil between the slide and the condenser (exactly as if the condenser were an objective of corresponding NA). Such condensers have a flat upper surface, to be oiled to the slide. There are two possible procedures: either the condenser is racked down to receive the oil and then gently racked up again, or else the oil is placed on the condenser top before the slide is placed on the stage.

Dia-illumination, darkground condenser

The use of a darkground condenser is described in Appendix 3A. The darkground condenser may be either a low-power (dry) type or a high-power (oil) type, depending upon the objectives with which it is to be used. The NA of the dark central cone of the condenser must not be less than that of the objective. The low-power type gives brighter illumination, because less of the centre is blacked out. A special objective with an aperture iris diaphragm may be used; the iris is closed down until the NA of the objective is less than that of the central dark cone of the condenser. This technique has been largely supplanted by epi-illumination, but may still be useful for scanning at low magnification (under which conditions epi-illumination is inefficient). The microscope is set up as follows:

1. Set up the microscope for conventional transmitted-light microscopy, as above, with a conventional condenser (see below for procedure for setting up lamp).
2. Insert the barrier and exciter filters.
3. Insert a fluorescent test slide and focus on the specimen.
4. Exchange the condenser for a darkground one. The condenser, if its NA exceeds 0·9, may require oiling to the slide (see previous section).
5. Insert the barrier and excitation filters.
6. Select the arc lamp; or, given a tungsten-halogen lamp only, turn up the power supply to the lamp.
7. Adjust the focus of the condenser for best results. If much too low, the condenser will give no illumination; if too high, or somewhat too low, the illumination is usually uneven. In principle, the condenser is focussed like any other substage condenser (with the objective focussed on a specimen, partially close the field diaphragm, and adjust focus of the condenser until the image of the field diaphragm is sharp). Sometimes a little variation from this position gives more even illumination.
8. Check centration of the condenser (this requires temporary partial closure of the field diaphragm). Readjust focus of the condenser if necessary.
9. If the result is not satisfactory, see Appendix 3B.

Setting up a fluorescence microscope with epi-illumination

This is now the most usual form of illumination. It is particularly suitable for use at high magnification, and can easily be combined with simultaneous or alternate phase-contrast or darkground illumination. The required steps are as follows:

1. Attach the epi-illuminator. If this is the kind that replaces the normal nosepiece (as in the Leitz Orthoplan, for example) the stage will have to be lowered first.

Ensure that the shutter in the illuminator unit is closed, or that the mirror-house is set to block the light from the arc lamp.

2. Set up the microscope for conventional transmitted-light microscopy. Use, if possible, a test slide which is also fluorescent (not necessarily at the wavelength to be used finally).
3. Ensure that appropriate excitation and barrier filters are present, if separate, or that an appropriate filter block is in place.
4. Ensure that the aperture and field diaphragms of the epi-illuminator are fully open.
5. Open the shutter, and look at the fluorescent test slide.
6. Adjust the epi-illuminator as necessary, including the field diaphragm and its centration, and the aperture diaphragm (which is normally left fully open).
7. Adjust the lamphouse controls (centration of lamp, mirror, collector lens) as necessary (see below).
8. Unless simultaneous dia-illumination is required, place a light-trap under the stage, to prevent light being reflected back. If a light-trap is not fitted to the microscope, lower the condenser and place a piece of black cardboard or cloth under the stage; or, close the condenser aperture diaphragm and place a black object below the condenser.
9. Close the shutter, change the filter combination if necessary, exchange the test slide for the specimen, open the shutter. After a change of wavelength of excitation, the focussing of the collector lens may require a minor adjustment.
10. If the result is not satisfactory, see Appendix 3B.

Light sources (lamps)

Gas-discharge lamps

The arc lamp should be installed in the lamphouse, following the instructions of the microscope manufacturer. One is always recommended to read the instructions supplied with the lamp! Special care must be taken not to touch the quartz-glass envelope of the lamp with the fingers, as grease is transferred and becomes burnt onto the surface when the lamp heats up. If necessary, the envelope can be cleaned with pure alcohol and then air-dried before it is first switched on. After installation, ensure that the lamp socket is securely in place. The lamphouse door must be securely closed; it should never be opened while the lamp is lit nor for at least 10 min afterwards, because of the risk of explosion, as well as the dazzling brilliance of the arc.

Ensure that there is adequate ventilation for the lamphouse, without an actual draught. Xenon arcs require special ventilation to remove ozone, which is mildly toxic to the operator (see below).

Starting the arc of gas-discharge lamps requires an initial very high voltage to start ('strike') the arc, after which the arc is maintained with a low voltage. Power units for arc lamps contain provision for supplying this initial high voltage. Most present-day power units incorporate an automatic arrangement to strike the arc by applying the initial high voltage immediately the power unit is switched on. Should the arc not strike, it is best to switch off and try again a few minutes later. Excessive use of ignition voltage reduces the life of the lamp.

Another kind of power unit has a rotating switch which will rotate only in one direction (e.g. clockwise). Turned from the OFF position towards the ON position, the switch passes through an intermediate position (without a detent) in which the ignition voltage is applied. To start the arc, the switch should be turned steadily, neither too quickly nor too slowly, from OFF through the intermediate position to the ON position. To switch off, the switch is turned in the same direction so as to reach the OFF position without passing through the starting position.

Mercury arc lamps require several minutes to warm up; xenon and tungsten-halogen lamps can be used immediately. The time of first switching on of an arc lamp should be chosen such that the lamp can subsequently be left lit and undisturbed for at least several hours; during this period the arc will stabilize, and flickering during subsequent use will be minimized.

Centring the lamp is carried out by adjustments in the lamphouse, moving the lamp horizontally and vertically. With the collector focussed (see below), images of the arc should be formed in the centre of the aperture diaphragm of the condenser (substage or epi-illuminator). In practice, it is sometimes recommended that a piece of frosted glass be placed on the dust-cover above the dia-illumination field diaphragm (assuming that the lamp can be used in this mode). Alternatively, in some cases an image of the arc can be arranged to be seen in the specimen plane, and centred that way (the substage condenser or epi-illuminator must already have been centred). Any diffusing disc in the lamphouse should if possible be removed, so that a good image of the lamp can be obtained. It may help to insert a grey (neutral) filter to reduce the intensity of the light.

Using epi-illumination, the following technique is useful. Remove one objective, and place a grey or white card in the specimen plane. A circular area of light should appear on the card. Adjust the lamphouse controls (centration of lamp, reflector position, and collector focus) until the circle of light appears even all over. However, the illumination beam will now be unnecessarily large in diameter, since not all the beam will enter the pupil of the objective. It is therefore necessary to re-focus the collector lens with the objective in place, while examining a specimen. Because the diameter of the exit pupil of the objective, through which light must pass, varies according to the NA and magnification, for optimal results it is necessary to readjust the collector lens for each objective. High-power objectives have much smaller exit pupils than low-power ones.

If an adjustable reflector is provided, as for example in the Leitz and Olympus lamphouses, the centring controls of the lamp and the reflector must be adjusted so that the image of the arc produced by the reflector is adjacent to the arc but not superimposed on it; if the image is superimposed on the arc, overheating will greatly reduce the life of the lamp. Also, the image of the arc must not lie on the quartz envelope; the quartz will recrystallize at this spot, resulting in an explosion of the lamp. The collective images of the arc, and of its image produced by the reflector, are imaged side by side into the aperture of the condenser by the collector lens. If no reflector is provided, the arc is centred onto the optical axis, and focussed into the aperture diaphragm of the condenser.

Cooling is important. A heat radiator can be attached to the upper terminal of the lamp, and is particularly recommended for xenon arc lamps. Adequate ventilation must be provided, both for cooling and for removal of ozone produced by the ultraviolet (UV). However, the lamp should not be allowed to stand in a draught. Philips, in respect of their CS series of mercury arc lamps, recommend maximal permissible temperatures of 700 °C for the envelope and 200 °C for the metal cap.

Mercury and xenon lamps have already been described (Chapter 3). Different excitation filters, including different heat filters, may be required for the two types of lamp, particularly in the case of dia-illumination. Also, different collector lenses (of different focal length) are required for different lamp types, according to the size of the arc. Check therefore that the microscope manufacturer's recommendations are followed in matching lamp, collector lens, lamphouse, and filters.

The life of the arc lamp is determined in part by the number of times the lamp is switched on and off; hence, the lamp should not be switched off if it will be required again within a short period – say up to an hour. The rated mean life of about 200 h for 100 W and 200 W mercury lamps refers to a duty cycle including about 70 switchings on and off, with about 3 h of operation on each occasion. Shorter operating times result in a reduced expectation of life. The life of a lamp is greatly reduced by burning for less than 20 min at a time. The life of the lamp is also reduced by operation at the wrong voltage, due to incorrect settings for either mains voltage or lamp type (L1 or L2 types of mercury arc lamp).

As the lamp becomes older, the inside of the bulb becomes coated with metal evaporated from the electrodes. This metallic coating absorbs both light and heat. In general, the lamp should be replaced when the light becomes too dim for satisfactory use, or when a sharp image of the arc can no longer be obtained, whichever comes first. Remember that a blackened lamp is more likely to explode, and that the effects on the lamphouse of an explosion are expensive to repair (see Fig. 5.3).

Tungsten-halogen lamps

The principle of operation of these lamps is based on the following features of the lamps:

1. In order to obtain maximal brightness, the filament should be run at or slightly above its rated voltage.
2. The life of the lamp is greatly reduced by running above its rated voltage.
3. The lamp's emission spectrum includes mainly long wavelengths, particularly infrared (which heats the specimen), and virtually no UV.
4. The lamp is designed for operation at high temperature. If the lamp is run at lower voltage, a black deposit will tend to form on the inside of the quartz envelope. This is a condensation of tungsten, which has evaporated from the filament during

operation. Ordinarily, the halogen (iodine) tends to prevent this deposition and to encourage redeposition on the filament instead.

A new lamp must be placed into its socket carefully. The lamp should be held in its sleeve, or a piece of paper, or with clean gloves, to avoid grease from the fingers getting on the envelope, where it will burn on and leave an opacity. Usually the lamp is supplied in a plastic or cardboard sleeve for this purpose. Care must be taken, when inserting the lamp into its socket, not to strain the pins sideways – this may crack the base. The lamp can normally be set up in any orientation (unlike gas-discharge lamps). The lamp should be centred, if provision is made for this. The procedure is identical with that outlined above for arc lamps, except that no grey filter is ever required as the lamp can be run at a lower voltage when full brightness is not appropriate.

Excitation filters for tungsten lamps are basically similar to those for xenon gas-discharge lamps; however, no precautions need be taken to filter out UV. Unlike mercury lamps, no spectral lines are available for selection; however, since the emission spectrum is continuous, filters with any peak wavelength from blue to red can be used. Adequate heat filtration is important.

One of the advantages of the tungsten lamps is that they can easily be turned off and on, provided that the current is limited when first switching on; such limitation can be achieved by means of a rheostat (for alternating current (AC) or direct current (DC) operation) or a thyristor dimmer (for AC operation only). Reducing the current (or voltage) when not in use increases the life of the lamp.

Selection of filters

Initially, an empirical approach is often best, trying different filter combinations to see which gives the best results. Logical selection of filter combinations requires a knowledge of the excitation and emission spectra of the fluorophore, together with a consideration of the aims or objectives of the microscopy. Some important considerations in the choice of excitation and barrier filters are discussed in Chapter 7. A record should always be kept of the *wavelength band of excitation* and the *barrier filter cut-off wavelength*, and these data included in any publication.

The excitation filter should in general be a bandpass filter transmitting only in the region of the excitation peak of the fluorophore. This is 'narrow-band' excitation. Sometimes, particularly if two or more fluorophores are to be visualized simultaneously, a wide-band excitation is used; this usually involves all wavelengths from UV up to the maximum suitable for the barrier filter.

A bandpass filter for specific excitation can be made up in three ways:

1. A glass filter with a narrow transmission peak. At present only the combination of Schott UG1 and BG38 glasses appears to fit in this category; UG1 transmits the 365 nm mercury lines, plus red (which is blocked by the BG38).
2. An interference bandpass filter, together with a glass filter to eliminate higher-order transmission.

3. A glass filter with a broad bandpass, together with a temperature-coloured glass of the barrier filter type to block short wavelengths. Thus for example BG12 can be supplemented with a longpass filters such as GG475 (and the obligatory BG38) to give reasonably selective blue transmission.

The barrier filter should always be chosen to transmit as much as possible, particularly as much as possible of the emission peak of the fluorophore(s), considering the nature of the excitation. Special bandpass filters to select the emission peak of the desired fluorophore, such as those for FITC fluorescence, should never be used routinely, because they may make extraneous fluorescence appear the specific colour (in this case, like FITC). Their proper purpose is to check on the specificity of fluorescence by comparison of the appearance with and without the specific filter. They are also useful to achieve maximal contrast, e.g. for photometry, and may be optimal for black-and-white photography.

The first step in selecting filter combinations is to identify sets of filters, each consisting of an excitation filter, a barrier filter, and a dichromatic mirror (if applicable). These sets may be identified by the manufacturer, and even linked together physically as in the filter blocks of Leitz and Reichert. For dia-illumination, separate excitation and barrier filters may be supplied by the manufacturers. The procedure to be followed to identify matching sets of excitation and barrier filters is as follows:

1. Insert an excitation filter; this may consist of more than one piece of glass, e.g. UG1 + BG38, or BG3 + K390 + BG38; it is assumed that a heat filter is already present.
2. Insert the longest-wavelength barrier filter (usually red).
3. Insert an appropriate test specimen.
4. Change the barrier filter progressively to the shortest-wavelength filter which still gives a dark background. This is the barrier filter which should be used with that excitation filter.

After the sets of matching filters have been identified, each set can be tried in turn and the best chosen on an empirical basis.

If the excitation maximum of the fluorophore is known, an appropriate excitation filter can be chosen. Data on excitation maxima are given in Volume II, Chapters 9 to 13 and Appendix 7. Commonly used systems, based mainly on the lines of the mercury arc spectrum, are listed in Table 5.2.

The arrangement of filters

The heat filter must be the nearest filter to the lamp, to protect the others from overheating. It is usually mounted permanently in the lamphouse. It must be loose in its mount, to allow for expansion and contraction as it heats and cools. Probably the most commonly used heat filter is KG1 glass (Schott), typically 2 mm in thickness. KG3, in my experience, is as good or better.

Table 5.2. *Common excitation systems*

Fluorescence excitation is most commonly carried out with a mercury arc lamp. Because of its line spectrum, excitation filters are commonly designed to transmit one or more of these lines. This table lists commonly used systems. It should be possible to identify, for each commercial microscope, a corresponding filter combination which can be selected. Only the approximate predominant wavelengths (mercury lines) are listed.

Name	Wavelengths of excitation (nm)	Barrier filter (cutoff, nm)
UV	364–366	410–430
UV-violet	365 + 406	430–460
Violet	405–406	430–460
Wide-band violet	365, 406, 439	480
Violet blue	439	480
Blue	475	515
Green	546	610

TEST OBJECTS

There is nothing quite as daunting as trying to set up a fluorescence microscope with a specimen of weak or doubtful fluorescence. Some standard preparation, with bright to brilliant fluorescence, is essential. A commercial test slide is supplied by BBL (Fluorescence Demonstration Slide, cat. no. BBL 40814 (X09–209); BBL division of BioQuest, PO Box 175, Cockneysville, MD 21030, USA; in UK, Beckton, Dickinson UK Ltd, York House, Empire Way, Wembley, Middx). This preparation consists of scattered particles of fluorescent material mounted under a coverslip on a standard glass slide. The particles are of three types, fluorescing respectively blue, yellow-green and orange-red. The average particle size range for each of the three is from 3·5 to 7 μm, making them very suitable for microscopy. In my experience, the fluorophores are reasonably stable and slides last indefinitely. A similar test slide can be made by embedding in DPX (or other non-fluorescent medium) a variety of pigments which are used in fluorescent paints.

In use, the test slide is placed on the stage of the microscope (right way up!) and examined. At low power, the particles are sufficiently closely and uniformly spaced for the alignment of the excitation optics to be carried out. Some idea of the specificity of the excitation can be obtained by noting the relative brightness of the three fluorophores. Useful though the BBL slide is, final adjustment of the microscope must always be carried out with a specimen of the specific fluorophore under consideration. Another kind of specimen suitable for general use is a tissue section stained with a reasonably permanent fluorochrome. Suitable dyes are listed in Table 5.3.

Sometimes a uniformly fluorescent test object is required. A piece of white paper is suitable for low-power use – most modern papers are impregnated with fluorescent

Table 5.3. *Suitable fluorescent dyes for test preparations*

These are recommended as giving relatively stable preparations which can be stored and used when required. See also Vol. II, Chapter 11 and Appendix 7.

Excitation wavelength	Dye	Fluorescence colour	Comment
UV	Calcofluor White	Blue	
	Thioflavine T	White	On amyloid
	Xylenol Orange	Red	
Violet	FITC	Green	
	Acriflavine	Yellow	
	Lucifer Yellow	Yellow	
Blue	Acriflavine	Yellow	Direct or Schiff
	FITC	Green	
	Eosin	Green	
Green	Basic Fuchsin	Red	Direct or Schiff
	Congo Red	Red	

brightening agents (see Volume II, Appendix 7) which fluoresce strongly under UV irradiation. Alternatively, a piece of an X-ray intensification screen may be cut and stuck onto a slide. Uranium glass (see Schott catalogue) is relatively resistant to fading. Fluorochromes can be incorporated into methacrylate and other clear plastics; I have several such pieces made by Koch-Light.

ADJUSTMENTS FOR OPTIMAL VISUALIZATION

The adjustments which are usually required are minor adjustments to the focus of the collector lens to give brighter or more even illumination, reduction of the field diaphragm opening, and sometimes a change of filters.

Maximizing brightness

For reasons which may be intuitively evident, and will be explained in Chapter 7, it is usually desirable or even mandatory to make the final microscope image as bright as possible; or at least, to avoid wasting any of the precious emitted photons.

The brightness of the image is related to three factors:

1. The effectiveness of the excitation.
2. The characteristics of the specimen.
3. The efficiency and magnification of the viewing system.

The effectiveness of the excitation depends upon the brightness of the illumination at the appropriate wavelengths (any other wavelengths are ineffective). Thus the brilliance of the fluorescence is increased both by increasing the overall intensity of the illumination and by illuminating at the most appropriate wavelength.

The brightness of the image is maximized by the following:

1. Using the most powerful light source available.
2. Using the most intense light source available, always provided that the source is large enough for its image to fill the aperture of the condenser system.
3. Correct adjustment of the lamphouse mirror.
4. Correct adjustment of the lamphouse collector lens, to focus the entire illuminating beam into the exit pupil of the objective (epi-illumination) or condenser aperture (dia-illumination).
5. Use of a collector lens of appropriate focal length.
6. Avoidance of frosted glass in the illumination system, e.g. on the collector lens or special diffusion discs.
7. Appropriate selection of excitation filter, which involves:
   (a) correct wavelength, to select light of the wavelength corresponding to the absorption peak of the fluorophore;
   (b) a wide-band filter, with as wide a bandwidth as is consistent with other requirements;
   (c) avoidance of an unnecessarily optically dense (thick) filter.
8. Appropriate selection of dichromatic mirror, if applicable.
9. Maximizing the effective NA of the condenser system. With a substage condenser, this will entail fully opening the aperture diaphragm, and, if the NA is greater than 0·9, oiling the condenser to the slide. In epi-illumination systems, the NA of the objective should be as high as possible. The illumination must be correctly focussed into the aperture of the condenser, which in turn must be correctly focussed.
10. Of substage condensers, the brightfield variety has a higher transmission than the darkground kind, because of the latter's central stop.
11. Of substage darkground condensers, those with the smallest central stop are to be preferred; this means the low-power variety if possible.
12. The image of the illumination source (Köhler diaphragm) should be demagnified as much as possible, consistent with filling the field. This means an appropriate focal length for the condenser. In epi-illumination systems, this means an objective with as high magnification as possible.
13. The objective's NA should be as high as possible.
14. The barrier filter should have as high a transmission as possible in the relevant wavelength (colour) range.
15. All unnecessary air–glass interfaces should be eliminated, as these cause loss by reflection at the surfaces. Items which may be dispensable include lenses in the nosepiece and intermediate magnification changers. Some microscopists believe that binocular heads should be avoided for the same reason; however, I have never found reduction of light to be a great problem in binocular heads, and this aspect is less important than the advantages of using both eyes – except of course for observers who prefer to use only one eye.

16. All air-glass interfaces in the viewing system should have anti-glare coatings.
17. The overall magnification should be as low as possible. This means using low-power oculars, and no supplementary magnifying system.

Evenness of illumination

The illumination of the microscope field should be uniform, especially for photomicrography. Such uniformity is notoriously difficult to obtain. It requires careful adjustment of the positions of the arc lamp, the reflector, and the collector lens. If these adjustments fail, a diffusing 'filter' can be used, at the expense of some loss of brightness.

The aperture of the condenser, i.e. the aperture iris of a substage condenser for dia-illumination or the exit pupil of the objective for epi-illumination, must be reasonably fully and evenly filled with light. In the case of epi-illumination, the matter is complicated by variation in diameter of the exit pupil of objectives when changing from low to high power and vice versa. In principle, the collector lens needs to be readjusted each time the objective is changed.

Flickering is a major problem with mercury arc lamps. It can be minimized by careful burning-in on first use (see above). The power supply unit should be protected from major variations in supply voltage – a voltage stabilizing unit may be required.

Reducing fading

The majority of fluorescent specimens fade during excitation, some quite rapidly. Fading (photobleaching) was mentioned in Chapter 2. In summary, photobleaching can be reduced by using as little irradiation as possible, choosing optimal wavelengths for excitation, and appropriate choice of mounting medium. Photobleaching is discussed in *Quantitative fluorescence microscopy* (Rost, 1991).

The intensity of irradiation may be reduced by closing the aperture diaphragm of the illumination system. An alternative is the use of a neutral-density (ND) filter: however, the latter becomes extremely hot, as under these conditions most of the illumination goes to heating the filter.

COMBINED ILLUMINATION TECHNIQUES

Double excitation by simultaneous epi- and dia-illumination

This technique is used for the simultaneous demonstration of two different aspects of the specimen, using two different fluorophores (usually two different fluorochromes, as in double labelling by immunofluorescence or neuronal tracing).

Simultaneous examination of two fluorophores with widely different excitation maxima requires excitation either over a broad bandwidth to include the excitation maxima of both fluorophores, or else simultaneous illumination at two wavelengths, to suit the excitation maxima of the two fluorophores. The former method is usually simpler and less expensive, but may give greater fading and less optical specificity than two-wavelength excitation.

Simultaneous excitation at two wavelengths is most easily carried out by using epi-illumination for one and dia-illumination for the other (Fig. 5.1). This technique appears to have been proposed first by Barer (1968). It makes no difference in principle which of the two wavelengths is directed by epi-illumination or dia-illumination; however, it is usually more convenient to provide the longer wavelength by dia-illumination, because of placement of the barrier filters (see below); also, if the longer wavelength can be supplied by a tungsten-halogen lamp, since this lamp will normally be placed in the transmitted-light path it is easiest to keep it there.

The barrier filter must be chosen to suit the longer wavelength of excitation. It may be helpful to have readily available a barrier filter to suit the shorter wavelength, so that the specimen can be examined under optimal conditions for each fluorophore separately if required. This arrangement is most easily achieved by the following procedures:

1. Epi-illumination at the shorter wavelength, with all appropriate filters (excitation, dichromatic mirror, and barrier) in a block.
2. Dia-illumination at the longer wavelength, with separate excitation and barrier filters; the barrier filter being in a slide in the body tube.

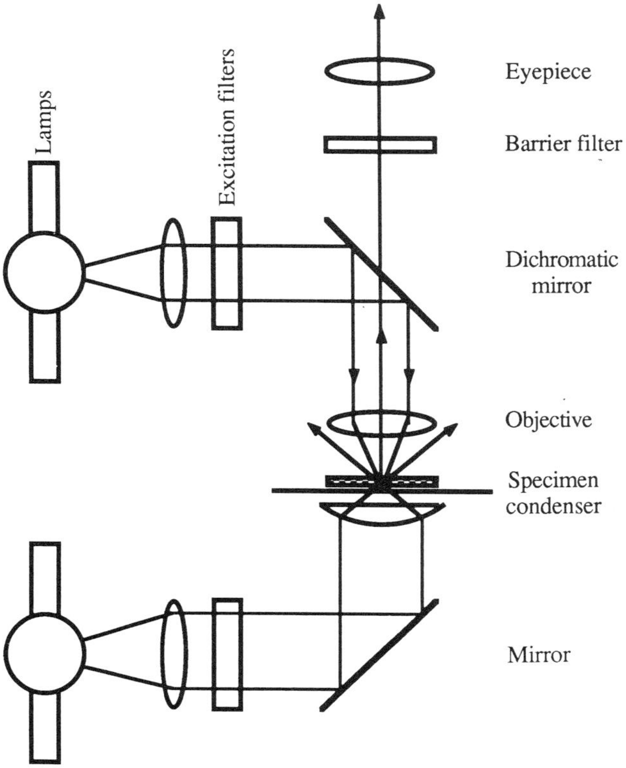

Fig. 5.1. Optical arrangement for simultaneous two-wavelength excitation of fluorescence (simultaneous incident- and transmitted-light fluorescence). The two excitation filters are different, e.g. one giving blue light and the other UV or green. The barrier filter must suit the longer excitation wavelength.

3. Any convenient arrangement to block either light path as required – for example, the usual shutter in the epi-illuminator, and a shutter or movable mirror in the mirror-house supporting the dia-illumination lamp.

This arrangement, with all filters in place and both light sources used, provides simultaneous excitation at both wavelengths and an appropriate barrier filter for the longer wavelength.

Instead of simultaneous excitation at two wavelengths, alternate excitation at each wavelength, together with the appropriate barrier filter for each, can be employed. With the above arrangement, examination at the longer wavelength (only) requires merely blocking the epi-illumination pathway (the additional barrier filter and dichromatic mirror cause some light loss but not much); examination at the shorter wavelength requires blocking the dia-illumination pathway and removal of the longer-wavelength barrier filter. However, the more usual method for alternate use of two wavelengths involves shutting off the dia-illumination pathway and changing the epi-illuminator from one wavelength setting to another – some manufacturers make alternation between different filter systems particularly simple and quick.

Fluorescent preparations typically consist of a few bright spots or small areas against a background which is black or nearly so (I call that kind of background a 'blackground'!). This means that it can be quite difficult to identify the various parts of the preparation. Recognition of parts of the specimen can be made much easier by simultaneous or alternate illumination by other means appropriate to the preparation, using brightfield, darkground, phase-contrast or interference-contrast, or polarization. In order not to mask the fluorescence, I find it essential to tint this additional illumination to a colour which contrasts with the fluorescence. Because of the barrier filter, the contrasting colour has to have a longer wavelength than the fluorescence. With blue, green or yellow fluorescence I have usually employed red for contrast. Red has the great advantage that, even if the light is quite bright, it does not spoil dark adaptation of the observer's eye. With red fluorescence no great colour contrast appears to be possible. The intensity of the background illumination must be adjusted so as to balance the brightness of the fluorescence. The filter for the contrast colour need not be elaborate; suitable red filters can be obtained from a photographic supply house. The gelatine filters of Cokin and Kodak (see Chapter 3) are very suitable; they should be mounted in glass to protect them, e.g. by mounting in a 'Super Slide' mount. At a pinch, coloured Cellophane may serve.

Simultaneous fluorescence/darkground illumination

For some purposes it is convenient to excite fluorescence by epi-illumination, while using simultaneous or alternate darkground illumination to demonstrate the specimen generally (and particularly non-fluorescent areas).

The system is set up in a manner very similar to that for combined epi- and darkground dia-illumination for fluorescence, the difference being that the dia-illumination light, when not used to excite fluorescence, need be less bright, and should usually be coloured red or other contrasting colour by a filter; the barrier filter is that appropriate to the excitation (Fig. 5.2).

With fluorescence dia-illumination, a beamsplitter is placed in the excitation pathway (usually in a mirror-house) to permit simultaneous use of two lamps. The beamsplitter consists of a sheet of plane glass, placed at an angle of 45° to the optic axis. This gives strong transmission for the excitation light, and weak reflection of a tungsten-filament light for the darkfield illumination. Appropriate filters are placed in the two lamphouses, and the balance of illumination is adjusted by varying the current in the tungsten-filament lamp.

As an alternative to the above, if the excitation filter is a combination of a glass such as UG1 and the red-stopping filter BG38, the latter can simply be removed. However, in such a case, the degree of red light available for darkground illumination cannot be controlled.

Applications. This combined technique is particularly suitable for examination of specimens containing particles, some of which are fluorescent and others non-fluorescent. The method can also be used for simultaneous demonstration of fluorescent objects and refractile, uncoloured or coloured structures, and can sometimes be used for the demonstration of the structure of tissue surrounding fluorescent material and the localization of the fluorescent components in tissues. Simultaneous phase-contrast may be better for the latter purposes. Simultaneous darkground is usually very simple to set up, requires relatively inexpensive equipment (see Appendix 3A), and can

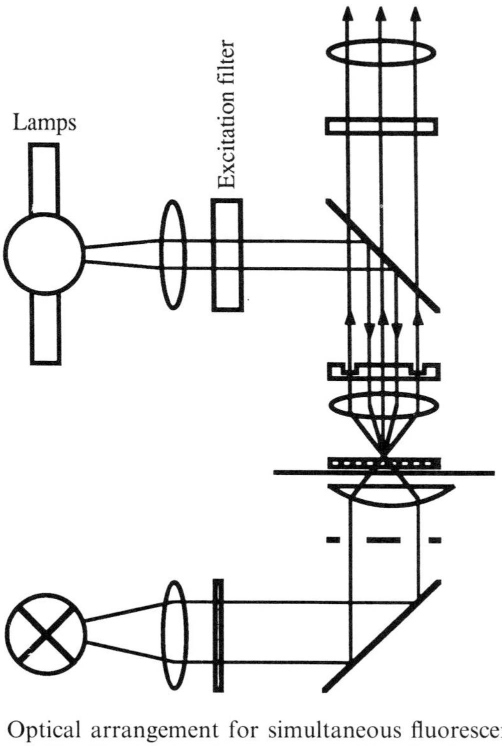

Fig. 5.2. Optical arrangement for simultaneous fluorescence and phase-contrast microscopy. The filter in the phase-contrast illumination pathway is optional, and should contrast with the fluorescence; it must be of a wavelength transmitted by the barrier filter (red is usually suitable).

be used with objectives of any type provided that their NA does not exceed that of the condenser's dark cone.

Combined fluorescence and non-fluorescence incident-light

A method for double incident-light illumination, for combined fluorescence and non-fluorescence, was described by Bagge (1981) for examination of living tissue by vital fluorochromy and reflectance microscopy. The method, on a Leitz microscope, employs a combination of the PloemOpak and the Ultropak illuminators. In the PloemOpak, the objective serves both for illumination and for observation: the backscatter of light from the objective makes this unsuitable for reflectance microscopy of low-reflecting living tissue. In the Ultropak, the illumination system surrounds the objective, so that the illumination does not pass through the objective. Bagge (1981) mounted the Ultropak below, i.e. in series with the PloemOpak.

Alternatively, a beamsplitter system similar to that described for dia-illumination with simultaneous darkground illumination can in principle be used with epi-illumination, but requires some modification to the filter block (the excitation filter being removed to the lamphouse) and the dichromatic mirror produces loss of illumination intensity at the longer wavelength. The system has the advantages that it can be used with any objective (however high the NA) and the intensity of the secondary illumination varies in proportion to the intensity of the fluorescence automatically when objectives are changed.

In principle, either of these techniques can be extended to simultaneous fluorescence excitation at two wavelengths.

Simultaneous brightfield illumination

The description given above concerning simultaneous fluorescence/darkground illumination can be adapted to brightfield illumination, the only difference being that a conventional brightfield condenser is used instead of a darkfield condenser. This is usually very easy to set up, requiring only the usual epi-illumination fluorescence arrangement and that the lamp for conventional brightfield viewing is turned on instead of off. A filter, of contrasting colour to the fluorescence, should usually be placed below the substage condenser or in front of the lamp. This is a simple technique to assist in localizing the fluorescence in the specimen; however, unless the specimen is coloured, other techniques are usually much more effective.

Simultaneous phase-contrast

Phase-contrast is a method for converting differences in phase in light passing through objects, which the eye is unable to appreciate, into visible changes in brightness. For details, see Beyer (1977).

Phase-contrast requires a substage phase-contrast condenser and special phase-contrast objectives (except in the Reichert system and the Zeiss Axiomat, where the phase-rings normally sited in the objectives are placed above them). Phase-contrast objectives are not at present designed for fluorescence work and some loss of light is to be expected, and indeed is inevitable because of the phase ring.

Some manufacturers offer more than one type of phase-contrast objective, varying in sign and degree of contrast. In *positive phase-contrast*, relatively dense parts appear dark against a light background; in *negative phase-contrast*, relatively dense components appear bright against a dark background, rather like fluorescence (I use the word 'dense' rather loosely here, to refer really to a longer optical path length; also, the general statements above apply only to fairly small differences in optical path length). Using epi-illumination, there are no special problems in setting up – the phase-contrast optics are used instead of ordinary objectives and condenser.

Special phase-contrast/fluorescence condensers were used for a few years from 1957, before epi-illumination became common. In these condensers, special phase-rings were provided in which the normally opaque regions were replaced by excitation filter glass (e.g. UG1). This allowed normal phase-contrast (by light passing through the transparent region) and excitation by dia-illumination (by light passing through the filter glass). Although workable, the method was cumbersome, the phase-rings expensive and non-permanent (the filter glass being subject to UV damage) and the condensers required numerous rings – a set being needed for each wavelength of excitation. The illumination system required also a corresponding excitation filter set in the arc lamphouse, and a beamsplitter to admit light from a separate tungsten-filament lamp. The voltage on the latter was varied to adjust the brightness of the phase-contrast system.

The technique of combined fluorescence/phase-contrast is usually the method of first choice for unstained specimens, the phase-contrast enabling the observer to determine the location of the fluorescent portions of the specimen (see Plate 1). In my experience, it gives a more uniform illumination than darkground, and usually gives a better demonstration of the structure of the tissue.

Choice of illumination mode

As in all microscopy, optimal illumination is very important. The mode of illumination must be adapted to suit the subject. The majority of fluorescent subjects can be classified according to their needs as regards illumination, into the following categories:

1. Solid, opaque objects require epi-illumination for excitation and also for normal (non-fluorescence) viewing.
2. Dispersed, microscopic opaque objects require epi-illumination for excitation. Non-fluorescent viewing by darkground illumination (simultaneous or alternate) or alternate epi-illumination.
3. Opaque, partly perforated or partly transparent objects (e.g. cloth, metal screens, some minerals) require epi-illumination for excitation. Non-fluorescent viewing by transmitted light, either alone or in combination with epi-illumination.
4. Solid, translucent, non-birefringent objects (e.g. sections of minerals or biological tissue). For excitation, epi-illumination preferred at medium and high powers; dia-illumination (darkground condenser) preferred at low magnifications. Combined or alternate excitation with epi- and dia-illumination at different wave-

lengths possible. Non-fluorescent viewing by any transmitted-light technique (including phase-contrast, interference-contrast).

5. Solid, translucent objects with birefringent components may be observed with excitation by epi-illumination, and non-fluorescent viewing by transmitted-light polarization, or any of the techniques listed above (see (4), above).
6. Dispersed, microscopic translucent objects are best seen with excitation by epi-illumination, and non-fluorescent viewing with darkground transmitted-light illumination. Also, where appropriate, any of the techniques listed under (4) and (5), above.

COMBINED FLUORESCENCE AND ELECTRON MICROSCOPY

There are obvious advantages in being able to carry out both fluorescence microscopy and electron microscopy (EM) on the same section or block. If it is feasible to use separate stains, the matter resolves to a technical matter of section preparation, such that thin sections can be cut for EM and 'thick' sections from the same block examined for fluorescence. Some specialized methods already exist for section preparation in such a manner as to enable examination of the same section by both fluorescence and EM, using different stains to study elastic tissue (Albert, 1973) or neurons (Furness, Costa & Wilson, 1977).

The ideal is a single reagent which is visible by both fluorescence microscopy and EM; such a reagent could be either a fluorochrome containing heavy atoms, or a fluorochrome conjugated to electron-dense material (Rost, 1972, 1980). For example, mercury is found in Merbromin (Mercurochrome 220, a fluorescein derivative), Fluorescein mercuric acetate and Fluorescein dimercuric acetate. The first success with a combined reagent appears to have been that of Molday *et al.* (1975), who visualized latex microspheres with attached fluorochromes by fluorescence microscopy and scanning electron microscopy (SEM); fluorescent microspheres were later demonstrated in transmission EM by Egensperger & Holländer (1988) using negative staining. Lectins were detected on cell surfaces with fluorescent derivatives of glycosylated ferritin (Kieda *et al.*, 1979). Horisberger & Vonlanthen (1979) introduced a combined label consisting of colloidal gold granules, 12 mm diameter, linked to the fluorochrome Tetramethyl Rhodamine isothiocyanate (TRITC) and thence to a lectin (Concanavalin A) either directly or through an avidin-biotin link. The double label gold-TRITC-avidin should be widely applicable. Lucifer Yellow, a dye used as a fluorescent tracer for neurons, was visualized also by electron microscopy by Maranto (1981). Thorball (1982) described the electron microscopy of fluorescent dextrans. Ferrer, Garrigues & Stockert (1983) were successful in using Merbromin as a combined stain for fluorescence and EM; however, Novello & Brauer (1986) obtained substantially different results in EM, finding that Merbromin was a specific stain for condensed chromatin. Fluorescein mercuric acetate has been used as a combined fluorescence/EM stain for -SH groups by Biggiogera & Pellicciari (1988).

Combined fluorescence microscopy and SEM is possible with an instrument with both functions (Hartmann *et al.*, 1978; Ploem, 1981; Ploem & Thaer, 1981; Wouters & Koerten, 1982; Wouters & Ploem, 1987).

SPECIAL FLUORESCENCE TECHNIQUES

Ultraviolet excitation below 365 nm

This is a difficult technique. Because ordinary optical glass does not transmit below about 350 nm, all optical elements in the illumination system must be made of quartz, fluorite or other UV-transmitting materials.

Xenon arc lamps are usually used, as they have relatively more UV below 360 nm than mercury arcs. The lamp bulb is made of quartz in any case, so this is not a problem. The condenser lens must be made of quartz. The normal heat filter (Schott KG1) cannot be used; it should therefore not be mounted permanently in the lamphouse but arranged to be swung or otherwise moved in and out as required. For epi-illumination, a special epi-illuminator with quartz optics, and quartz objectives (e.g. Zeiss Ultrafluar) are required. For dia-illumination, the substage mirror must have its reflective coating on the front surface (aluminium is better than silver for the coating) or be made of quartz, and a quartz condenser is required. In principle I suppose that a UV darkground condenser of the mirror type could be constructed with front-surface reflecting surfaces.

Excitation below 350 nm also brings problems with the specimen slide and coverslip. Both are available in quartz, at a price. To save expense, the slides are made much smaller than standard, just large enough for the specimen, and are themselves mounted in a metal holder. Naturally, slides and even coverslips are usually too expensive to use only once, and are cleaned for re-use. Strictly, since only the excitation is below 350 nm, only one or other need be quartz; for dia-illumination, only the slide need be of quartz, and for epi-illumination only a quartz coverslip and a non-fluorescent slide are required. The latter arrangement is less expensive.

The mounting medium and immersion fluid (if used) also need to be transparent to UV. Useful substances include water, glycerol and liquid paraffin, which are transparent to UV and non-fluorescent (see Table 5.1 and Volume II, Chapter 17).

Fluorescence microscopy with UV excitation and emission has not yet, as far as I know, been carried out. However, the technology exists: UV-transmitting optics, and a UV-sensitive video camera to observe the image (see Chapter 8).

Infrared fluorescence

Fluorescence emission in the infrared is widely known in chlorophylls, and has also been observed from blue and green dyes such as Indocyanine Green. The fluorescence is invisible to the eye, but can be recorded on infrared-sensitive photographic film (Dhéré & Biermacher, 1936; Elkin & Park, 1975; Puchtler, Meloan & Paschal, 1980) or examined with an infrared-sensitive video camera (Moneta *et al.*, 1987).

Long working distances

For some purposes, it is necessary to be able to maintain a longer than usual distance between the specimen and the front of the objective. For example, studying the surface of an isolated organ in a chamber (Moravec *et al.*, 1972) or tissue in a cooling chamber or cells in a culture chamber. While fluorescence microscopy is still possible, special problems arise. The main problem is that long working distance for an objective or condenser is incompatible with high NA. In general, the longer the distance required, the lower the power possible for the objective (and condenser, if required). Objectives of the same magnification and NA vary in working distance according to the design, and it is worth while shopping around to locate the most suitable optical equipment. Unfortunately, the procedure of using a high-magnification ocular to compensate for an unduly low magnification objective leads to low image brightness. With epi-illumination, the brightness of the image depends on the fourth power of the NA of the objective, and is independent of its magnification.

Substage condensers with long working distance are available. Leitz supply long working distance caps for the Series 600 condensers. These caps have a lower NA than the standard caps; the longer the working distance, the lower the NA (because the diameter of the condenser top is limited). For phase-contrast, special rings are available for use in conjunction with these caps.

Fluorescence in a dissecting microscope

A fluorescence epi-illuminator, made by Leitz, is available to fit certain of the Wild stereo microscopes. Besides providing stereo images, this arrangement extends the range of fluorescence microscopy into low magnifications, for examination and photography of larger objects.

For other stereo microscopes, the additional equipment required is a separate source of appropriate light, focussed onto the specimen from the side and above the stage. Note that tilting a high-pressure arc lamp may not be permissible – manufacturer's recommendations must be followed (see Chapter 4). Some mercury arcs can be used in a horizontal position, implying a conventional lamphouse turned on its side and rotated about the axis of the lamp to direct light onto the stage. Possibly quartz-fibre optics would be suitable to carry the light to the right place. Using present-day technology, a stand is required to accept a standard lamphouse, and a separate lens to act as a condenser. Such lenses were used in former days under the name of 'bull's eye' lenses. A photographic enlarging lens of appropriate diameter and focal length would be ideal. In addition, barrier filters would need to be fitted, either in the body tube or in pairs in the eyepieces.

In this service, 'black-light' lamps such as the Philips HPW may be appropriate; I have used a pair of HPW 125 lamps, set horizontally, with a low-power stereo microscope. See also Macrophotography in Chapter 6.

Fluorescence with an inverted microscope

In principle, the process of setting up fluorescence for an inverted microscope is exactly the same as for an 'ordinary' one.

Underwater fluorescence

There is no theoretical difficulty in the study by fluorescence of surfaces of objects under water, e.g. in a small tank using epi-illumination. Water is an excellent transmitter of

UV. Water-immersion objectives can be racked down into the water to the required distance from the specimen. For examination from underneath, an inverted microscope is appropriate.

Polarization

Polarization fluorescence microscopy presents a number of technical problems, and has not so far been widely used. For information on normal polarization microscopy, see Patzelt (1986), Ganse (1977), James (1976), Missmahl (1966) and Oster (1955). The conversion of a normal fluorescence microscope for polarization studies requires the addition of an analyser in the observation system, and possibly a polarizer in the excitation system; the latter is not always required. Insertion of an analyser presents no particular problem; it is commonly placed in the eyepiece, and can thereby be rotated. A straight monocular body is preferable in this context, if only to avoid complications due to the beamsplitter prisms in a binocular head. If an epi-illuminator is used, containing a dichromatic mirror or other plate beamsplitter, this will produce some polarization, the extent depending on how closely the Brewster angle of the beamsplitter approaches the 45° angle at which it is set. Because the orientation of this polarization is determined by the geometry of the beamsplitter, the polarizer and analyser must be oriented correspondingly. Under these circumstances, neither the analyser nor the polarizer can appropriately be rotated, at least for quantitative studies. Therefore, probably a metallurgical microscope with a prism illuminator in part of the objective aperture should be adapted. A fluorescence epi-illuminator is made for some Leitz polarizing microscopes (see Chapter 4); see also Eisert & Beisker (1980). Alternatively, a substage condenser can be used, thereby avoiding the problems of the epi-illumination mirror. A centrable rotating stage is required to rotate the specimen.

The production of plane-polarized UV or blue light for fluorescence excitation at the intensities normally required for fluorescence microscopy requires a polarizing device relatively resistant to high-density illumination. Suitable Nicol prisms of large size are available, unfortunately at rather high cost. In my experience Polaroid™ is not normally suitable for this service because it cannot withstand strong irradiation, but it has been used by dos Remedios *et al.* (1972).

Quantitative and scanning techniques

Microfluorometry (quantification) and microspectrofluorometry (measurement of fluorescence excitation and emission spectra) are described briefly in Chapter 8.

Video intensification microscopy. A video camera can be attached to a fluorescence microscope, for observation, recording and quantification (see Chapter 8).

LOCATION OF THE MICROSCOPE

Ideally, the microscope should be located in a room which can be darkened, and is well ventilated. The object of darkening the room is to allow the eyes to adapt to the low level of light. It is rarely necessary to have complete darkness, and neither is this convenient for working. Generally, an overall soft, dim light is best. It is best to avoid having a brilliant, dazzling spotlight to illuminate a brilliant white notebook. Generally, the most suitable level of lighting is dim enough for the iris of the eye to be fully dilated (to accept the exit pupil of the microscope) and bright enough to preserve scotopic vision (i.e. full colour vision; see Chapter 7). If colour is not significant, further darkening of the room to permit photopic vision will allow very much dimmer fluorescence to be seen. Red ambient lighting can be helpful, as visual adaptation is less affected. Red goggles are available (used by medical radiologists) to preserve adaptation during visits to the brighter world outside.

CARE OF THE FLUORESCENCE MICROSCOPE

Maintenance procedures and general care of a fluorescence microscope are basically those of any optical microscope, but there are a few additional points to which attention should be given.

The lamp used for excitation will need to be replaced from time to time. It is recommended that a record be kept of use of each lamp. Advice on extending the life of gas-discharge lamps has already been given. The lamp must be replaced when its bulb blackens significantly. Caution: examine it only when cold, and never while switched on!

Dust is a particular nuisance in fluorescence microscopes because it is often fluorescent. Therefore, dust should be kept off the microscope as much as possible, and the optical surfaces cleaned from time to time with a blower, mini-vacuum and/or brush. Dust accumulation can be minimized by keeping the microscope under a cover when not in use (caution: if the cover is plastic, it must not be allowed to touch a hot lamphouse, or the cover will melt). It may be possible to keep the microscope in a dark box with doors, like a cupboard.

Optical surfaces need cleaning from time to time. Dust must always be removed first, as otherwise it may grind into the surface. Coated surfaces must be treated with particular care. Optical surfaces are best cleaned, in the first instance, with a blower or a jet of compressed gas to remove dust. A soft brush may be needed to dislodge adherent particles. Any bad contamination will require cleaning with a soft linen rag, chamois leather, or paper tissue. Personally I prefer soft paper tissue to 'lens tissue', because it is softer, even though it tends to shed paper fibres which have to be blown off. Dust-free material is essential. A solvent is frequently required: distilled water, xylene, or a mixture of equal parts of ethanol and ethyl ether (caution: highly flammable!) or a commercial window-cleaning fluid. Ethanol/ether has the advantage that it evaporates quickly. However, a greasy film usually remains, resistant even to the above. The complete cleaning of small, concave lens surfaces such as the front lenses of some objectives is particularly difficult. Final cleaning is best accomplished with a freshly broken surface of polystyrene foam (Styrofoam, Styropore, Poresta), using a coaxial rotating motion (Pabst, 1974): any xylene used for preliminary cleaning *must* be allowed to evaporate completely first, so as not to dissolve the polystyrene.

Filters should be inspected regularly – at least annually – to see if they need replacement. The life of excitation filters is limited, particularly if they are subjected to overheating. The effect of excessive irradiation is usually to render the filter denser or more opaque, possibly in an irregular fashion, giving reduced transmission and an uneven field of illumination. Barrier filters are commonly made of several layers cemented together; sometimes the cement becomes partially unstuck, degrading the image.

Faults, if reasonably simple, can best be identified by checking the microscope systematically, commencing at the light source and ending at the oculars (or camera, if there is one). A list of some faults and their likely causes is provided in Appendix 3B.

General servicing of the microscope should be carried out regularly by a trained technician; this usually means having the instrument serviced by the manufacturer or manufacturer's representative.

DANGERS OF FLUORESCENCE MICROSCOPY

Because of the brightness of the lamp, there is a risk of dazzle if the barrier and/or excitation filter is removed while the observer is looking down the microscope.

Ozone is generated by the action of UV light on atmospheric oxygen. This ozone is mildly toxic, leading mainly to headache. Ozone is generated particularly by xenon arcs, because of their relatively high output of short-wavelength UV. If the microscope is situated in a confined space, e.g. a darkroom, a special ventilation system is required for these lamps. In such circumstances, I have used an extractor fan in the adjacent wall, about half a metre or less from the lamp; better, especially if the distance is greater or for the more powerful lamps, is a length of air trunking to collect the gases emanating from the top of the lamphouse. A strong draught, however, should be avoided as it is dangerous for the lamp.

Stray UV light is a bit of a nuisance. There is a general fear of UV light being dangerous to the eyes; however, the stray light from a well-designed microscope is very unlikely to do any harm, and the risk has probably been exaggerated. Eyes can be shielded from the bright light seen in the region of the specimen by a glass shield of the barrier-filter type. Such shields tend to be expensive and are not usually essential, except for inverted microscopes.

Explosions of high-pressure arc lamps sometimes occur. For this reason, the lamphouse must always be kept closed except for access. The lamphouse should be kept closed until at least 10 min after the current is switched off. If it is absolutely necessary to open the lamphouse while the lamp is lit, protective goggles and clothing must be worn. Although explosion is not common, the explosion is powerful – usually powerful enough to shatter the collector lens and filters! An example of the debris is shown in Fig. 5.3. Explosions are usually caused by weakening of the quartz envelope as a result of recrystallization of the quartz due to overheating. This overheating is likely to be caused by:

1. The reflector producing an image of the arc on the quartz envelope.
2. Blackening of the inside of the envelope, causing excessive heat absorption.

3. Mercury deposition inside the envelope, due to operation of the lamp at less than the rated voltage (the deposit absorbs heat).
4. Blackening of the envelope due to excessive ignition.
5. Deviation from the vertical. A tolerance of about 15° is usually allowed. Some mercury arc lamps can in principle be used in a horizontal position, but so far as I am aware these have not been used for fluorescence microscopy.

Take care when cleaning up after an explosion. First, switch off and allow the lamphouse and its contents to cool. Then remove (with care) all the broken glass and remnants of the lamp. Remember that mercury is toxic.

Chemical dangers

Microscopists should be aware also of the chemical dangers of substances used in preparation and mounting of biological tissues and other objects. For information, see Lenga (1988) and Bretherick (1981). Particularly hazardous substances are *organic solvents*: (flammable, toxic vapours), *immersion oil* (may be carcinogenic, due to trichlorophenol), and fixatives, e.g. *formaldehyde* (toxic vapour).

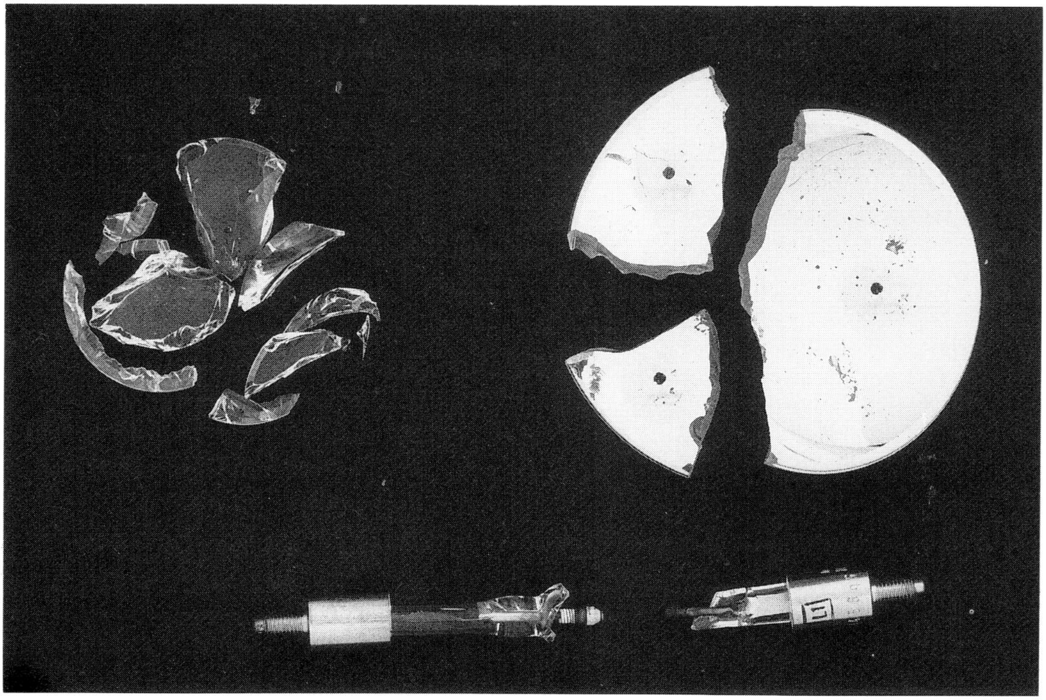

Fig. 5.3. The debris collected after a 200 W mercury arc exploded, showing pieces of the lamp, collector lens, and reflector.

6

Photography of fluorescence

This chapter deals predominantly with photomicrography of fluorescence, i.e. photography using a fluorescence microscope; there are also short sections on photomacrography of fluorescence and on ultraviolet (UV) photography.

PHOTOMICROGRAPHY OF FLUORESCENCE

Photomicrography of fluorescent specimens presents special problems, mainly because of the low light levels involved and fading of the fluorescent material. The solution of these problems is of particular importance because many fluorescent preparations are evanescent, and photomicrography is essential to obtain a permanent record. Because of the dimness of the fluorescence microscopic image, sometimes a photograph may reveal details which were not recognized when viewed in the microscope.

Not very much has been written specifically on fluorescence photomicrography: the literature includes brief reviews by Koch (1975) and Reynen (1967), and specific discussions in relation to immunofluorescence (Schuit, 1970) and porphyrins (Cripps, Hawgood & Magnus, 1966). Photomicrography was treated comprehensively by Loveland (1970); see also James (1976), Lawson (1972), and Croy (1968). Commercial fully automatic cameras for photomicrography were reviewed by Cowen (1969); although now out of date, this article forms a useful basis for assessment of current equipment. For information on photography in general, see Jacobson, Ray & Attridge (1988) and Blaker (1977).

Good photomicrography is nine-tenths good microscopy and one-tenth good photography, as I have remarked previously (Rost, 1980). It is essential to have a good-quality optical system and to adjust the microscope as finely as possible.

The low light intensity of the fluorescence image leads to two major difficulties in photography: measurement of the exposure (requiring a sensitive photometer) and the length of the exposure time. The former is a merely technical problem, solved by appropriate equipment which is commercially available. The long exposure, however, usually results in fading of the fluorescence under the prolonged irradiation. Fading of the specific fluorescence is often faster than that of the background, so that the photograph suffers from loss of contrast. There is a serious complication, due to the properties of the photographic emulsion: as the length of the exposure increases, the sensitivity of the film decreases, so that a still longer exposure is required to compensate, and during this additional exposure there is more fading, giving a vicious circle of fading, longer exposure required, loss of film sensitivity, longer exposure required . . .

PHOTOGRAPHIC MATERIALS

Photosensitive materials are central to any photography, and therefore will be discussed first. Lack of space precludes more than a brief outline of the relevant characteristics of photographic film. For further details, the reader should consult books on photography, and manufacturers' literature concerning individual films.

Black-and-white film

The actual photosensitive material, which reacts to light and ultimately forms the picture, is known as the *photographic emulsion*. This forms a thin layer on the *support* – film, glass or paper. The emulsion consists basically of a suspension of silver halide (mainly bromide and iodide) particles in a solution of gelatine, which is used to coat the support and allowed to dry. Optical sensitizers are also present; these are organic dyes which absorb light and transfer the energy of the photons to the silver halide. The silver salts in an emulsion are photosensitive; that is, they undergo a change when exposed to light. The precise nature of this change is not well understood; however, see James (1986). It appears that some of the the microscopic particles of silver halide acquire a minute black speck of silver, know as a sensitivity speck or centre. The total of such specks constitutes the *latent image*, which is not visible to the naked eye but is intensified by the process of development to produce a visible image. Strictly speaking, a visible image can be produced given sufficient exposure; however, the effect of development is to decrease the required exposure by an enormous factor, estimated at a thousand million times!

Negative material (film or plate) consists of a photographic emulsion coated onto a transparent support (film or glass). In addition, to avoid light not absorbed by the emulsion being reflected off the rear surface of the support and returning to the emulsion (causing halation around bright areas), the support is usually coated at the rear with an *anti-halation backing* which is water-soluble and dissolved during subsequent processing.

Black-and-white (B&W) film may be characterized in three main ways: sensitivity to different parts of the spectrum, speed, and grain. An emulsion containing only silver halides, without optical sensitizers, is sensitive only to the shorter wavelengths (higher quantum energies) of radiation: UV, violet and blue. Optical sensitizers present in the emulsion extend the sensitivity to include green and yellow (*orthochromatic* films) and also red (*panchromatic* films). *Speed* or *sensitivity* refers to the capacity of the film to record dim light; the faster or more sensitive the film, the weaker the light with which a photograph can be taken in a given exposure time. Numerical values for film speed

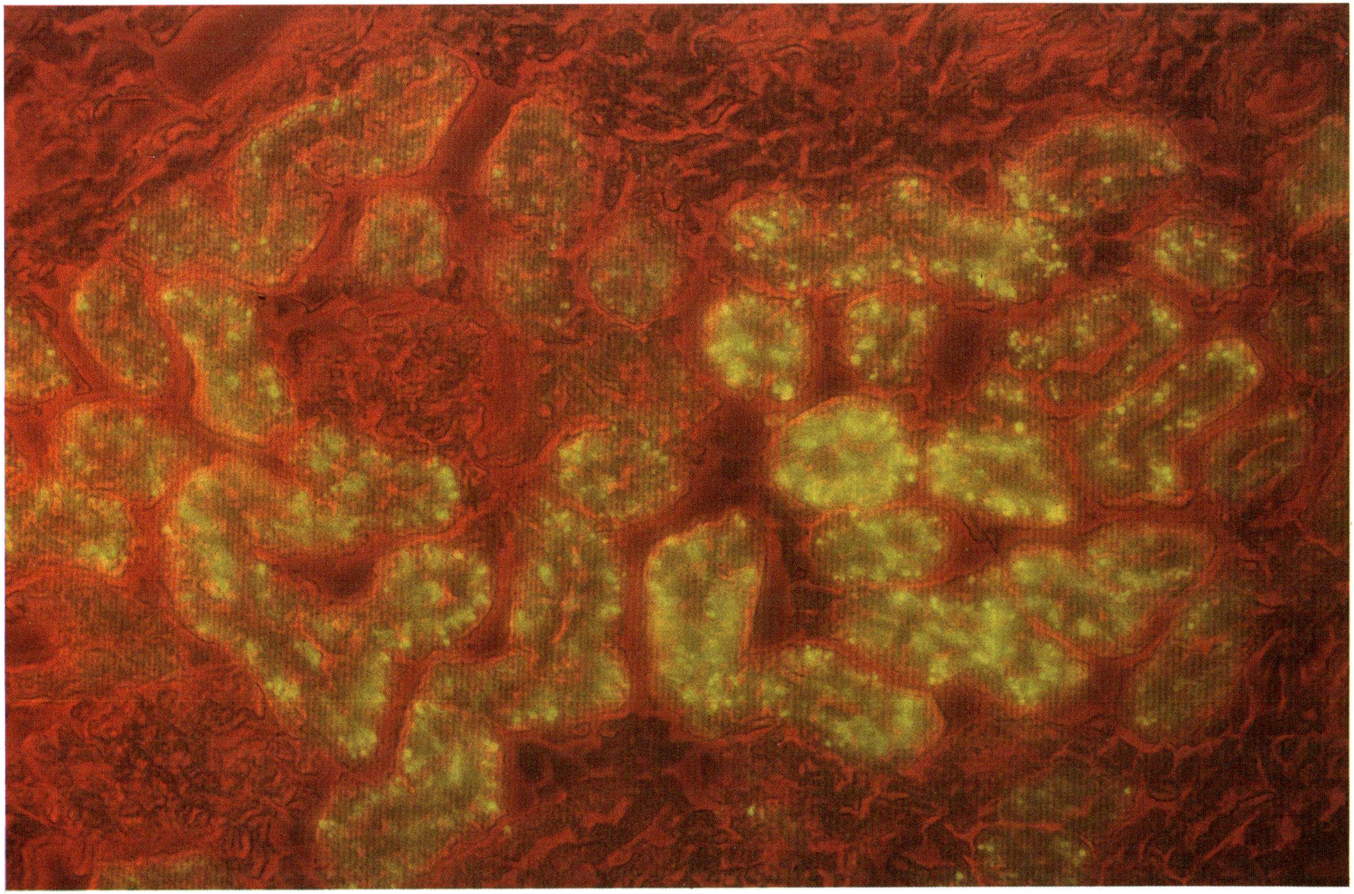

Plate 1. Combined fluorescence and phase-contrast microscopy. The phase-contrast image is red. Autofluorescence of lysosomes in renal tubular cells in rat. Original magnification $\times 130$, field size $1{\cdot}8 \times 2{\cdot}5$ mm.

(a)

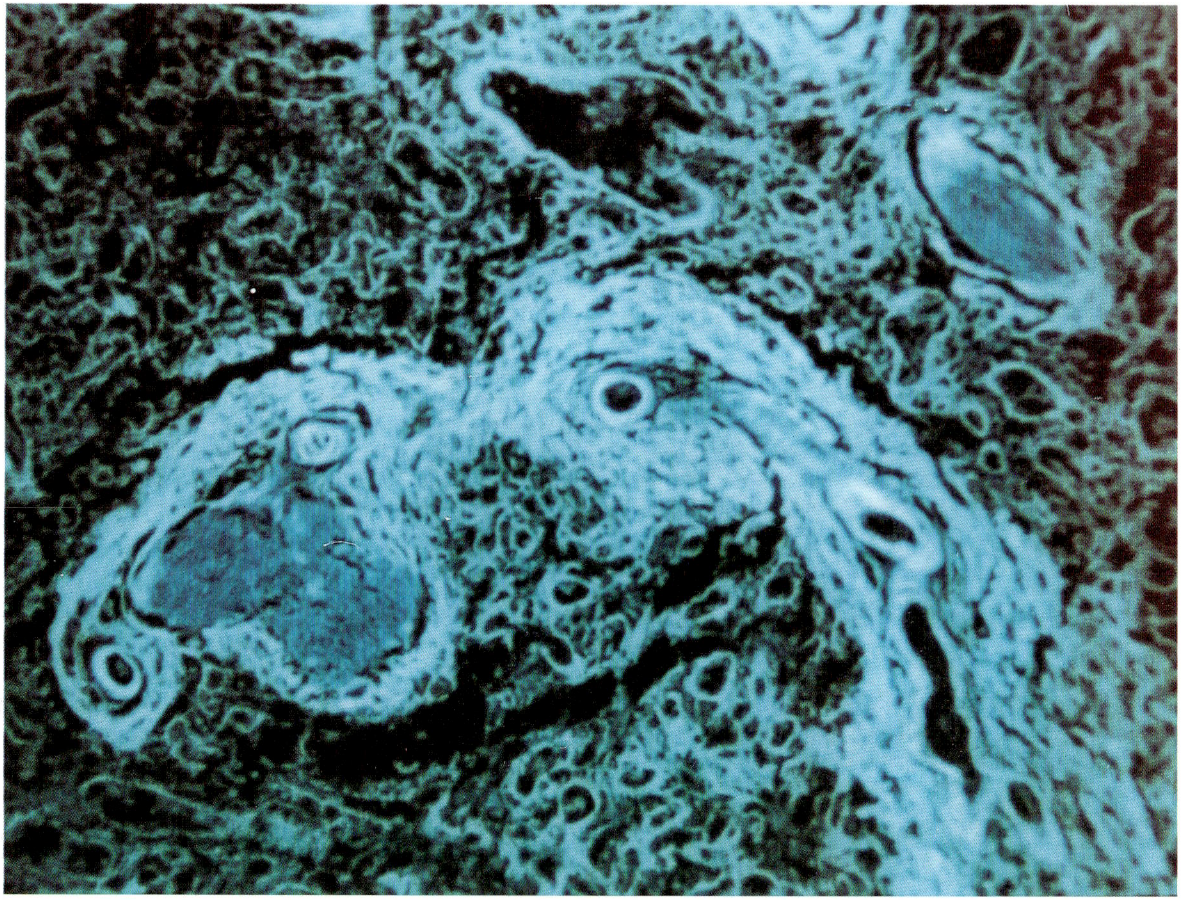

(b)

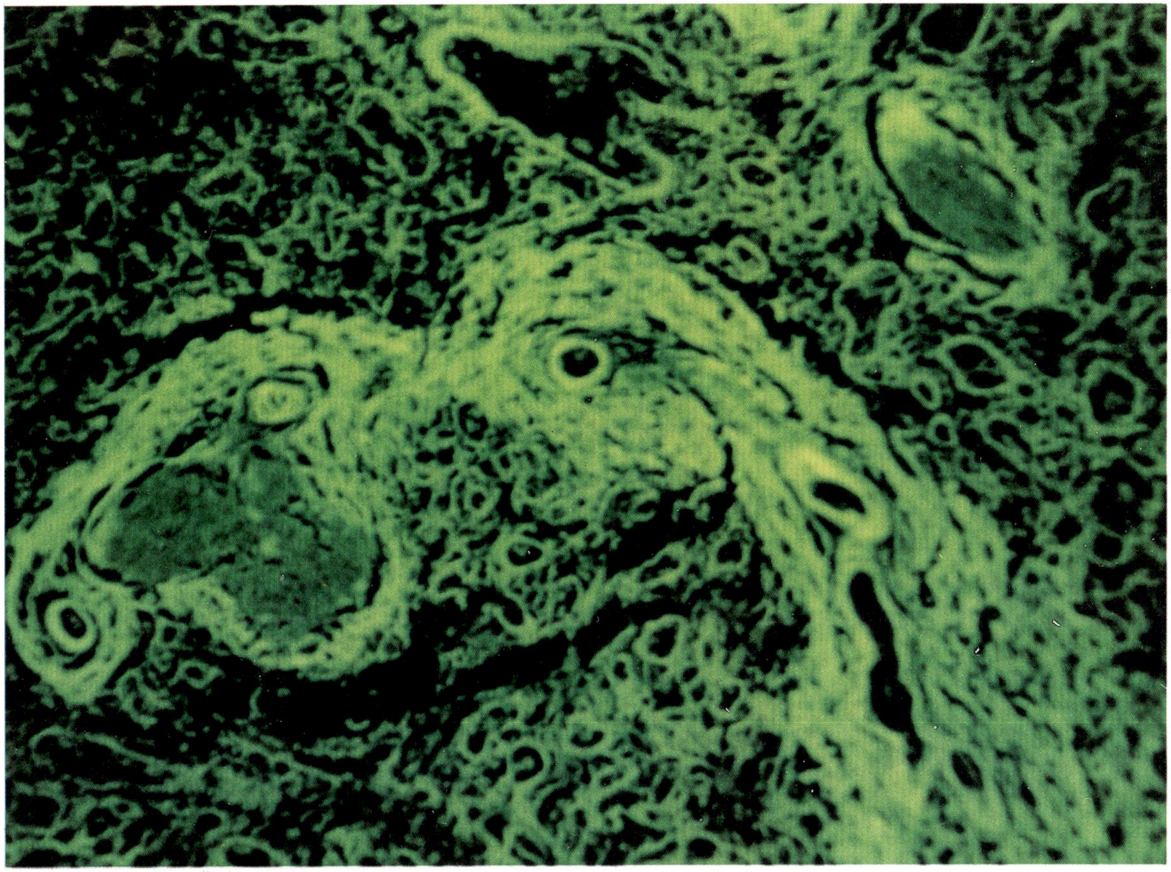

Plate 2. (a, b) Effects of different barrier filters. Amyloid in kidney, stained with the fluorochrome Thioflavine T (C.I. 49005). Leitz Orthoplan microscope, UV excitation. Original magnification ×32, field size 0·7 × 1·0 mm. (a) K430 (colourless) barrier filter; note that at least three colours can be distinguished, namely blue, silvery-white, and green. (b) K530 (yellow) barrier filter; everything appears green. See also Chapters 3 and 7.

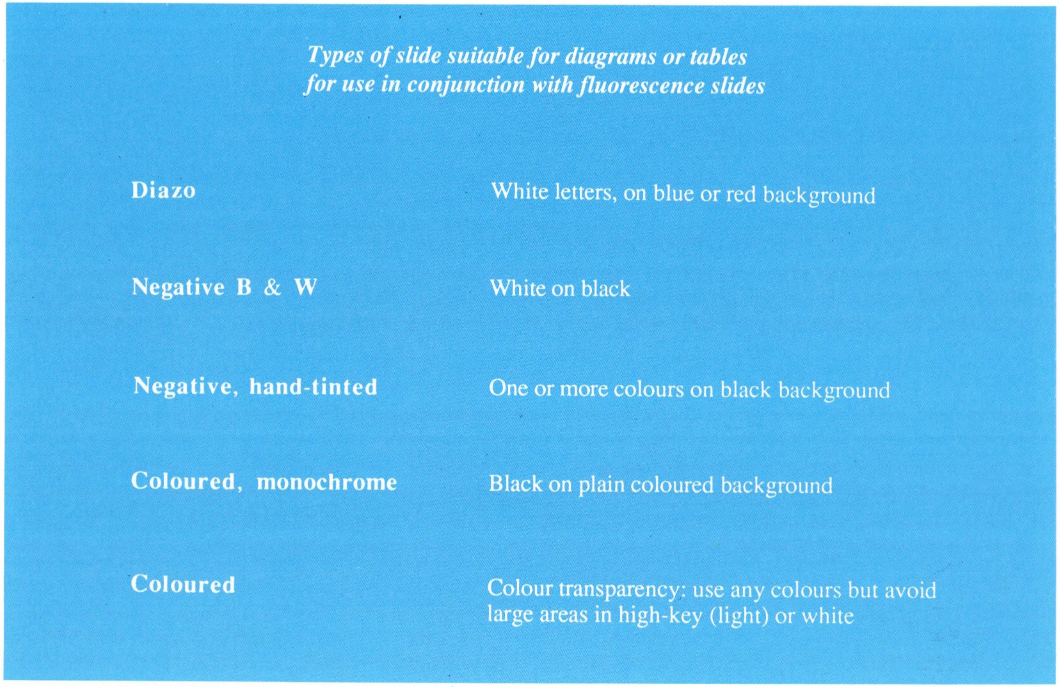

Plate 3. Type of slide suitable for diagrams or tables for use in conjunction with fluorescence slides.

ratings, called the *exposure index* (EI), are now usually given in the ISO (International Standards Organization) system, which is similar to a combination of two systems previously used, ASA (American Standards Association) and DIN (Deutsches Industrie Norm). In the ASA system, the number is directly proportional to the speed; so that a film rated at 64 ASA is twice as sensitive as one rated at 32 ASA. The DIN system uses a logarithmic scale; an increase of 3 DIN degrees represents an increase in sensitivity by a factor of 2. The ISO rating has an arithmetic rating (similar to the ASA) followed by a logarithmic rating expressed in degrees, similar to the DIN system; however, the ISO rating is based on criteria which differ slightly from those of the ASA and DIN. An ISO rating of, for example, 100/21° is one-quarter the speed of ISO 400/27°. The ISO rating is based on certain standard conditions of exposure and development. Since, in a practical situation, these conditions may not be met, a different EI is often set on the camera. In photography, a factor of 2 in exposure or sensitivity is referred to as 1 *stop*, this being the usual relationship between successive graduations of dials for setting lens aperture. Multiples or fractions of a stop are in geometric proportion, e.g. 2 stops corresponds to a ratio of 1:4, 1/2 stop corresponds to 1:1·4, 1/3 stop to 1:1·26, and 2/3 stop to 1:1·59.

The sensitivity of the film, used with a specified development, is described by the *characteristic curve* of the film; a hypothetical example is shown in Fig. 6.1. This shows the relation between the length of the exposure (expressed logarithmically) and the optical density of the film. The characteristic curve is altered by a change in developer and/or development time, so that each film has a series or family of characteristic curves. The curve may be regarded as having several regions: the toe, the approximately straight-line portion, the shoulder and the region of solarization. It is only on the straight-line portion of the curve that density differences in the negative correspond proportionally to differences in brightness in the original image. The slope of the curve

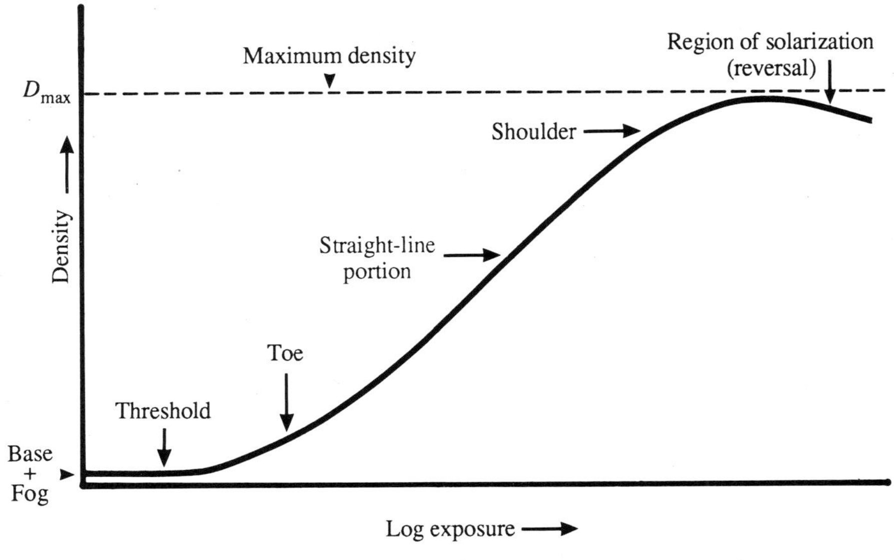

Fig. 6.1. Characteristic curve of a typical photographic negative.

is a measure of the *contrast factor* or *gamma* (γ) of the emulsion; this is the amount of increase in density resulting from unit increase in the logarithm of the exposure. Gamma can also be regarded as the ratio of the contrast in the photographic image to that of the original subject. Gamma usually has a value somewhat less than 1·0, typically about 0·8. Below the toe, the curve is parallel to the log exposure axis and a little way above it. The value of density in this region, normally very small except in colour negative film, is known as *base+fog* or D_{min}. Fog is due to the development of unexposed grains. The point on the curve corresponding to the first perceptible increase in density is called the *threshold*; the need to overcome this threshold is the basis of the pre-exposure technique (see below). The region of solarization, or reversal, is not employed in ordinary photography; in this region an increase in exposure leads to a decrease in density. The exposure required for this effect is of the order of 1000 times greater than normal exposures.

The sensitivity of the film is affected by several factors. The ISO rating quoted by the manufacturer refers to exposures, usually, of the order of 1/100 s; at the much longer exposures required for fluorescence photomicrography (possibly 10 or 100 s) the sensitivity is appreciably less, enough to invalidate calculated exposures (see below, reciprocity failure). The speed at which film should be rated is also affected by variations in development; and underexposure can to some extent be compensated by prolonging the development time (see below, Pushing). Sensitivity to very weak light can be increased by prior controlled exposure to weak light; this was applied to fluorescence photomicrography by van der Ploeg, Bosman & van Duijn (1967). Because of the the number of variables involved, the photographer should take some test shots, with several pictures of similar material over a range of different exposures, in order to calibrate the exposure meter (as described later in this chapter).

Reciprocity failure

All photographic material is subject to an effect known as reciprocity-law failure, reciprocity failure, or the Schwarzschild effect. The reciprocity law states that for constant image density, the required exposure time is inversely proportional to the light intensity (i.e. there is a reciprocal relationship between light intensity and the required exposure time). The word 'failure' in this context implies that the reciprocity law does not hold under certain conditions, involving very long or very short exposures, where the emulsion is less sensitive.

Reciprocity failure can be explained in practical terms as follows. A particular number of photons (say a million) falling on a photosensitive surface (photographic emulsion) during a period of 1/100 s have more effect on the emulsion than the same number of photons falling on the emulsion during 100 seconds. At the other extreme, the same million photons falling on the emulsion during a very short period (say 1 μs) would again give less effect, manifested as lower density of the photographic image. In other words, if the light is very weak, increasing the exposure time in inverse proportion to the light intensity is insufficient; an additional increase in exposure is needed to compensate for loss of emulsion sensitivity in dim light. The range of exposure times during which greatest sensitivity exists varies according to the type of emulsion, as does the degree of reciprocity failure outside that range. There is of course still an inverse relationship between the intensity of the light and the required exposure time; the relationship takes the form:

$$\log(\text{exposure time}) = p/\log(\text{light intensity})$$

where p has a value greater than 1 instead of exactly 1. For exposures of the order of 10 to 100 s, p typically has a value of about 1·2 to 1·5. Data on the reciprocity characteristics of some films are given in Appendix 4B. A recent practical study of films is that of Dragesco (1988).

Automatic correction for reciprocity failure in a camera with automatic exposure determination uses the formula $T_c = (T_m)^p$, where T_c is the corrected time, T_m is the uncorrected time, and p is a constant for which values are set by a selector switch. I have calculated values of p for a few films (see Appendix 4B). The calculation is simple, given a known uncorrected exposure time T_c and the corresponding corrected time T_m:

$$p = \log(T_c)/\log(T_m)$$

Grain and resolution

Fast films have a coarser grain structure than slow films, in general, and are therefore unable to record details as fine as can be recorded with a slower film. However, some graininess is usually acceptable. In any case, the resolution of the film measured in line pairs per millimetre (LP/mm) need not be more than twice the theoretical resolution of the optical system.

Development

Exposure of the film to light results in the production of a 'latent image', which is converted to a visible image by the process of development. Development essentially involves the reduction of silver halides to metallic silver, which is deposited as grains of silver. The process of development increases the effective sensitivity of the film by an enormous factor, estimated to be one thousand million times. The processing is completed by *fixation*, in which unreduced silver halides are dissolved away, leaving a negative stable to further exposure to light, *washing* to remove remaining developer and fixer, and *drying*.

Black-and-white films are best processed by the user. Developing these is quite easy and requires minimal equipment. Details of film processing are given in standard books on photography (e.g. Jacobson *et al.*, 1978).

My favourite developer is Ilford's *Microphen* (this of course is purely a personal preference). Microphen gives a useful increase in film speed, and a fine grain. I prefer to use the developer diluted 1 + 3 (that is, 1 part Microphen stock solution plus 3 parts water); the diluted solution is thrown away after developing one film. Development time is increased, to about 15 min at 20 °C (9–30 according to the film and the amount of contrast desired). This method offers the advantages that, firstly, the developer is always fresh, so no correction to developing time has to be made for previous use, and, secondly, the development time is not critical.

It is very important to find a dust-free place to hang the film to dry. A commercial drying cabinet is fine, provided that dust is not allowed to accumulate inside and that the air in the room is not dusty.

Pushing (push-processing) is a procedure by which the film is underexposed and developed for a longer time to compensate for this underexposure. This strengthens what would otherwise have been a very pale negative, and allows the speed rating of a film to be increased. Push-processed B&W films give negatives which have increased contrast and a larger grain size than film processed normally. Some detail will be lost, due both to overall reduction of image quality (due to increased graininess) and to underexposure in the darker regions of the optical image.

Although pushing does produce a more acceptable negative than would be obtained by underexposure alone, the film does not really undergo a speed increase. Since the ISO speed rating of negative film is based on the minimal exposure required to produce an excellent photograph, underexposing the film will excessively underexpose the areas corresponding to darker areas of the optical image. Overdeveloping the negative will darken the areas which were adequately exposed, but will not develop the excessively underexposed areas and it is for this reason that negative film cannot have its true sensitivity increased, and also why contrast is increased.

As a rule of thumb, the required increase in development time for push-processing is an increase of 50% for the first stop, and an additional 25% of the basic development time for each additional stop. For example, suppose that a particular film is normally rated at ISO 100/21° and requires a development of 8 min. The required development times are set forth in Table 6.1.

Different film/developer combinations have different development times and an appropriate table is easily constructed to suit. Of course, there is little point in pushing an ISO 100/21° film to 400/27° when there exist several films of the higher rating; the above example was chosen for clarity of the numbers. Push-processing of colour film is discussed under Colour film, below.

Chromogenic film

Chromogenic film is film in which the final image is formed by a deposit of organic dyes. The dye replaces the silver, which is removed by a fixer during processing, leaving the dye behind to form the image on the film. Chromogenic B&W film is similar to colour negative film (described below) but with only a single dye layer.

The only current chromogenic B&W film is Ilford's XP1-400. This film can be developed by the standard C-41 colour negative process and all compatible processes, but for optimal results XP1-400 should be processed in the special developer (Ilford's XP1 developer). In principle, chromogenic film can also be processed as a conventional B&W film, or reversal-processed to a transparency (E-6 process; reduce the speed rating to ISO 100–200 or push-process). XP1-400 film can be rated at any speed from ISO 50/18° to ISO 1600/33° and yield quite acceptable negatives; moreover, the film

Table 6.1. *Development times needed for push-processing at different speed ratings*

Speed rating set on camera (ISO)		Development time (min)
100/21°	(standard)	8 (standard)
200/24°	(pushed 1 stop)	12
400/27°	(pushed 2 stops)	14
800/30°	(pushed 3 stops)	16
1600/33°	(pushed 4 stops)	18

speed dial on the camera can be changed from shot to shot within this speed range. Push-processing is possible, particularly for the range ISO 800/30° to 3200/36°. The latitude of the film gives substantial tolerance for incorrectly estimated exposure. Underexposed XP1-400 negatives actually have a finer grain, the converse of that expected and obtained with standard negatives. In any case, the chromogenic films have quite small grain size in relation to their speed. Because of their potential high speed, wide latitude, capability to cope with a high-contrast subject, and comparatively fine grain, these chromogenic films may be preferred to conventional B&W films for fluorescence photomicrography. The processing procedures are more complicated and inconvenient for a small laboratory not normally concerned with processing colour negatives, but on the other hand (at least here in Australia) commercial processing is quick and widely obtainable.

Colour film

Colour film differs from B&W film in having three photosensitive layers, each arranged to record a different range of wavelengths and each, on development, producing a different coloured dye. The final apparent colour of the film is due to combination of the colours of the three layers. Colour film is made in two main types: negative (print film) and positive (reversal). Negative colour film is used in a manner similar to B&W film; it gives a negative image, with colours complementary to the original colours but usually with an overall yellow added which is compensated for in printing. Positive (reversal) film gives a positive image, with true colours (within the limitations of the film); such film is used for projection (as a transparency or dia-positive).

Colour negative film, in principle, should be ideal for general use, since colour or B&W prints or slides can be made from it. Unfortunately, in practice, colour negative film is little used for scientific purposes, owing to difficulty in producing prints with correct colour rendition. In any case, colour prints can be made from colour transparencies, and their colour easily compared with that of the original transparency. If a B&W print is required from a colour transparency, an intermediate B&W negative is made.

Colour positive film is made in two main varieties, 'daylight' and 'artificial light'. Only the daylight type is suitable for fluorescence, because it reproduces the colours correctly (in so far as film can). The artificial light variety is intended for taking photographs by tungsten lamps, which have, compared to daylight, too much red and too little blue light; in compensation, artificial light-type film has increased sensitivity at the blue end of the spectrum, and less sensitivity to red. For ordinary photography, artificial-light film can be used in daylight if a conversion filter is used; however, this reduces the sensitivity of the film. Colour films are adjusted to give natural-looking results when the original subject is illuminated with light having a particular spectral distribution, normally defined by the colour temperature – the temperature which a perfect radiator ('black body') would have to have to give approximately the same emission spectrum. Do not be misled by the names 'artificial light' and 'daylight'; these refer only to the colour correction of each. Fluorescence photomicrography may involve artificial light, but use the 'daylight' type of film always!

The filters sold for photography in 'fluorescent light' have of course nothing to do with the subject of this book: they are designed for correcting the green cast of fluorescent lamps used for general illumination.

Reciprocity failure of colour film is similar to that of B&W film. However, in this case reciprocity failure can affect each of the three photosensitive layers differently, so that reproduction of colour may be distorted after long exposures. This can be compensated for, if required, by the use of appropriate filters, as specified by the film manufacturer. Unfortunately these filters reduce the intensity of the light, aggravating the reciprocity problem. For photomicrography, the colour changes are unlikely to be significant and no colour correction filters need be used. If changes are to be made in the colour rendition of the picture, it would be better (probably) to copy the picture and to incorporate the colour corrections during the copying process. Colour correction filters of coloured gelatine can be obtained for this purpose.

Film processing

'Amateur' Kodachrome film is usually processed by the manufacturer, and should be sent to Kodak as soon as possible after exposure. Here in Australia, my experience is that it is best to hand the film to my friendly pharmacy for transmission and return by courier – this is quicker and more reliable than the postal service. Film is normally returned mounted in plastic mounts; on request, the film will be returned as a strip without the mounts; if the pictures are predominantly black, the Kodak technician may have difficulty deciding where the individual frames (pictures) begin and end, and it may be preferable to cut and mount them oneself.

Most other colour films can be processed by the user, or by an independent laboratory. In my experience, setting up to process colour film is only worth while if large numbers of films are to be processed and plenty of technical assistance is available, otherwise inconsistent results are likely.

Pushing of colour positive film is possible and effective; the increased developing takes place in the first development stage of the processing. Unlike negative film, pushing colour positive film gives an actual increase in sensitivity. With properly (normally) exposed colour reversal films, the full potential of the image density is not utilized because the slide would appear too dark. If therefore the film is underexposed by say 1 stop, since the film has the potential to record detail at higher densities, detail in dark regions of the optical image is recorded even though the slide appears too dark. Compensation of the overexposure by longer development of the film will lighten the transparency and restore detail in dark areas. Therefore a real increase in effective speed results. In the E-6 process, the first development time should be increased by 2·5 min to push 1 stop, 6 min for 2 stops and 12 min for 3 stops. Greater pushing than 3 stops is not recommended. Push-processing of professional Kodachrome film and of films requiring the E-6 process is available commercially, for an additional charge; information as to the amount of pushing required must be sent with the film. Some special varieties of film are specifically designed for push-processing (see Appendix 4A).

Special-purpose films

Infrared film

Special films are available with extended sensitivity in the far red and infrared. It should be explained that silver halide emulsions are basically sensitive only to the high-energy (blue) end of the spectrum: the construction of red-sensitive ('panchromatic') films require the addition of an organic dye which absorbs in the red and transfers the photon energy to the silver halide. Infrared colour film has been used for recording the fluorescence of chlorophyll (Elkin & Park, 1975).

Astronomical and hypersensitized films

The twin problems of dim light and reciprocity failure are even better known to astronomers than to fluorescence microscopists. Astronomical photography generally employs special emulsions whose reciprocity failure is minimal. In addition, astronomers utilize a variety of 'hypersensitization' techniques to increase the speed of the film (for reviews, see Anon., 1989; Smith & Hoag, 1979). Generally, the principle is that of complete dehydration of the emulsion (see James, 1972); the film is baked in dry hydrogen or hydrogen/nitrogen mixture ('forming gas') at 60–70 °C for 2–3 h. Hypersensitized film can be kept for 2–3 months in nitrogen in a freezer. The process can be applied to a wide variety of B&W and colour films. Recent practical studies of hypersensitized films were made by Dragesco (1988) and Quirk (1990). The application in microscopy of hypersensitized Kodak Technical Pan Film 2415 at up to ISO 1600/33° was described by Smith (1986). Commercially hypersensitized film is available, at least here in Australia (from K. Ryan, PO Box V105, Mt Druitt Village, NSW 2770).

'Instant' films

Polaroid (TM) films, available in both B&W and colour, require separate discussion. There are at least two potential advantages in using 'instant' photography for fluorescence. Firstly, if each picture is developed separately, the exposure can be checked and subsequent exposure times modified as necessary. Secondly, any other error (such as a wrong filter combination, or a mechanical malfunction) will show up immediately. The main difficulty is in obtaining sufficient speed; the relatively coarse grain of Polaroid film is not likely to be a problem in this context.

Of current 35 mm Polaroid films, the colour film (Polachrome CS) and the B&W continuous-tone transparency film (Polachrome CT) are slow (ISO 40/17° and 125/22°, respectively). Polachrome CS suffers badly from reciprocity failure; however, I have successfully photographed autofluorescence in plant cell walls with it.

The print films are in larger formats. One can, however, calculate the relative speed compared to the 35 mm format (see Appendix 4C). For example: an image on 'quarter-plate' format (73 mm × 95 mm) is 7·3 times dimmer than the corresponding image on 35 mm format. Therefore, Polacolor Type 108, speed ISO 80/20°, is equivalent in speed to a 35 mm film of ISO 10/11°; B&W print film Type 47, ISO 3000/36°, is equivalent to 35 mm film of ISO 400/27°. The former is too slow, the latter is satisfactory. The above are theoretical design considerations. In practice, the situation depends on the relative magnifications of the 35 mm and Polaroid camera for the particular microscope in use. For example, I have used an Olympus Vanox (old series) microscope in which the Polaroid attachment gives a magnification 3 times greater than the 35 mm camera; the ISO value of the Polaroid film should therefore be divided by 9 for comparison with the smaller format. I have used Type 667 film (for B&W prints) and found it quite satisfactory and convenient.

Choice of films

One must compromise between high speed (for a short exposure) and fine grain (associated with slow films).

For B&W, the chromogenic films (e.g. XP1-400) are possibly the first choice for fluorescence photomicrography. However, any moderately fast film (ISO 125/22° to 400/27°) should give satisfactory results with appropriate development. Some types of film have bad reciprocity characteristics, and these should be avoided – a slower film with better reciprocity may require little more exposure, and show finer grain.

In colour film, the principle is to select a daylight type film with as high a sensitivity as possible consistent with an acceptable degree of graininess, and as little subject to reciprocity failure as possible. Sometimes a medium-speed film with good reciprocity characteristics may give as good results as a fast film subject to severe reciprocity failure. Among present-day 35 mm colour films, in my experience Kodachrome 200 film is by far the best for general fluorescence photomicrography. There are several faster colour films which can be used if necessary, but with some loss of picture quality. Of colour negative films, Ektar 1000 can be recommended for its high speed and relatively fine grain.

CAMERAS

The camera for photomicrography of fluorescence is usually a specially constructed camera, made to suit the particular microscope, and mounted on the vertical tube of a trinocular head (see Fig. 4.12). In microscopes specially designed for photomicrography, the photographic system may be incorporated into the microscope body or form a substantial module attached to the top of the stand (see Figs. 4.5 and 4.14). An ordinary camera can be adapted, as will be described, but this cannot be expected to be as effective.

An ordinary photographic camera normally consists of the following components: a mechanism to hold the photographic film and to transport the film frame by frame, a lens to form an image of the subject on the film, a light-tight box to keep light away from the film except for the light passing through the lens, and a shutter to keep the light entering the lens away from the film except during a definite period (the exposure time). The amount of light reaching the film is determined jointly by the opening of the shutter and by the lens aperture – which latter, corresponding to the NA of the objective in a microscope, is normally made variable by means of an iris diaphragm.

A photomicrographic camera is basically similar to an ordinary camera, the microscope system serving as the lens. There are two main kinds of system used for forming an image on the film. In all cases, objectives are utilized in the normal manner. However, whereas for observation the ocular produces a virtual image and the lens of the eye converts this to a real image on the retina, in the photomicrographic system a real image is required on the film plane and therefore the normal ocular must alternatively (a) be supplemented by a lens equivalent to the lens of the eye, or (b) be replaced by lenses which produce a real image on the photographic film, or (c) be re-focussed to produce the required real image in the film plane. This latter can be achieved either by increasing the tube length (between the objective and ocular) or by re-focussing the microscope as a whole. In some photomicrographic cameras, the camera contains lenses which completely replace the ocular; an example is the Leitz Orthomat E, which is fitted at the top of the photographic body tube in place of the ocular. Some manufacturers employ a system in which an ocular resides at the top of the phototube, and the camera sits above that with additional lenses to form a real image. A similar effect can be obtained by using a conventional camera, complete with lens, arranged to look through the normal eyepiece; the object will be in focus if the lens of the camera is set to infinity and provided also that the lens of the camera is not too close to the ocular, i.e. is held at the eyepiece exit pupil. However, with a 'standard' lens the full field of the microscope will give a circular image only partially filling the film frame. Method (b) above is employed in the Olympus adapter for OM cameras; the ocular projects an image directly onto the film; a special projection ocular is normally used, and of course for fluorescence the lowest power available ($\times 2{\cdot}5$) is preferred.

The objective should normally be of the flat-field (e.g. planachromatic or planapochromatic) type in order that the whole field be in focus. It should have a high NA in order to collect as much light as possible, to reduce the time required for photographic exposure; if a substantial depth of field is required, a lower NA may have to be employed. As in all fluorescence work, the projection ocular should have as low a magnification as possible.

The shutter mechanism may be either a leaf shutter or a focal plane shutter. The former is usually placed in an aperture-limiting plane in or near the lens, and resembles an iris diaphragm which closes completely. This kind of shutter has very light moving parts, symmetrically arranged, so that it generates hardly any vibration when opening and closing. A focal plane shutter usually consists of two roller blinds just in front of the film; these are arranged on either side, so that one opens to expose the film and is followed by the other (moving in the same direction) to close the shutter. The length of the exposure is determined by the interval between the opening and closing; for exposures shorter than that represented by the full opening of the first blind before the other moves, the second blind follows closely behind the first so that in effect a slit between the two blinds passes in front of the film, each part of the film receiving an exposure whose length depends on the width of the slit as well as on the speed with which the slit traverses the film. The blinds may move either longitudinally or transversely with respect to the axis of the film. In some cameras, the cloth blinds are replaced by one or more thin sheets of titanium. Focal plane shutters are very reliable, but may generate a substantial amount of vibration as the first blind is brought to a halt.

Microscope body-tube for photomicrography

A microscope designed for photomicrography should be equipped with both a binocular or monocular tube for observation and a separate monocular tube to accommodate the camera. A sliding mirror or beamsplitting device allows the light from the objective to pass alternatively to the observation tube or to the photographic tube or to both. For photomicrography of fluorescence, it is important that the body tube have provision for all (100%) of the light to pass up to the camera. Some makes of microscope have only a beamsplitter, so that some of the light is always diverted into the observation tubes, and the full amount of available light cannot be utilized for photomicrography. This is unsatisfactory, having the effect of unnecessarily prolonging the exposure required. Also, if this kind of tube must be used, the eyepieces should be covered when not in use to prevent stray light from the room reaching the photometer and film.

The ideal specifications of a photomicrographic camera for fluorescence

In my view, the ideal photomicrographic camera for fluorescence should have the following characteristics. Some of these features may not be practicable for a particular application, on economic grounds.

1. High light transmission: no beamsplitters, zooms, etc.
2. Low camera magnification. The film format should cover only a little less than the objective's field of view; indeed, for B&W, the format need not even be fully filled. The format should be as small as practicable (normally 24 mm × 36 mm on 35 mm film).
3. Illuminated graticule for focussing and composing the picture, for visibility against a dark background, preferably visible in both normal eyepieces. The illumination should preferably be variable in intensity, possibly red in colour (to

avoid spoiling the eye's adaptation), and with an on/off switch so that the graticule does not disturb normal observation.

4. Revolving stage, camera system or image for composition of the picture. At least 180° of rotation is required.
5. Variable magnification of the focussing graticule – high magnification for focussing, low for composing the picture. Possibly an automatic focussing mechanism, operative even at low light levels, with manual override, would be helpful.
6. Photometer of adequate sensitivity, optional integrated measurement or spot measurement from a defined moveable spot.
7. Exposure control system with provision for setting ISO film speed over a wide range (especially high ISO values), compensation for % bright/dark fields, and compensation settings for reciprocity failure characteristics. Bracketing should be easy, at about 1/2 or 2/3 stop values.
8. An appropriate shutter mechanism, preferably optionally automatic or manual, with long exposure capability to at least 10 min.
9. The control and exposure information of the exposure control unit should be clearly visible under low illumination levels. Any lamps or illuminated displays should not dazzle the eye, and preferably be in red colour to maintain dark adaptation.

Commercial cameras

Commercial cameras are described in Chapter 4, with the corresponding microscopes. See also Cowen (1969).

PHOTOMICROGRAPHIC TECHNIQUE

The procedures required in taking photomicrographs are described below in approximately the order in which they are usually carried out.

Setting up the microscope

It is of major importance that the microscope be set up optimally for the specimen (see Chapter 5). Every effort must be made to obtain the brightest possible image on the film. The objective's NA should be as high as possible; this may be limited by a requirement for depth of field. Objectives of high NA, and whose effective NA can be varied by an iris diaphragm, are convenient. An efficient barrier filter, transmitting most of the fluorescence, is important; if the cut-off of the filter is at too long a wavelength, not only will the colour be distorted (as will be explained in Chapter 7), but also the intensity of the image will be reduced because only the long-wavelength tail of the fluorescence is transmitted by the filter. The most efficient type of body-tube sends 100% of the light to the camera, and none is diverted to the eyepieces. Beamsplitters of any kind, and supplementary magnifying systems, should be avoided because of light loss.

The camera must be loaded with the correct film, and wound on ready to shoot. In loading a 35 mm camera, it is important to ensure that the film is firmly attached to the take-up spool and that the driving sprockets mesh properly with the perforations of the film, so that the film will transport properly.

Filters

For photomicrography in colour, the correct filter combination is the same as for visual observation. However, because film is sensitive to UV and relatively more sensitive to violet than the eye, it is particularly important that the barrier filter should block short wavelengths adequately. If in doubt, add an additional UV-blocking filter to the barrier filter; if the main filter is yellow, orange or red, the UV filter should be placed before the main filter so as to block UV which may otherwise cause autofluorescence in the main filter.

For B&W photography, the barrier filter may with advantage be a narrow-band filter with peak transmission corresponding to the emission peak of the relevant fluorophore, to increase the specificity of the resulting photograph.

Focussing and composing the picture

The object to be photographed is first selected, taking care not to fade it unnecessarily. This is best accomplished by placing a neutral-density filter in the excitation light path, so that the specimen is observed with dim light. Preliminary focussing and location of the specimen without fluorescence excitation can be undertaken by phase-contrast, interference-contrast, dark-ground or even ordinary transmitted light (possibly with the aperture diaphragm of the substage condenser partially closed to increase contrast). The size of the field of view to be photographed should be selected by choosing the appropriate objective (set any magnification changer or zoom to minimal magnification; better, remove it from the system). The specimen can be oriented in the field by turning it (e.g. on a rotating stage) or by rotating the camera; the latter is usually best accomplished by rotating the entire body tube with camera assembly, if this is possible.

Focussing is ideally done through the microscope eyepieces. For this purpose, some microscopes have an arrangement whereby the image, if in focus in the eyepiece, is automatically in focus in the camera. Using a binocular body tube, the length of the body tube varies according to the distance apart of the oculars, varied to suit different interpupillary distances. This variation in tube length can be compensated for either automatically or manually; in the latter case, a scale indicates the distance apart of the oculars, and the eyepieces should be adjusted individually (by turning the tubes) until index marks on the eyepiece tubes are set to positions on their scales corresponding to the interpupillary distance. A graticule defines the plane of focus in the eyepiece. This graticule usually contains two circles or lines very close together, which can be distinguished as separate only when accurately focussed. The graticule also includes marks to indicate the boundary of the photographic field. Ideally, for fluorescence photomicrography the graticule should be illuminated, so as to be visible against the dark background. This arrangement is provided in the latest Olympus PM-10AS camera (Fig 4.12), which offers a choice of red or yellow; red has the advantage of not spoiling the dark adaptation of the observer's eye, while the yellow provides contrast against red fluorescence or if simultaneous red phase-contrast is used. The graticule can alternatively be reflected into the eyepiece by a prism, so that the effect of rotating the camera or changing the magnification can be observed (as in the Leitz Orthomat-E). To assist precise focussing, the image and graticule can be examined through a telescope designed to be applied to the eyepiece (Olympus make a good one).

Direct ground-glass screen focussing, in which the film is temporarily replaced by a ground glass, is only useful for large formats and is therefore not normally relevant to fluorescence photomicrography.

Single-lens reflex focussing is employed in some camera bodies used for general-purpose photography, as for example in the Nikon F3 (which I use for general photography). Here, an image is reflected by a mirror onto a ground-glass screen and examined through a magnifier; when the shutter release button is pressed, the mirror swings aside to allow the light to pass to the film through a focal-plane shutter. The screen in this case is not necessarily ground glass; moreover the screen normally supplied for general photography is not suitable for photomicrography, with fluorescence or otherwise. The grain of the screen is usually too coarse, masking the detail of the image and rendering accurate focussing impossible. Moreover the stigmometer (split-image or microprism focussing device) blacks out. For photomicrography, Nikon provide a special focussing screen (Type C), which has a fine-ground screen with a central clear area with cross-lines for 'aerial image' focussing. In the latter procedure, the observer moves his/her eye from side to side; if the image is focussed in a plane different from that of the crosslines, they will appear to move relatively – if the image is focussed further from the eye than the cross, the image appears to move in the same direction as the eye; if the image is focussed nearer to the eye than the cross, the image appears to move in the opposite direction to the eye. The reverse applies to the cross. A magnifying finder (such as the Nikon DW-4) or an eyepiece magnifier assists greatly in focussing.

The mirror of a single-lens reflex camera may generate substantial vibration when swinging aside. Some cameras are better than others in this respect; the best have provision to raise the mirror before the shutter is released, as in the Nikon F3. In some cameras, operation of the self-timer releases the mirror, allowing time for vibrations to die away before the exposure is taken.

Depth of field. There is only a limited thickness of the specimen which can be in focus at any one position of the focussing adjustment. The thickness or depth which can be in focus at any one time is known as the *depth of field*. There is a corresponding term, *depth of focus*, for the range over which the film plane can be moved while maintaining a given object plane in focus. Both are primarily dependent upon the NA of the objective and the degree of blurring acceptable at the film plane. Increasing the NA decreases the depth of field. This means that objectives of high NA, giving excellent light-gathering power, have only a shallow depth of field. Some objectives incorporate an iris diaphragm to limit the NA; if necessary, a circular stop (fixed diaphragm) can sometimes be placed in the rear focal plane of an objective to reduce its NA. Such a diaphragm or stop may have three applications: increasing the depth of field, reducing the NA to permit use with a dark-ground condenser, and reducing flare sometimes seen with very brightly fluorescing objects (particularly if the objective is of medium quality only). See also Chapter 3.

Composing the picture is accomplished primarily by moving the stage carrying the specimen. Rotating the picture is usually achieved by rotating either the stage carrying the specimen, or the body tube plus camera, or the camera alone, in about that order of preference. A special optical system to rotate the image is also possible, as in the Leitz Aristoplan, and is even more convenient. Since the photograph can always be turned end-to-end, only 180° of rotation (90° in each direction) is strictly necessary.

Aesthetic considerations

Photomicrographs of fluorescence can be very beautiful, particularly those with several fluorescent colours or with simultaneous phase-contrast in an harmonizing colour. Even for a photograph intended only as a record of a specimen, not for publication, a little additional time and care in composition may yield rich dividends in beauty. Artistic considerations are of even greater importance in the production of pictures intended for publication or public presentation – the latter including lectures to students.

Often, a composition can be improved by no more than rotating the specimen (or the camera) to obtain the most harmonious arrangement within the field. The most prominent object in the field should not normally be placed exactly in the middle of the field, and neither should any division in the specimen divide the field into two precisely equal halves. When using a spot-measuring photometer operative only in the centre of the field, measure the exposure first if possible, then adjust the composition. A good place for the most prominent object is approximately one-third of the distance from side to side and top to bottom (strictly, at the point of the Golden Section, which is more exactly 0·384, or slightly more than one-third). Sufficient magnification should be used to fill the frame with the relevant object area.

Many books on the practical aspects of art and photography have sections on composition. The most detailed treatise on composition to my knowledge (and a most excellent book) is that of Bouleau (1980); see also Jonas (1976).

Exposure time

The correct exposure time is determined by a combination of many factors. They may be summarized as: the brightness of the fluorescence, the extent of fading during exposure, the magnification of the image onto the film, the amount of light losses in the optical system, and the sensitivity of the film under the prevailing circumstances.

The brightness of the fluorescence is maximized, and fading minimized, by methods which have been described in Chapter 5. Obviously these techniques are of particular importance for photomicrography.

The magnification of the image at the film plane affects the intensity of the image; the latter is inversely proportional to the magnification. In order to minimize the magnification required to fill the picture with the desired object, the film format should be as small as practicable. Cameras with 35 mm format are almost universally used for fluorescence photomicrography. Larger formats should be avoided (but see comments above on Polaroid films). Still smaller formats, e.g. 24 mm × 24 mm on 35 mm film, or even the half-frame 35 mm (17 mm × 24 mm) would give shorter exposure times. Appendix 4C indicates the relative exposure times required by various formats; it will be seen that the tiny formats might, if suitable equipment were available, be useful for photographing fluorescence.

Strictly, the above remarks concerning the effect of magnification on image brightness apply only to images of areas, and do not apply in the special case of point sources of light, i.e. an object consisting of unresolvable minute particles. In such case, the images of the particles should, if sharply focussed, be diffraction-limited and their dimensions independent of the magnification. The situation is analogous to that of astrophotography, where the stars constitute unresolvably small sources of light.

Loss of light in the optical system, always unfortunate in fluorescence microscopy, is to be avoided at all costs for fluorescence photomicrography. Apart from sources of loss mentioned in Chapter 5, those peculiar to photomicrography are: a beamsplitter in the phototube of the microscope, which reflects part of the light into the oculars (use only a trinocular tube with provision to transmit 100% of the light to the camera), a beamsplitter in the camera to reflect light to an exposure-measuring device, and multiple air–glass interfaces in the system. Usually, the simpler the design of the camera optics, the better.

Exposure measurement

Determination of the correct exposure is essentially a matter of relating the brightness of the object to the sensitivity of the film. The matter is complicated by the fact that the range of brightness within a fluorescent object is usually very much greater than that which can be recorded on film. In colour transparency film, overexposure of the highlights results in loss of colour saturation (i.e. the colour tends towards white), while underexposure gives a dark, grainy image. The background is less critical – particularly as regards underexposure. Exposure determination ideally should be done with a video camera and computer; this may become an economically viable proposition in the future. Exposure could then be based on either the peak brightness or the mean peak brightness.

There are three ways in which the length of exposure can be determined: trial and error, brightness measurement before exposure, and continuous measurement during exposure. Photomicrographic cameras are usually provided with means to measure the brightness of the field. For this purpose, light is diverted either by a sliding mirror which can be inserted into the light path, reflecting light to a photometer; or alternatively a permanent beamsplitter reflects part of the light to a photometer. The latter arrangement is more convenient and allows measurement during the exposure itself (to allow for fading), but has the great disadvantage that the beamsplitter reduces the brightness of the image on the film, thereby necessitating a longer exposure. In ordinary cameras, another method has comparatively recently been introduced: measurement of light reflected from the film during long exposures, or from the blind of a focal-plane shutter before short exposures. This, although a Minolta patent, was first introduced in the Olympus OM2 camera. I have no experience with it. This method of photometry, applied to a photomicrographic camera, should be ideal for fluorescence because measurement could take place during exposure but without the loss of light associated with a beamsplitter. The photometer has to have a much higher sensitivity than is required for ordinary photomicrography by transmitted light.

Trial-and-error, as the sole method of exposure determination, may be acceptable for routine photography under highly standardized conditions, but otherwise is not satisfactory in my experience. The method involves first taking a series of trial exposures at various exposure times (each exposure say twice the previous); the exposure giving the best result is noted and made the basis of further photomicrography. Exposures using this method should always be bracketed – that is, additional pictures should be taken, e.g. at double and half the estimated exposure, or treble and one-third. The method may be effective under highly standardized conditions. If the magnification is changed, corrections to the estimated exposure can be calculated or determined empirically. It is necessary to allow the lamp to warm up fully, and long-term diminution in the light output from the lamp must be watched and compensated for, e.g. by measurement of the fluorescence from a standard object. To improve accuracy of measurement, the standard object can be a brilliantly fluorescing object, more brilliant than the fluorescence to be photographed.

Measurement before the exposure is similar to the method usually adopted for conventional photography. The brightness of the image is measured, the exposure estimated and given. Again, the exposure should be bracketed, although usually a smaller range of exposures is required (e.g. a factor of 1·5 each way, above and below the estimated exposure). This method is much more accurate than the previous, but is still subject to error due to fading during the exposure.

Measurement during the exposure has already been referred to. In principle it should compensate for fading of the specimen during exposure. Use of a beamsplitter, however, results in loss of light which would be better employed on the photographic film; however, the method of measurement from light reflected from the film overcomes this problem.

Spot and integrated measurements

Spot measurements, as their name implies, are made from a small area, usually in the centre of the field. If the specimen consists of a small fluorescent area in a generally dark field, spot measurement is helpful, provided that the fluorescent object is big enough to fill the spot. If the measuring spot is fixed in the middle of the field (see Fig. 6.2), necessitating moving the specimen to get a suitable area into the spot and then again to compose the picture, this method is suitable only for those cameras which measure the exposure before opening the shutter. In a fully automatic camera measuring during the exposure, it is essential to be able to move the spot to a relevant area.

Integrated measurements are made over the entire field. If the fluorescence occupies only a proportion of the field, the calculated exposure must be corrected accordingly (inversely in proportion to the proportional area of the specimen). Figure 6·3 illustrates the situation. The correction consists of a reduction in the exposure below that indicated *prima facie* by the exposure meter – if for example the fluorescence occupies 25% of the field, the exposure should be reduced to 25% of that indicated. A similar correction can be applied to spot measurements if the object does not fill the spot. The proportion is difficult to estimate by eye, partly due to optical illusion – the bright objects on a dark background tend to appear larger than they really are. Figure 6.3 provides some standards for comparison. These diagrams were produced by a computer, programmed to draw rectangles consisting of square blocks of white on

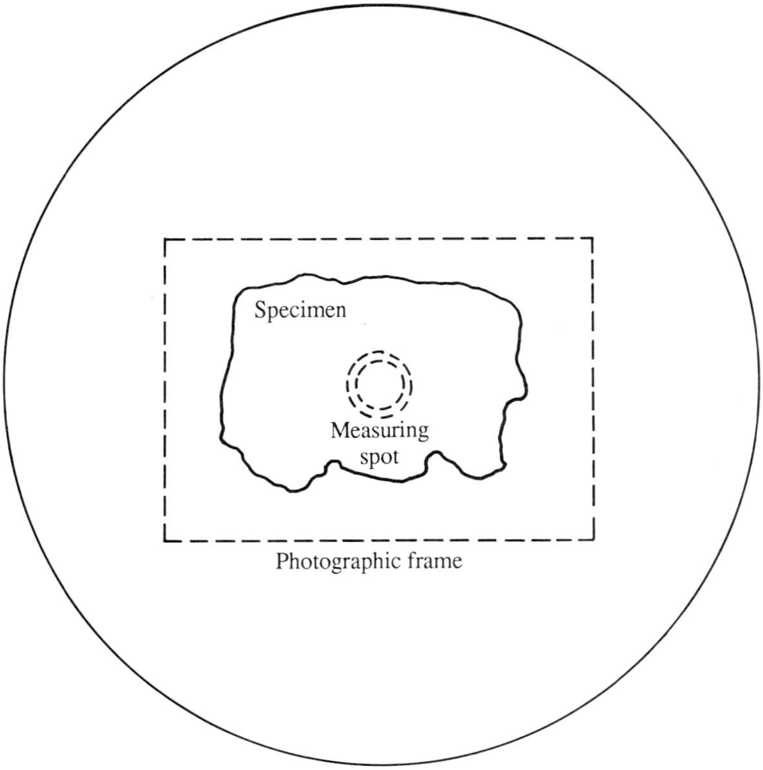

Fig. 6.2. Use of spot meter for exposure measurement: the measuring spot is represented by the small circle in the middle. Note that if the arrangement of the object in the graticule is fixed, it may not be appropriate for the measuring spot to be in the centre of the field. On such occasions, either the spot must be movable or else the measurement be made before moving the object into the right position for photography.

(a) (b) (c) (d)

Fig. 6.3. All diagrams represent objects of equal brightness; however, they differ in the proportion of the field occupied by the objects. An integrating photometer, measuring from the whole field, would give readings proportional to the areas occupied. A correction for the latter must be estimated by eye, and the photometer reading divided by the proportion occupied by the specimen. These diagrams can be

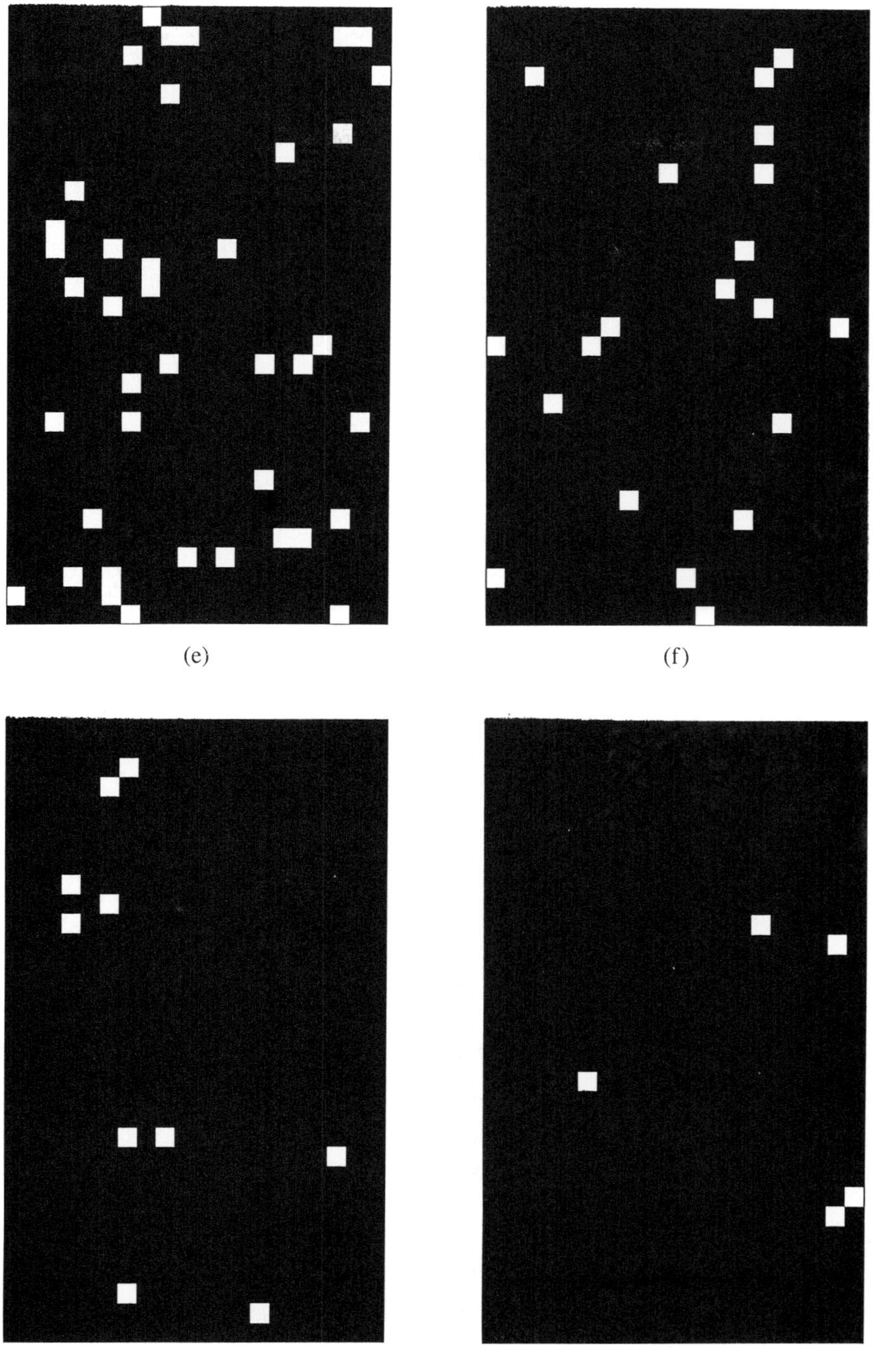

(e) (f)

(g) (h)

Caption for Fig. 6.3 (*cont.*)
used for comparison; the proportions occupied by the light areas are given by the following list:

Figure:	a	b	c	d	e	f	g	h
Proportion:	1/2	1/4	1/5·3	1/8	1/16	1/32	1/64	1/128
	50%	25%	18%	13%	6%	3%	1·5%	0·8%

Table 6.2. *Compensating factor for proportion of area fluorescent*

This table gives the compensating factor (F) by which the exposure must be multiplied, given the proportion (P) of the field which is fluorescent and the ratio (R) of the intensity of the background to that of the specific fluorescence. The figures are given to somewhat more decimal places than are required for photomicrography, so that the significance of the numbers is clearer to the reader.

Contrast ratio (R)		**1/2** **0·500**	**1/4** **0·250**	**1/8** **0·125**	**1/16** **0·063**	**1/32** **0·031**	**1/∞** **0·000**
Proportion (P)		Compensating factor (F)					
1	1·000	1·000	1·000	1·000	1·000	1·000	1·000
1/2	0·500	0·750	0·625	0·563	0·531	0·516	0·500
1/4	0·250	0·625	0·438	0·344	0·297	0·273	0·250
1/8	0·125	0·563	0·344	0·234	0·180	0·152	0·125
1/16	0·063	0·531	0·297	0·180	0·121	0·092	0·063
1/32	0·031	0·516	0·273	0·152	0·092	0·062	0·031
1/64	0·016	0·508	0·262	0·139	0·077	0·046	0·016
1/128	0·008	0·504	0·256	0·132	0·070	0·039	0·008
1/∞	0·000	0·500	0·250	0·125	0·063	0·031	

black with specified proportions of white to black. The arrangement of the white blocks is random.

A further factor to be taken into account in this case is the brightness of the background. If the proportion of the field which shows the specific fluorescence is P, where $P \leq 1$, and if the ratio of the brightness of the background to that of the fluorescence is R, then the required compensating factor F is given by:

$$F = P + R(1 - P)$$

Values of F for various values of P and R are given in Table 6.2. For cameras such as the Leitz Orthomat E which have a compensating setting for the proportion of field fluorescing, the appropriate setting is given by $F \times 100\%$ light-on-dark.

Centre-weighted measurement is a form of integrated measurement in which greater emphasis is given to the brightness of the centre portion of the field. The calculation for correction of proportion of field fluorescing is more difficult in this case.

Fully automatic cameras

Fully automatic systems in which the exposure is controlled automatically by a photometer may be convenient in some respects, and may be excellent for general purpose photomicrography, but four problems may arise: (1) some light is lost by the beamsplitter reflecting light onto the photometer; (2) identical exposures cannot conveniently be given for test and control preparations; (3) with spot measuring systems, the area of the field from which the measurement should be taken may not be in the middle of the field where the measuring spot usually is; and (4) appropriate corrections may not be made automatically for reciprocity failure. It follows that such equipment may not be suitable for fluorescence, or at least must be operated with special care. Some modern photomicrographic systems provide a locking system for exposure times, to overcome problem (2) above.

Automatic electronic compensation for reciprocity failure is provided in some models; the better ones have a choice of several settings for use according to the film and the estimated length of the exposure. If there is only a single correction provided, it may not be appropriate to the film/developer combination in use, and calibration by the user is essential, followed if necessary by the setting of a film speed rating appropriate to the exposure time. The nature of the correction applied by the system should be ascertained from the manufacturer, and compared with the reciprocity failure curve of the film type in use (see also Appendices 4A and 4B).

Automatic correction for reciprocity failure uses the formula $T_c = (T_m)^p$, where T_c is the corrected time, T_m is the uncorrected time, and p is a constant for which values are set by the selector switch (see Appendix 4B). It would be helpful if manufacturers indicated the actual values of p on the selector switch; advanced instruments with digital entry could even allow the desired value of p to be entered. Such an arrangement would encourage film manufacturers to quote appropriate values of p (see Appendix 4B for examples), which could then be applied to any equipment.

Corrections by trial-and-error

Despite having an accurate photometer, some manual compensation is almost always required, based on experience. I recommend keeping a detailed record of all pictures taken. The record should include the ISO film speed set, the photometer reading, the compensation settings (if any), and the length of the exposure given (my detailed recommendations are given below under Record keeping). Afterwards, when the pictures have been developed, note the ones which were correctly exposed. I have found it invaluable to plot a graph, for all successful pictures, of photometer reading against exposure (or, with an automatic camera, ISO setting against exposure time). A separate graph is required for each film type. An example of such a graph is shown in Fig. 6.4. From this graph, one can estimate from a given photometer reading what the optimal exposure is likely to be and also estimate the amount of error likely, from which the bracketing exposures can be calculated. For an automatic camera, a similar plot of ISO speed against exposure time allows estimation of the ISO value(s) to be set.

It is normal to expect up to a half-stop variation in exposure calibration due to changes in film batches and processing runs of colour films.

Coping with reciprocity failure

There are several ways to tackle this problem, namely:

1. Measure the intensity of the image, and consult your graph (referred to above), from which the correct exposure can be estimated. The graph will automatically correct for reciprocity failure.
2. Measure the intensity of the image, and calculate the exposure in the ordinary

way. Then use reciprocity data (see Appendix 4B) of the film type in use to calculate the exposure to be actually given.

3. Some automatic systems have electronic compensation for reciprocity failure. The correct functioning of these systems for the film/developer combination in use must be checked by test photographs.
4. Automatic systems without automatic compensation for reciprocity failure must be pre-set for a suitable compensation factor. It will be necessary to have some idea first of the approximate length of exposure to be expected. The relevant sensitivity rating (ISO, ASA, DIN) for that film type at that approximate length of exposure is then set on the camera control. Data is given in Appendices 4A and 4B, or can be calculated from manufacturer's data.

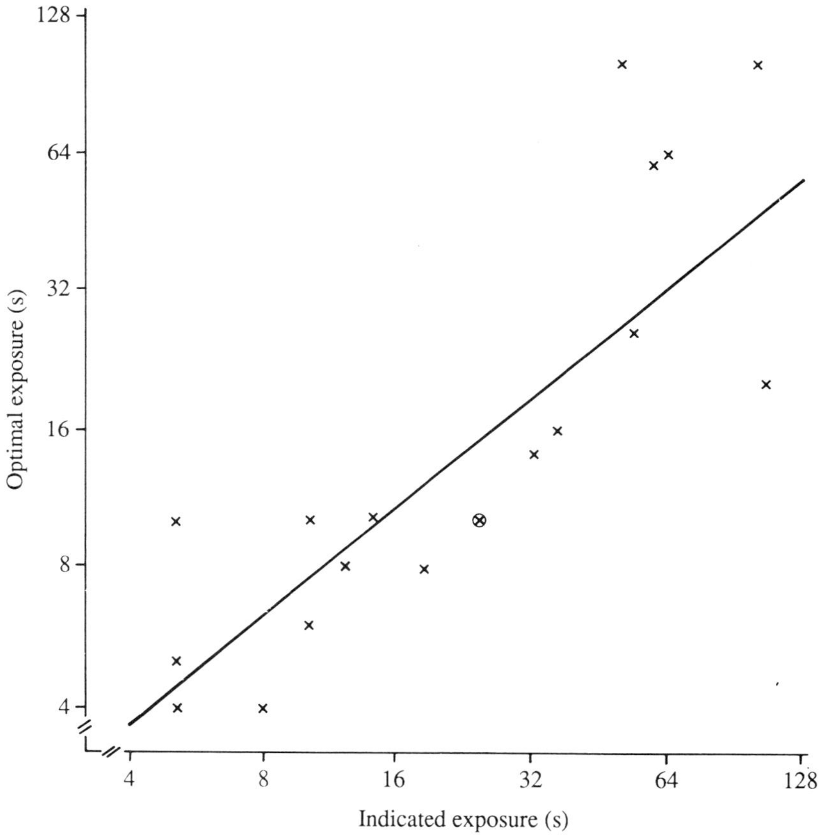

Fig. 6.4. Example of graph used for calibration of a photomicrographic photometer. Scales are logarithmic. Each cross on the graph represents one correctly exposed picture; the circled cross represents two crosses on the same point. Data from actual photographs on Ektachrome 200, of plant autofluorescence. The camera was an Olympus PM10-A, with the reciprocity failure correction set on '5' in accordance with the manufacturer's recommendations. The diagonal line is the calculated regression line (line of best fit).

While these calculations, mental or otherwise, are proceeding, block off the excitation to avoid fading the specimen.

Taking the pictures

To take a picture, if the camera has shutter speeds set manually: with the full excitation, operate the photometer to determine the brightness of the image. Block off the excitation while calculating the exposure to be given. Set the exposure to be given, open the excitation shutter, and shoot. Record the exposure in your notebook.

It is usually necessary to bracket the exposures, i.e. take additional exposures at shorter and longer exposure times, in order to ensure the best results. With colour film, take two additional pictures, one each about 0·5 to 0·7 and 1·4 to 2·0 times the calculated optimal exposure (i.e. 1 to 1/2 stop; 2/3 stop is convenient for instruments calibrated in 1/3 stops). For B&W give 3 to 4 times less and more; or take a series of pictures at 1–stop intervals. Because of the fading problem, I usually give these exposures in the following order: middle (expected to be correct), shorter, longer; or alternatively, I start with the shortest and then take the progressively longer ones. Thus, suppose the estimated exposure on HP5 film (B&W) is 12 s at ISO 400/27°; from the reciprocity data supplied, the corrected exposure is 30 s; so give 30 (or 32) s, 8 s, and 2 min. If fading is severe, re-measure after each exposure. Do not forget to keep a record of the exposure meter reading and the exposures given.

Control pictures must always be taken at the same exposure time as the ‘test’ pictures. This is particularly important in immunofluorescence, if ‘positive’ and ‘negative’ preparations are to be photographed. The correct procedure is to photograph the positive preparation first, noting the exposure given, and then to photograph the control specimen under exactly the same conditions and with the same exposure time. A fully automatic camera (determining the exposure for itself) can be used; but for the control (negative) preparation, the automatic system must be overridden and the same exposure time given as for the positive preparation. Otherwise, the automatic camera will give the control preparation a much longer exposure, in an attempt to make it appear as bright as the positive one!

Troubleshooting

If the pictures do not come out well, or cannot be taken at all because of technical difficulties, the fault may lie in the microscope, the camera, the film, or the processing of the film. Appendix 4D offers some hints for identifying the cause or causes of the problem.

Summary of strategy

The camera must be loaded with a suitable film; this will have a high speed (sensitivity) and may be B&W or colour; if the latter, the film should be of the ‘daylight’ type. The optical system (microscope and camera) must be set up in such a way as to give greatest possible efficiency of transfer of fluorescence from the specimen to the film, in order to minimize the exposure time. The exposure time should be calculated from readings of a photometer incorporated into the camera system, either manual or automatic (in the latter case, the photometer is connected to the shutter).

Efforts must be concentrated on minimizing the exposure of the specimen to irradiation both before and during photography. The following steps may be helpful:

1. Preliminary viewing of the specimen, focussing and composition of the picture with a neutral density filter in the excitation light path, or with phase contrast or other technique requiring lower light intensity or longer wavelength.
2. Selection of a high-speed film. It may be possible to raise the speed at which the film is rated, by a suitable developer (e.g. Microphen) or increased development time ('pushing') or pre-exposure.
3. Optimization of the optical system to give greatest efficiency in transfer of light to the film. This includes using a low overall magnification, objective of high NA, and avoidance of unnecessary optical elements.
4. Multiple pictures of the same field should be taken with different exposures (usually most easily arranged by adjusting the ISO film speed setting). The shortest exposures should be taken first, or else the 'best estimate' first followed by the remainder, shortest first.
5. If successive pictures are to be taken of different fluorophores in the same field, the first photograph should be that requiring the longer excitation wavelength (e.g. green before blue or UV, blue before UV) so as not to fade the second fluorophore.

Control photographs, if taken, must be exposed under the same conditions and for the same exposure time.

Flash-fluorescence photomicrography

Photomicrographs of fluorescence can be made using an electronic flash as light source (Young, 1969). The electronic flash has spectral characteristics that are essentially those of a xenon arc, but giving only a brief flash rather than continuous output. Flash duration of conventional flash units is of the order of 0·1 to 1 ms (1/10 000 to 1/1000 s). Young (1969) employed flash to record the fluorescence of motile organisms vitally stained with Euchrysine, which could not be photographed by ordinary means because they moved during the exposure. The flash, the Zeiss (Oberkochen, West Germany) photomicrographic unit, had an output of 60 J, with a duration of 0·5 ms. Similar flash units are available to fit other brands of microscope. In the ordinary circumstances of a non-motile specimen, more than one flash can be given as multiple exposures on one frame of film to achieve the required exposure. Note that because of an effect similar to reciprocity failure, an intermittent exposure of this type may not equal in effect one longer flash of energy equal to the sum of the shorter flashes. Also, reciprocity failure occurs at very short exposures such as may be given by a flash unit. On the other hand, use of flash for fluorescence should assist in minimizing fading. Photomicrographs have also been taken with pulsed dye lasers (Bergquist, 1973).

Videointensification photomicrography

If fluorescence is examined with a television camera, photographs can be taken from the monitor screen. Resolution is much impaired (compared to direct photography) but the exposure is determined only by the brightness of the monitor and therefore the exposure is short (by fluorescence standards) and consistent. Alternatively, images can be recorded on videotape.

Videointensification microscopy is discussed in detail in *Quantitative fluorescence microscopy* (Rost, 1991). Photography from a monitor screen is described there and in Kodak Publication L-10 (Anon., 1989). The following is a summary. Almost any black-and-white or colour film of normal contrast can be used. The shutter speed, about 1/8 to 1/2 s, should be long enough to cover several frames of video, otherwise the raster pattern will appear. This problem is particularly apparent with cameras with focal plane shutters (as in most SLR cameras), which record a diagonal stripe if a short exposure is used. The brightness setting of the monitor should be noted; the brightness setting, once set to a suitable level, will rarely need be changed, and the exposure required for the camera will also remain constant. It is preferable not to use the 'auto' setting of a camera shutter, since this may inappropriately vary the exposure from one experiment to the next.

FIDELITY OF REPRODUCTION

The reproduction of the fluorescence in the photograph can differ from the original in the following main ways: in colour, contrast, definition, and form. Colour reproduction cannot be expected to be exact, because of differing reciprocity failures of the three photosensitive layers of the colour film. However, the eye is not good at distinguishing colours at low intensities either! Therefore, it is a waste of time worrying about small variations in colour, and certainly there is no point in using the colour correction filters recommended by the manufacturers to compensate for reciprocity failure. Contrast will depend upon the film, the development process, and the exposure, as well as the specimen and the degree to which it fades during exposure of the film. For example, Kodachrome film has higher contrast than Ektachrome film; contrast is increased with increasing development, up to a point, then decreases; some developers give higher or lower contrast; and highest contrast is normally obtained with the 'correct' exposure. Definition or resolution in the photograph usually depends mainly on the optical system, and otherwise on the nature of the film and the manner of development.

Resolving power of the optical system is determined by the NA of the objective (in fluorescence microscopy, the specimen being self-luminous, the NA of the condenser does not affect the resolution of the microscope). The resolution of the film may also affect the resolution of the final photograph. In practice, the resolution of film is unlikely to be a limiting factor in the detail shown by the photomicrograph, although with high-speed film graininess may be apparent.

Certain important aspects of the interpretation of photomicrographs are discussed in Chapter 7.

PHOTOMACROGRAPHY OF FLUORESCENCE

Photography of specimens at a magnification too low for a microscope requires photomacrographic equipment and techniques. This may involve either a dedicated photomacrographic apparatus, or a normal 35 mm SLR (single lens reflex) camera with a macro lens, bellows or extension tubes. Appropriate lighting is crucial for satisfactory results.

Photographs of small fluorescent specimens can be made using conventional photomacrographic equipment with appropriate primary and barrier filters and either a UV lamp or electronic flash as a light source. An additional source of ordinary (visible) light may be required for focussing, particularly if the actual picture is to be taken with flash, or for photographing the specimen by ordinary light for comparison.

As sources of UV light, several kinds of lamp are available. The easiest to set up are the fluorescent tubes, available in several sizes to fit standard holders. I have had no experience of photographing with them. Probably better are the Philips HPW 'black light' high-pressure mercury arc lamps, which, however, require a special power supply unit.

Alternatively, an ordinary photographic flash unit can be used; or better, two flash units, one on each side of the specimen. An appropriate excitation filter must be placed in front of the flash unit; square filter glasses (Schott standard size is 50 mm square) are convenient and can be attached temporarily to the flash head with opaque black tape. The excitation filters appropriate for use with a flash unit are those specified for xenon arc lamps. For UV excitation, I have used a combination of UGI (2 mm) and BG38 (4 mm). For use with fluorescein, special filters have been developed for use by ophthalomologists for fluorescein angiography: for review see Myers (1985).

For a barrier filter, it may be possible to use a standard yellow photographic filter, provided that it is not strongly fluorescent. Otherwise, a 50 mm diameter unmounted glass filter may be used in a suitable holder. I have a holder made in Britain by BDB, and another made specially to accommodate thick or multiple filters and to accept a lenshood on top.

Equipment which I have used for fluorescence photomacrography is shown in Fig. 6.5. The illumination is provided by a pair of Philips HPW 125 W high-pressure mercury arc lamps, mounted horizontally in a metal reflector. The lamp bulbs have a coating to absorb most of the visible light, and no additional filtration is necessary. As a barrier filter, I currently use the combination of 2 mm Schott GG395 glass and a Nikkor Y44 yellow filter. The latter is strongly fluorescent, which is why the GG395 is needed, placed on the outside (nearer the lamp) to absorb sufficient UV to prevent fluorescence from the Y44 filter. A lenshood also keeps direct UV light off the filters. As shown, I usually use a Nikon F3 camera body, a Leica bellows (with appropriate adapter) and an enlarging or similar lens. An example of results obtainable is shown in Fig. 6.6. See also Blazek (1983) and Eastman Kodak (1972).

Exposure measurement tends to be a problem unless either the fluorescence is very strong or the camera's metering system is unusually sensitive. While one is using continuous lighting, if the meter is sensitive enough, the exposure can be measured with the lens aperture wide open (for greatest sensitivity) and then the exposure time calculated for the aperture to be used. Corrections need to be applied for the proportion of the field which is fluorescent, and reciprocity failure, exactly as for photomicrography. Through-the-lens flash metering is likewise possible, provided that corrections are made for the proportion of the field which is fluorescent. In all cases, bracketing is required. Automatic sensing devices on the flashgun itself cannot be used.

Specimens under water present special problems mainly due to the refraction of light at the air–water interfaces for both illumination and photographic light paths. Excellent examples of photographs of fluorescent marine organisms in aquaria are to be found in the works of Catala (1964) and Gillett & Yaldwyn (1969).

My own limited experience suggests that the following considerations are particularly important. The smaller the aquarium, the closer the specimen can be to both the camera and the lights. While water transmits UV reasonably over short distances, it is

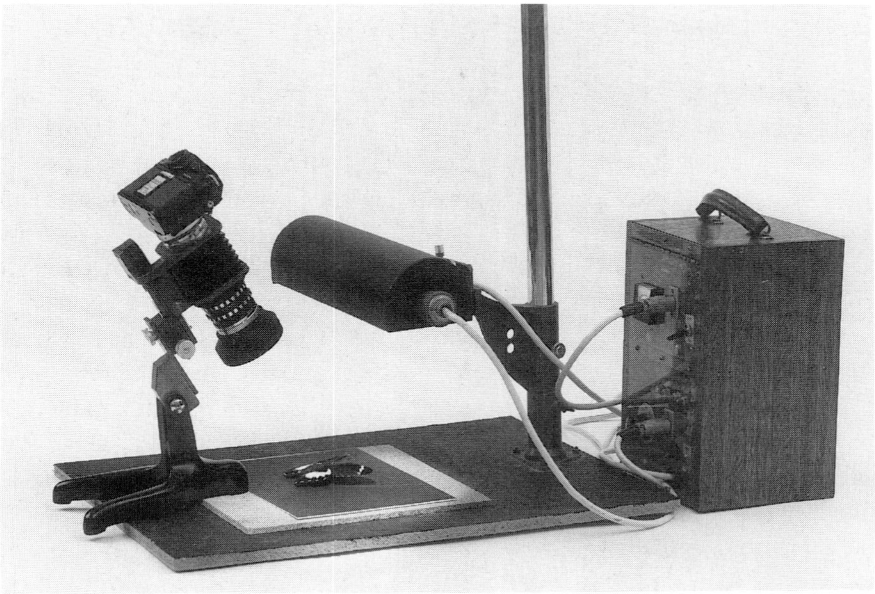

Fig. 6.5. Equipment for photomacrography of fluorescence. The illumination is provided by a pair of Philips HPW 125W high-pressure mercury lamps in a reflector held on a copying stand (centre), with a power supply unit (right). The camera is a Nikon F3 camera, mounted on a Leica Bellows II (by an adapter) which has a Schacht 100 mm f/3·3 lenshead, barrier filters, and lenshood. The bellows unit is supported on a stand derived from the foot of an old Watson microscope. Five countries contributed to the manufacture: Australia, Britain, Germany, Japan, and the Netherlands. Illumination unit on loan courtesy of Mr K. Gillett.

essential that the water be kept clear: any turbidity scatters the illumination, particularly at short wavelengths. Because the surface of the water cannot be kept flat, both illumination and photography should take place either through the walls of the aquarium or through special devices projecting into the water. The design of air–water interfaces (ports) was reviewed by Mertens (1970).

Ultraviolet photography

I mention this apparently irrelevant topic for two reasons. Firstly, sometimes people are a little confused between photography of fluorescence using UV excitation, and photography in ultraviolet light. As we know, the former requires illumination with UV, without contamination with visible light, and photography of fluorescence through a barrier filter; whereas the latter likewise requires UV illumination which, however, can include any amount of visible light, and photography at UV wavelength through a UV-passing filter similar to the excitation filter used in fluorescence.

Fig. 6.6. Fluorescence of scorpion, *Urodacus hoplurus*. Specimen from Armidale, New South Wales, photographed with the apparatus shown in the previous figure. Agfapan Vario XL film (a chromogenic film), C-41 development. Field 32 mm × 48 mm. Original magnification ×0·75, exposure 15 s at f/28 (lens set to f/16, effective aperture reduced to f/28 due to bellows extension). For filters, see text.

My second reason for mentioning the matter is that equipment designed for fluorescence photography may be adaptable, relatively easily, to UV photography. As in photomicrography, ordinary optics can generally be used at 365 nm, except Leica lenses which since about 1962 have incorporated a UV filter. There are special macro lenses for UV: those which have come my way are the 105 mm *f*/4 UV-Nikkor for Nikon cameras, recently introduced (there was a much earlier model, discontinued long ago), and the 85 mm *f*/4·5 Ultra-Achromatic-Takumar which was made for the screw (M42) Pentax. There is also a Zeiss UV-Sonnar CF 105 mm *f*/4·3 for the Hasselblad.

For photomacrography, I have used the same illumination system as described above for fluorescence, with a 365 nm interference filter over the lens. The filter may, alternatively, be a glass filter combination of UGI + BG38, or Kodak Wratten filter 18A. Do not use a filter which has previously been used as an excitation filter, because its optical quality will have been spoilt by the heat.

A high-contrast B&W film is the most appropriate, e.g. Kodak Technical Pan film. Many films, especially daylight-balanced slide films, have a barrier overcoating to absorb UV. This may make them unsuitable for some UV photographing techniques. If you use a colour transparency film, use one balanced for tungsten light, because these films are designed for full sensitivity at the blue end of the spectrum and do not have a UV-absorbing overcoat. Exposure has to be determined by trial-and-error.

GENERAL

Record keeping

It is important to keep a full record of every photographic exposure made. Such a record provides data for construction of a graph such as that shown in Fig. 6·5, is an invaluable guide to further photography, and is also valuable for reference if a picture is later to be published.

My own practice is to use a notebook of A4 format. Details of the microscope should be written down at the beginning: make, model, light source, illuminator, filter specifications, objectives, camera ocular (projective) (if any), camera type. If any of these are changed later, an annotation should be made in the book.

Each time I load a film into the camera, I start a new page, and write down the *type of film*, its speed rating (ISO), the reciprocity setting (automatic cameras only), the *camera body* used (more than one body may be in use simultaneously, for different film types), and a *serial number* for the roll of film. At the beginning of each photographic session, the *date* and *subject* are noted. For each exposure, the following information is required:

1. Objective identification (type, magnification, NA).
2. The overall magnification.
3. Exposure meter (photometer) reading.
4. Film speed setting (if different from the manufacturer's rating).

5. Compensation factor for area of fluorescence (if required).
6. Duration of exposure.

Usually it is only necessary to identify the objectives fully once, and thereafter to specify either the overall magnification or the objective magnification, assuming that the overall magnification can be calculated from the objective magnification. Similarly, using an automatic camera, compensation for the proportion of the area fluorescent can be made either as such, if the camera allows, or by fudging the speed rating (ISO/ASA/DIN), and a record only needs to be kept of whichever parameter is used. With a non-automatic camera, neither (4) nor (5) need be recorded, but (6) becomes vital. It is rarely necessary actually to write all this information for each frame of the film; it is usually sufficient to specify those aspects which have changed from the previous line of the notebook.

Later, when the pictures have been developed, they should be compared with the notes, and the latter annotated as to the success or otherwise of the result: satisfactory (I make a tick), underexposed, or overexposed.

Over the years, one builds up a priceless mine of information concerning photomicrographic technique as well as details of one's pictures. The notebooks (I am into my sixth volume) deserve to be kept in a safe, fireproof place, such as in the same filing cabinet as the pictures.

Duplication of transparencies

Duplication of pictures is particularly relevant to fluorescence photomicrography, for two reasons: the best photographs may be unique and unrepeatable, and poor ones which are too dark but otherwise valuable can be reproduced somewhat lighter and thereby made more suitable for projection. The 35 mm transparencies can easily be duplicated, using a reflex camera and one of the many duplicating devices (which are of variable quality) or with a photomacrographic apparatus. Some loss of detail and colour balance is to be expected. In general, high-contrast films such as Kodachrome film should usually be duplicated onto relatively low-contrast films, e.g. Ektachrome film, otherwise the resulting contrast will be extreme. A special duplicating Ektachrome film is available. Commercial duplicating services exist, but may not be able to carry out special procedures such as making a transparency lighter.

Care of photographs

All photographs should be clearly labelled. Negatives are usually stored in loose-leaf transparent sheets which should be labelled at least with the serial number of the roll. Transparencies are usually numbered when returned from processing; the roll number should be added – I find that blue ball-point ink is best.

Transparencies can be stored in boxes, projector magazines, or plastic files in filing-cabinets. It is important to store them in a dry, well-ventilated area to discourage growth of fungi (moulds). If you have many slides, you will need an index; these days indexing is best done on a computer, if available, or file cards. For detailed information on the care of photographs, see the book by the Editors of Time-Life Books (1972), and Anon. (1988).

THE PRESENTATION OF PROJECTION SLIDES (DIAPOSITIVES)

At a scientific or technical meeting, when slides (diapositives) of fluorescent specimens are shown, there are certain steps that can be taken to enhance the presentation.

First, precious slides must be properly protected, preferably in a glass mount. Some projectionists, both professional and amateur, seem unable to avoid putting their fingerprints on slides, and only glass mounts give sufficient protection. I remember one projectionist who actually put butter on a slide – she was eating some of her lunch early! The orientation of slides should be indicated by a prominent dot in the bottom left-hand corner of the slide as seen looking at the picture.

The audience will always appreciate a clear indication of the scale of the picture. This is best shown by a scale bar on the picture itself. With the Nikon Microphot-FXA it is possible to photograph a scale bar directly on the original film. Otherwise, it is possible (but not usually practicable) to mark the bar directly on to the transparency (on the emulsion side): only a duplicate transparency should be risked in this procedure. A dark bar may be added in black ink using a fibre-nib pen; a light-coloured bar can be made by scratching a line in the emulsion. To avoid damaging the original slide, a scale bar can be photographed on a second film and the two films mounted together, emulsion sides towards each other so that both will be in focus simultaneously in the projector. Alternatively, a print from the original slide can have a bar drawn on it and then be re-photographed; unfortunately this procedure loses much of the tonal range of the original, due to limitations of the print.

Any temptation to block out unwanted fluorescence should be resisted. It always shows, and will make the audience doubt your intellectual honesty.

Bearing in mind that fluorescence pictures are by their nature dark, and are commonly shown in a dark auditorium, the audience's eyes should be given an opportunity to adapt both before and after showing slides of fluorescence. The process is analogous to the preparation and resolution of a musical discord. First, prepare your audience by showing a relatively dark slide: a coloured photograph, or a diazo slide (white on blue or red; they look like Plate 3). Then, after your fluorescence slide, again use a relatively dark slide to enable your audience to get accustomed to the light gradually. Avoid using a black-on-white diagram immediately before a fluorescence slide: your audience will not be able to appreciate the fluorescence slide which follows. If you cannot have a diazo (coloured) slide made, colour your slide; one way is to sandwich the slide with a sheet of coloured gelatine.

7

The interpretation of the fluorescence microscopic image

Because of its complexity, microscopy of fluorescence is fraught with more than the usual set of dangers in interpretation of the image. Microscopists who rely heavily on visual observation might with advantage contemplate the optical illusions described and illustrated by Tolansky (1964).

This chapter deals with the factors which may affect the appearance of the image in a fluorescence microscope. These factors are related to the specimen, to the microscope optical system, and to the observer's own optical and neurological characteristics. Even photography, apparently objective, may be misleading if not interpreted correctly.

THE SPECIMEN AND ITS ARTEFACTS

A microscope image is never even as good as the specimen; it can only be worse! It therefore behoves the microscopist to start with as good a specimen as possible. Specimen preparation is discussed in Volume II, Chapter 17. The fluorescence microscopist must select and prepare the tissue, perform the required staining or other reaction, and mount the preparation, all with care and professional skill.

Fluorescent preparations are of course subject to all the artefacts to be found in other types of preparations. A common confusing artefact is a bit of fluorescent rubbish which fell onto the specimen by accident, often a piece of vegetable fibre, debris from paper tissues used for cleaning. The foreign object may fluoresce any colour, depending upon its chemical composition, and may even mimic the specific fluorophore. I recall having seen such an object exactly mimic the fluorescence colour of fluorescein isothiocyanate (FITC) in an immunofluorescence preparation; only by changing the filters could it be shown to be not FITC. Holbrook & Tishler (1980) reported that searching for fluorescent Y-chromosomes (Y-bodies) stained with Quinacrine is often complicated by the presence of artefacts indistinguishable from Y-bodies by fluorescence microscopy. These artefacts could, however, be distinguished by darkground illumination with white light. They employed a darkground condenser for both forms of illumination, fluorescence being excited by an HBO200 mercury arc lamp and BG12 filter. A true Y-body, whose refractive index is apparently similar to that of the surrounding cytoplasm, is rendered invisible by the change to white light illumination, whereas the artefacts remain clearly visible.

Fluorescent dust on the top of the substage condenser may be confusing, but this stays put when the object is moved.

Autofluorescence of normal tissue may be difficult to distinguish from specific fluorescence under some conditions, particularly if the specific fluorescence is blue or if the barrier filters are badly chosen (see below).

Clean dehydrating and clearing agents must be used for mounting fluorescent specimens; those contaminated with Eosin, commonly found in histological laboratories, must be avoided as the fluorescent Eosin may stain the tissues.

Pseudofluorescence

Sometimes a phenomenon is observed which mimics fluorescence. This has been called pseudofluorescence (van der Ploeg, Bonnet & van Duijn, 1979). An example is the Wolman effect due to excitation light in the visible region being refracted or reflected from the specimen and not blocked by a barrier filter (Wolman, 1970).

Illumination by a darkground substage condenser should result in an image of the highest contrast because none of the excitation light is intended to enter the magnifying system. However, some light is refracted or reflected as if to form a normal darkground image at the excitation wavelength, which should be blocked by the barrier filter. However, if the excitation light is in the visible spectrum, and if, as is customary, the barrier filter is less opaque than would be provided for brightfield illumination, some of the diffracted/reflected light may appear as if it were fluorescence (Wolman, 1970). A similar effect in red colour can be seen if the excitation filter combination has inadequate blocking in the red (Tanke, Huikshoven & de Jonge, 1984).

An artefact similar to the Wolman effect can be observed in specimens freeze-dried, embedded in paraffin wax, and examined for formaldehyde-induced fluorescence (see Volume II, Chapter 12) after mounting in paraffin oil without removal of the wax; the crystalline structure of the wax can produce refraction or reflections, which may in some circumstances mimic fluorescence, unless eliminated by fully dissolving the wax or liquifying it by heating.

Another form of pseudofluorescence, due to anomalous dispersion, was described by van der Ploeg *et al.* (1979).

THE MICROSCOPE

There are three main parameters of the image which are of particular interest: the colour, the contrast, and the resolution.

Colour

Effect of barrier filter

The apparent colour of a fluorescent specimen depends on the transmission characteristics of the barrier filter, as well as on other factors to be discussed later. If the barrier filter is 'colourless', i.e. with uniform transmission throughout the visible range, the apparent colour of the fluorescence will not be affected. However, such barrier filters can be used only in conjunction with ultraviolet (UV) light. Barrier filters designed for use in conjunction with excitation in the violet, blue or green have cutoff wavelengths in the blue, yellow, or red, respectively. Held up to white light, they appear yellow, orange or red. Objects seen through these filters do not, of course, show their natural colours, as can be confirmed by looking through them at objects of various known colours. The same distortions of apparent colour take place in the fluorescence microscope; a yellow (minus blue) filter will render blue invisible, while a red filter will make everything appear in shades of red. The effects of different barrier filters are illustrated in Plate 2 and Fig. 7.1.

For example, the fluorescence colour of bisbenzimide has been variously reported as blue (Berkley *et al.*, 1980) and green (Kuypers *et al.*, 1980); this difference is probably due to the effect of a barrier filter. Similarly, the formaldehyde-induced fluorescence of noradrenaline should appear as a greenish-blue colour with a transparent barrier filter, but it is more commonly observed as green because of the use of a yellow filter, and may be photographed as blue by inappropriate use of 'artificial light' film (see below).

Care must be used in interpretation of green fluorescence seen with the special BP525

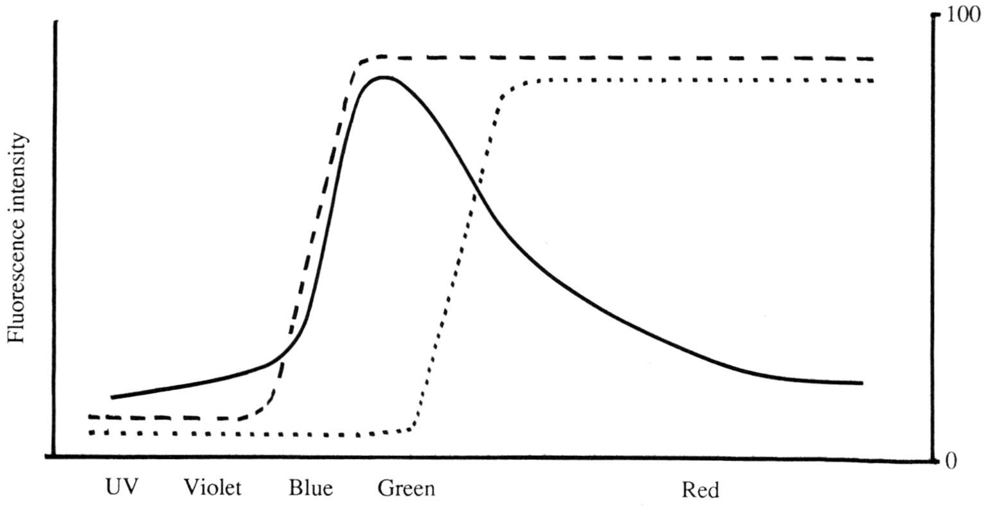

Fig. 7.1. Explanation of the colour distortion effect of barrier filters. The solid line (–) represents the emission spectrum of a typical fluorophore, with peak in the blue and a long tail extending into the red. The dashed line (---) shows the transmission of a barrier filter with cut-off in the violet (e.g. Leitz K430), which allows virtually all of the emission to pass, so that the fluorescence appears its true colour, probably greenish blue. The dotted line (·····) shows the transmission of a barrier filter with cut-off in the green (e.g. Leitz K530). This would make the fluorescence appear yellow. An even longer wavelength cut-off would result in the fluorescence appearing orange or red. Cf. Fig. 2.2. Diagram redrawn after Rost (1981).

emission filter for FITC. By design, the filter transmits *only* green light, and therefore *anything* seen through it appears green. Using this filter, it may reasonably be assumed that all FITC fluorescence is transmitted. However, not all apparently green fluorescence is FITC: it may be something else, with an emission peak either in the green (like FITC) or in the blue with a tail into the green. In the latter case, the BP525 filter will show only the green component of that fluorescence.

Instrumental contrast

The contrast of a microscopic image can be defined roughly as the difference (ratio of brightness) between the image of the object and the background. Contrast must be considered in two ways: real and apparent. The real contrast is that which can be measured by a photometer; the apparent contrast is that perceived by the eye, which is not necessarily the same thing. Contrast may exist in tone (that is, brightness) or in colour. In the former case, an objective measurement is the ratio of brightness of the image of the object to that of the background.

It is also necessary to distinguish the contrast between the object and empty field on the one hand, and between specific parts of the object and the remainder of the object. For the moment, we consider only the former, instrumental contrast, which relates only to characteristics of the fluorescence microscope itself.

The contrast of an image is affected by a number of instrumental factors, of which the most important consideration relevant to fluorescence microscopy is the matching of the excitation and barrier filters. However good the filter, there is always some transmission by the excitation filter at wavelengths transmitted by the barrier filter, and vice versa. This is particularly true of interference filters, which tend to have a higher transmission in the tail of their spectrum than dyed-glass filters; a practical interference filter therefore usually incorporates a layer of coloured glass filter. However, leakage of light past the filters can be kept to a very small proportion of the light available for excitation, if filters are properly chosen.

The most common mistakes in setting up the filters, whereby a bright background appears instead of the normal dark one, are (a) omission of the red-stopping filter from the excitation filters, giving a red background; (b) an insufficiently thick excitation filter, particularly in the case of brightfield transmitted-light illumination; and (c) wrong choice of barrier filter, with too short a short-wavelength cut-off. Even with a supposedly correctly matching filter set, it must be admitted that some commercial filters are better than others, and the less satisfactory ones should be avoided if possible.

Contrast is increased if excitation light is kept out of the optical system used for magnification of the specimen, because the barrier filter is then less critical and there is less autofluorescence in the optical system.

Autofluorescence in the optical system

Contrast of the image is invariably degraded if there is any autofluorescence within the optical system. It is due to the light intended for excitation inducing fluorescence in the optical elements; this fluorescence can occur in the optical glass, in cement used between adjacent components of a multi-element lens or filter, and in dust or other dirt on the optical system.

Objectives are particularly subject to autofluorescence in epi-illumination systems, in which the full blast of the excitation passes through the objective. Unwanted light at the exciting wavelength may also be diverted towards the ocular in epi-illumination systems as a result of light-scatter from dust on the objective lenses, and reflections from the surfaces of the lenses and from the inside of the lens mounts.

These problems can be minimized by appropriate design of the objectives and by keeping the lenses scrupulously clean. Several manufacturers offer special objectives for epi-illumination, in which autofluorescence has been reduced or eliminated.

Immersion oil from a reputable manufacturer should be non-fluorescent, as well as having the other required characteristics described in Chapter 5. If there is any doubt, there are several possible non-fluorescent alternatives, including glycerol, paraffin oil, and water.

Dichromatic mirrors, so far as I am aware, create problems only if they are dirty.

Oculars do not seem to suffer from autofluorescence problems, presumably because, firstly, the barrier filter stops the excitation light from reaching the ocular, and secondly, they are of simple construction with few elements.

Barrier filters ought not to fluoresce, but unfortunately may do so. Glasses otherwise suitable for barrier filters but having significant fluorescence include GG14, OG1, OG2 and OG5. A useful test of a barrier filter involves holding the filter in the path of the excitation beam, and examining the filter through another barrier filter. It is generally necessary to excite with a somewhat shorter wavelength than would be used with that particular filter, and to examine it though a barrier filter with a correspondingly shorter wavelength cutoff. If the filter is seen to fluoresce, it is not suitable, since its fluorescence will degrade the contrast of the image.

Specimen contrast

I refer now to contrast between specific fluorescence and fluorescence of surrounding tissue. Autofluorescence of surrounding tissue can be minimized by careful attention to specimen preparation, and if possible to selection of the longest possible excitation wavelength, as autofluorescence is normally excited most strongly by UV. Sometimes it is possible to stain the background with a substance which quenches fluorescence; the latter effect can also be achieved by counterstaining with a dye which is not excited by the wavelength used to excite the specific fluorescence.

Changes in pH have widely differing effects on different fluorescent compounds. This may make it possible to select a value for the pH of the mountant so as to maximize the fluorescence of the desired fluorophore and to reduce the fluorescence of others; it may even be possible to select a pH value which, while not necessarily giving optimum conditions for brightness of the particular fluorescence being studied, may eliminate obscuring fluorescence from other compounds present (O'Brien & McCully, 1981).

Counterstaining is a method of changing the contrast, and will be dealt with below.

Resolution

The resolution of the microscope is affected by essentially the same factors as in an ordinary microscope, except that the numerical aperture (NA) of the illumination system is not relevant. Resolution is determined by the NA of the objective, the wavelength of observation, the degree of correction of the objective (i.e. the degree to which it approaches the theoretical resolution), and the contrast of the image.

This is not the case in a confocal fluorescence microscope (see Chapter 8), for three reasons. Firstly, the resolution is governed by the wavelength of excitation, rather than that of emission. Secondly, the lateral resolution of the system is at least 1–4 times greater than in a conventional system. Finally, the resolution is very much less affected by the thickness of the specimen; in a conventional microscope the resolution is limited to approximately 1/10 of the specimen thickness.

THE HUMAN VISUAL SYSTEM

The eye: anatomy and physiology

The human eye (Fig. 7.2) is an approximately spherical organ, about 24 mm in diameter. On the outside, it has a tough fibrous coat called the *sclera* except at the front where it is transparent and called the *cornea*. The cornea serves both as a tough, transparent outer coating and, because of its curved front surface, as a powerful lens. The sclera is lined internally by an opaque pigmented layer; this is continued behind the cornea as the *iris*. The iris has a hole in the middle, the *pupil*. The iris is a muscular organ, which can dilate or constrict the pupil; the iris thereby constitutes a variable aperture diaphragm with a maximal aperture of about 7 mm. Light enters the eye through the cornea and iris. Behind the iris is the *lens*, which is transparent and biconvex in shape. The cornea and the lens collectively constitute a lens system which forms an image on the *retina*, which is a layer of nerve cells lining the inside of the back of the eye inside the opaque pigmented layer. Strictly speaking, the retina consists of several layers of nerve cells of different types. The image is focussed on to a mosaic of sensory cells, which detect light and initiate electrical impulses which are relayed by other nerve cells into the brain. The sensory cells of the retina are of two types, known as the *rods* and *cones*. The cones are concentrated mainly in a central region, the *fovea*, corresponding to the centre of the field of vision. Rods are found mainly in the periphery. The cones provide colour vision and the highest visual acuity; rods provide more sensitive vision in dim light, but with reduced acuity and no discrimination of colours.

The characteristics of vision in bright light and in dim light are quite different. The type of vision characteristic of bright light is called *photopic vision*; that in dim light is called *scotopic*. The first stage in the detection of light by the rods and cones is absorption of incident quanta by pigments, known as visual pigments, which then undergo a photochemical reaction. The best known visual pigment is *rhodopsin*, found in the rods. It is light-sensitive, being purple in the dark and becoming bleached in the light. The cones are believed to comprise three sub-populations, cells of each sub-population containing one of three similar pigments. These pigments differ in their spectral absorption characteristics. As might be expected, the spectral distribution of the eye's sensitivity to light is believed to correspond to the absorption curve of the visual pigments.

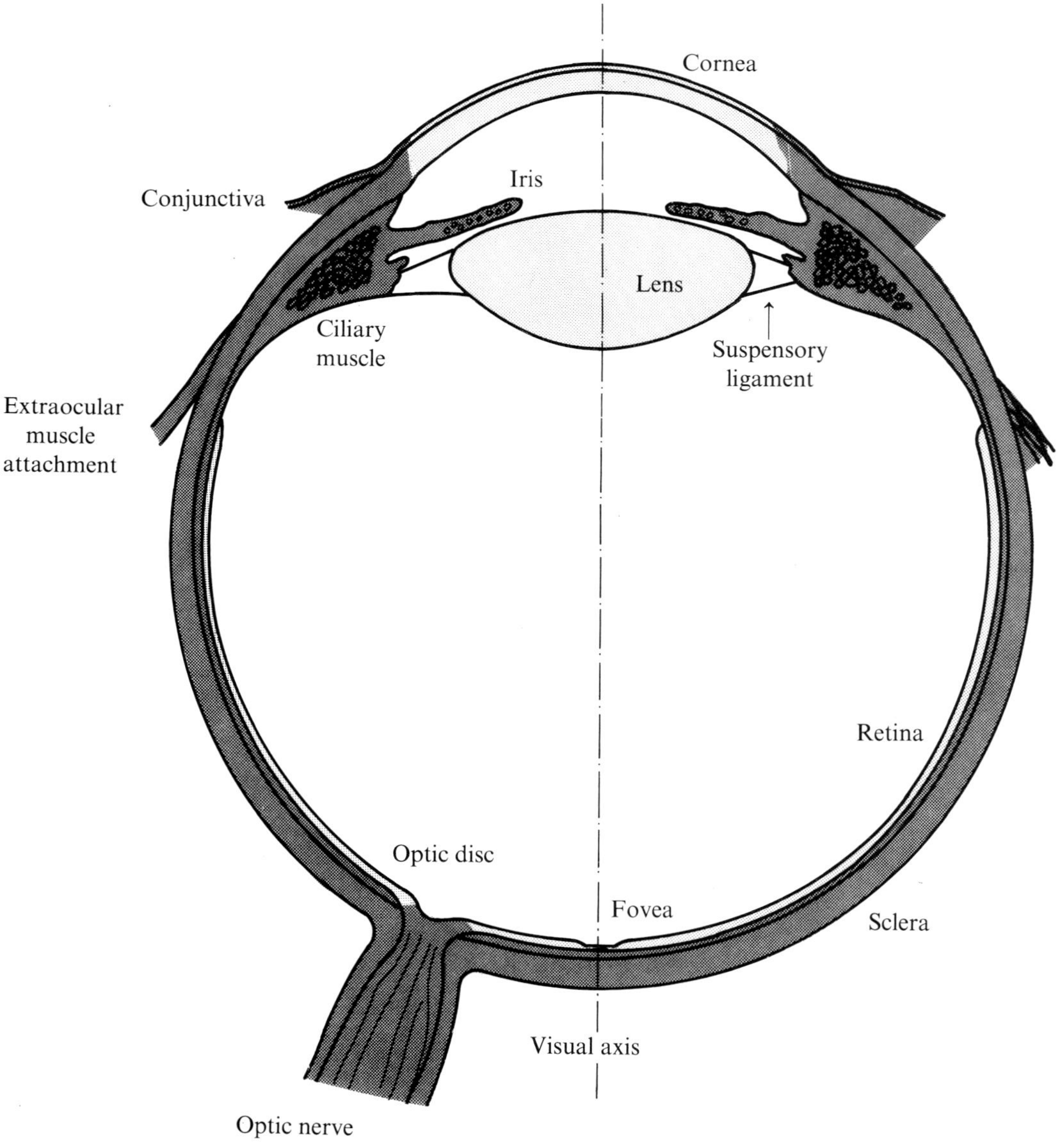

Fig. 7.2. The human eye (highly diagrammatic).

The absorption of rhodopsin is greatest for blue-green light, at a wavelength of 510 nm. This corresponds to the sensitivity of human scotopic (dim light) vision, which is greatest for blue-green light, although the colour of it cannot be recognized by the eye in such dim light. Scotopic vision is colour-blind, everything appearing in tones of grey.

It is believed that colour vision depends upon differential absorption of each wavelength by three pigments found in the cones. This corresponds approximately to the three-colour processes in colour photography. Because rods only have one pigment, scotopic colour vision is not possible. Figure 7.3 compares the spectral distribution of sensitivity of the eye under dim and bright light conditions, the *scotopic visibility curve* and the *photopic visibility curve*. The latter is similar to the former, but is displaced towards the red (low quantum energy) end of the spectrum, with peak sensitivity at about 555 nm.

The sensitivity of vision

Individual rods appear to be able to respond to single quanta of light; however, the dimmest light it is possible to see consists, probably, of a single quantum in each of several (2–14) rods. More light than this has to reach the eyeball, because of losses in those parts of the eyeball preceding the retina. This very high sensitivity cannot be achieved immediately after exposure to bright light. After exposure to bright light, the sensitivity of the eye is relatively low. When placed in the dark, or red light (to which the rods are insensitive), the sensitivity becomes greater as time passes, becoming maximal in about an hour. This is *dark adaptation*. Apart from an initial increase in sensitivity due to opening of the pupil, the increase in sensitivity appears to be due partly to regeneration of visual pigments previously bleached by light, and partly to adaptation of the nervous system. In the cones, dark adaptation gives only a relatively small

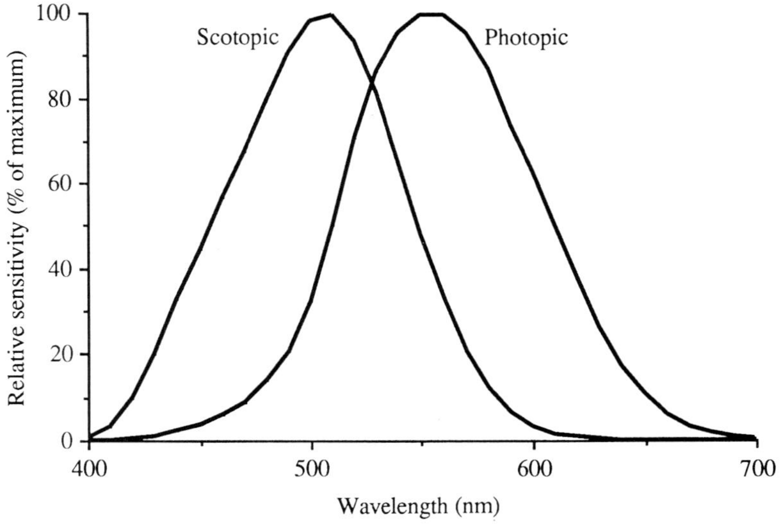

Fig. 7.3. Scotopic and photopic visibility curves: the approximate relative sensitivity of the eye to different wavelengths, in bright light (photopic vision) and dim light (scotopic vision).

increase in sensitivity, maximal after only about 7–10 min and detectable only by measurements with deep red light invisible to the rods. Full dark adaptation increases the sensitivity of the human eye 10 000-fold; this is accompanied by a change in spectral sensitivity as already indicated.

The sensitivity of the eye under bright-light conditions is determined also by the aperture of the iris. This aperture opens or closes as required, with a response time of about 2–3 s.

Colour blindness

Colour blindness (in photopic vision) is common, being found in 8% of all men and 0·4% of all women. Possible varieties can be explained on the basis of deficiency of one or more visual pigments in the cones. The most common form of colour blindness is *anomalous trichromatism*; sufferers, like people with normal vision, employ mixes of three colours to achieve matches, but the intensities in the mixtures are different. This is one of the reasons why abnormal colour vision is not always obvious: another reason is that colour-blind persons can often deduce a great deal from other, non-chromatic clues.

The possible effect of variations in the observer's colour vision has been discussed by Sander (1970) in relation to immunofluorescence. Sander concluded that only 50% of the observers tested were suitable individuals to scan immunofluorescence (FITC) slides, because of poor judgement of green colours.

The observer

The effect of the observer's visual characteristics on fluorescence microscopy has been discussed by Ritzén (1967) and Sander (1970). As explained above, the human eye is unable to appreciate colour correctly in dim light. In strong light, the eye has a maximal sensitivity at about 555 nm; at decreasing light intensities, the peak sensitivity of the eye shifts towards shorter wavelengths, and finally colour discrimination ceases. The first noticeable effect is a loss of vision of red.

The apparent contrast of a visual field depends in part on the overall brightness of the field, and possibly on the degree of dark adaptation of the eye, because dimmer parts may be below the limit of effective colour vision. This may be of particular importance if the intensity of background staining or the relative brightness of test and control preparations may appear to change because of alterations in viewing conditions.

Viewing of dim fluorescence is assisted by dark adaptation of the observer's eyes, which requires keeping the microscope in a dimly lit room (see Chapter 5). Complete darkness is not necessary and should be avoided, as it hinders working and renders the eye liable to being dazzled by any lamp. Bright illumination, if required anywhere, is best coloured red, as this will not spoil dark adaptation. Neon pilot bulbs can be used freely; the kind which is sold as a night light, or for darkroom use, is suitable. It is less well known that dark adaptation can be preserved during a subsequent period of brighter illumination (e.g. a visit to another laboratory) by wearing red goggles. The red light admitted by the goggles permits satisfactory vision for most purposes, while preserving dark adaptation.

To minimize the difficulties associated with vision in dim light, the optical conditions should normally be chosen to make the fluorescence microscopic image as bright as possible. The only accurate way to determine the colour of fluorescence emission is to measure the emission spectrum with a microspectrofluorometer, as will be described in Chapter 8.

There is an interesting discussion of vision in relation to astronomy (which has many of the same problems) by Sidgwick (1979).

PHOTOMICROGRAPHY

There is an old saying, 'The camera cannot lie'. I recall, at a meeting of the Royal Microscopical Society, having capped that with the addition '. . . but the photographer can!' What I meant was that the photographer can choose his film, field, filters, exposure, development, and so forth, and can produce a misleading photograph either by design or accident. I do not include here the retouched photographs which have been occasionally displayed at meetings of learned societies, with fluorescent junk obliterated with black ink. I always mistrust such photographs, on principle. Another artful dodge is to take a photograph in black-and-white, and show a transparency of it tinted with a coloured gelatine filter sandwiched in the mount, to imitate FITC or whatever; I think that this is allowable provided that one announces the picture as such.

Photographic film has its own characteristics, which differ from those of the human eye. The characteristics of photographic film were described in Chapter 6: the relevant facts may be summarized as follows. Black-and-white film is by definition unable to record differences in colour; the spectral sensitivity of the film may differ markedly from that of the human eye, particularly in increased sensitivity to blue light and in being sensitive also to UV. Colour film is made with two types of colour correction: 'daylight' and 'artificial light'. The former is designed to reproduce colours as closely as possible in appearance to the original image; the latter has increased sensitivity in the blue. Pictures taken on 'artificial light' film will therefore appear unnaturally blue. Both kinds of film tend to give an increase in contrast (dim parts of the image may not appear at all in the photograph) and deviate from normal colour balance when long exposure times are given to capture low light levels.

SPECIFICITY

When fluorescence microscopy is used to demonstrate a histochemical reaction, the specificity of the fluorescent image is never greater than that of the histochemical reaction, and may be adversely affected by lack of specificity in the fluorescence observation. We thus distinguish two components contributing to specificity, or lack of it: the histochemical (cytochemical) specificity of the staining or other reaction, and the optical specificity of the microscopic technique.

As an example: Congo Red is occasionally used as a stain for amyloid. The stained preparation can be examined perfectly well in an ordinary microscope; it can also be examined in a fluorescence microscope. There is nothing to be gained, apart from a possibly prettier appearance, by the latter technique.

Thioflavine T is another stain for amyloid. As a fluorochrome, it stains amyloid a silver-whitish colour, while other tissue components usually appear more yellowish. If an inappropriate barrier filter is used, everything stained by Thioflavine T appears yellow, so there is a loss of specificity.

In immunofluorescence with FITC, there is a strong temptation to assume that everything appearing green is FITC. This assumption can be checked by changing the barrier filter to one with narrow-band transmission around 515 nm, the emission peak of FITC, and also by examining with different excitation wavelengths.

Optical specificity in the fluorescence microscope can be maximized by proper choice of the excitation wavelength and barrier filter. The excitation wavelength should correspond to the absorption peak of the the desired fluorophore; while the barrier filter should transmit as much as possible so as to minimize colour distortion. A narrow-band interference filter, at the emission maximum of the presumptive fluorophore, can be used as a check. The only really certain way is to measure the excitation and emission spectra with a microspectrofluorometer (described in Chapter 8).

From the fact that the fluorescence is always associated with a transition from the first excited state (S_1), irrespective of the level to which the molecule was originally raised by absorption of a photon, it follows that the emission spectrum and therefore the colour of fluorescence cannot be altered by a change in excitation wavelength. However, a change in wavelength, particularly one towards the high-energy (UV) end of the spectrum, may result in excitation of some other fluorophore which may be present, e.g. a contaminant in a dye.

The specificity of reactions is usually discussed in works relating to the individual reactions. For a general discussion, see Horobin (1982).

LOCALIZATION

One of the problems in fluorescence microscopy is that the image tends to be that of some fluorescent areas or particles in an otherwise empty field. This makes it difficult for the observer to recognize parts of the specimen and to determine the identity of fluorescent structures. There are several ways in which this difficulty can be overcome: by using the autofluorescence of tissues, by a fluorescent counterstain, and by revealing the morphology of the specimen by simultaneous or alternate examination by a different form of microscopy, e.g. phase-contrast.

Autofluorescence has classically been used as a guide to the morphology of tissues. This method requires UV excitation, since longer wavelengths do not excite the general proteins of tissues. With UV excitation and an appropriate barrier filter, autofluorescence generally appears blue, although other colours occur. This blue can only be distinguished from, say, the green of FITC, if a colourless barrier filter is used; otherwise, the background and the FITC may be made to appear a similar colour (e.g. yellow) as an artefact of the barrier filter.

Where UV excitation is not optimal for the fluorophore, as in the case of FITC, UV has nevertheless been used by some workers in order specifically to excite autofluorescence. This can be accomplished either by just using UV instead of green light for excitation; or else by using simultaneous or alternate excitation at both wavelengths (Nairn & Ploem, 1974). This practice has two major disadvantages: there is a loss of optical specificity, and the unnecessarily energetic excitation leads to more rapid fading. It may be less expensive to implement than a simultaneous phase-contrast system, requiring special objectives and condenser.

Combined illumination methods

Fluorescence excitation by epi-illumination can easily be combined with substage illumination for simultaneous or alternate studies by phase-contrast or darkground illumination, or, with somewhat more trouble, polarization or interference-contrast. Examination by the secondary method can be either simultaneous with the fluorescence excitation, or used alternately; the latter is most easily effected by means of a system of movable mirrors, as in the Leitz Mirrorhouse 500, which enables quick changes between the various possible modes of illumination.

The secondary (non-excitation) illumination can be either white, or made a contrasting colour by the addition of a filter. Red is usually a very suitable colour for the purpose, particularly for phase-contrast and darkground illumination where the image is normally monochromatic in any case. Moreover, red illumination does not spoil the eyes' dark adaptation, required for observation of dim fluorescence. The only exceptions to the general usefulness of red are those cases where a coloured counterstain is employed, and in simultaneous non-fluorescence illumination of an object which fluoresces red, in which latter case differentiation would be lost. Red or other coloured filters, if not available from the microscope manufacturer, can be obtained easily from other sources. Probably the most convenient are the filter glasses supplied for ordinary photography, which may be placed on the foot of the microscope below the condenser; these may be had from any photographic supply shop. Filters of coloured gelatine may be obtained from Cokin and Kodak; to avoid scratching, these should be mounted between glass; cut to size, they can be fitted in the glass mounts supplied for Super Slide or larger transparencies. In an emergency, coloured Cellophane can be utilized, but its optical qualities are poor unless it is kept absolutely flat without wrinkles.

Darkground illumination can be used for the demonstration of the morphology of the tissue. It is particularly useful for assessment of the relative numbers of fluorescent and non-fluorescent particles. Moreover, the relatively dim image tends not to spoil the eyes' adaptation to dark.

Phase-contrast is particularly suitable for the visualization of non-fluorescent tissue. Simultaneous phase-contrast and fluorescence studies are best carried out with a phase system giving negative phase-contrast, in which denser tissues appear light against a dark background. This preserves the dark background of the fluorescence microscope, against which both fluorescent and non-fluorescent objects appear. Because the general tone of the field is kept dark, the eyes' dark adaptation is easier to maintain.

Interference-contrast gives a better view of the specimen than does phase-contrast; in particular, there is no halo around objects. On the other hand, the equipment is more expensive. Combined Nomarski differential interference-contrast and epi-illumination of fluorescence is particularly easy in those instruments which use infinity-corrected objectives.

Modulation contrast (Hoffman, 1977) by transmitted light can be combined with excitation by epi-illumination (Hoffman, 1978). Modulation contrast equipment comprises a modulator in the objective, a special slit aperture in the front focal plane of the condenser, and a contrast control polarizer before the slit.

Polarization systems, using transmitted white light, can be used for simultaneous demonstration of polarization and birefringence. A difficulty is that the analyser will usually absorb some 60% of the light, reducing the intensity of the fluorescent image.

Simultaneous incident and transmitted light excitation is normally suitable only for the simultaneous demonstration of two fluorophores with widely differing absorption maxima. In principle, one of the excitation wavelengths could be UV to demonstrate the autofluorescence of tissue; however, this is not recommended as a method for displaying the morphology of the tissues, for the reasons given above.

Counterstains

A counterstain is a second (or further) stain applied to the specimen, in addition to the specific stain. Fluorescence microscopy gives unrivalled possibilities for counterstaining. The possibilities for counterstains include: (1) a fluorescent dye, visible simultaneously with the specific fluorescence; (2) a fluorescent dye whose excitation peak differs sufficiently from that of the specific fluorescence that the two can be visualized alternately by changing the excitation wavelength; (3) a non-fluorescent dye, visible by normal transmitted light and invisible by fluorescence; and (4) a coloured fluorescent dye, visible by both fluorescence and normal transmitted light.

The basic requirements for the counterstain are that: (1) it must be of a colour clearly distinguishable from the specific fluorescence; (2) it must not mask the specific fluorescence; (3) it must not quench the specific fluorescence; (4) it must stain suitable tissue components; and (5) it must either have the same excitation maximum (approximately) so as to be visible simultaneously with the specific fluorescence, or have a quite different excitation peak so as not to be visible while the specific fluorescence is excited.

Counterstains are usually of one or other of two types: nuclear stains, and cytoplasmic stains. The best of the former are specific stains for deoxyribonucleic acid (DNA), unlikely to cause fluorescence outside nuclei.

A specific study on fluorescent counterstains, applied to neural tissue, is that of Schmued, Swanson & Sawchenko (1982); see also Alexander & Potter (1963). A counterstain can also be used to modify autofluorescence: Evans Blue has been used for this purpose, staining connective tissue and changing the tissue fluorescence to a dull red, contrasting with formaldehyde-induced fluorescence (de la Lande & Waterson, 1968).

8

Quantitative and scanning fluorescence microscopy

Quantitative fluorescence microscopy is concerned with the measurement of fluorescence emission from a defined area or areas of a specimen. The basic measuring technique is called microfluorometry (MFy).

The information to be gained by microfluorometry is of several types. Firstly, microfluorometry is most commonly applied to determining the amount of some specific substance, such as deoxyribonucleic acid (DNA), present in particular regions, such as cell nuclei, by comparison of the intensity of the fluorescence of the specified regions with that of a standard. Secondly, determination of the fluorescence spectra, and possibly of other characteristics described below, may enable the identification of specific fluorescent substances. Thirdly, determination of parameters of fluorescence of a known fluorophore introduced intracellularly as a probe can give information about the micro-environment of the probe, and thereby about the inside of the cell. Fourthly, scanning or counting devices can quantify the distribution of fluorescent components in tissue or other material.

This chapter is intended to give a brief overview of quantitative fluorescence microscopy, with emphasis on those aspects which are important to an understanding of fluorescence microscopy in general. The topic is dealt with in much greater detail in a companion volume, *Quantitative fluorescence microscopy* (Rost, 1991). Accordingly, I have limited the references to reviews and a few papers of particular importance or which report very recent work; more references are given in the other book. Scanning microscopy is dealt with in this Chapter because it involves measurements of fluorescence intensity.

Brief reviews of microfluorometry were given by Ploem & Tanke (1987), Rost (1980), and Fukuda, Böhm & Fujita (1978). Useful information is also found in the chemical and biochemical literature on fluorometry: see, for example, Miller (1984) and Wehry (1982). The standard text on microscope photometry in general is that of Piller (1977). Many topics related to quantitative cytochemistry are discussed in detail in the books edited by Thaer & Sernetz (1973), Wied & Bahr (1970), and Wied (1966).

Microfluorometry, basically, involves measuring the brightness of fluorescence emission from a defined area of the specimen under standardized conditions. This process can be extended to the measurement of other characteristics of the fluorescence. The parameters or characteristics of the fluorescence which can be measured from a given region of the specimen are:

1. The intensity of fluorescence.
2. The emission spectrum.
3. The excitation spectrum.
4. The quantum efficiency (ϕ) of the fluorophore.
5. The polarization of fluorescence.
6. The fluorescent lifetime (τ) of the fluorophore.
7. Structure-correlated information, e.g. the relative areas of fluorescence and non-fluorescence.

Changes in all of the above can be followed over a period of time, either short (comparable to the fluorescent life time) or long (e.g. rate of fading).

Of the above possible types of measurement, (1) is microfluorometry proper; (2) and (3) are microspectrofluorometry; (7) requires scanning, either mechanically or with a video camera. It may be possible to measure more than one parameter simultaneously. The measurement of cells *en passant* in a flow system is flow cytofluorometry.

MICROFLUOROMETRY

For the remainder of this section, microfluorometry may be said to be a technique whereby the intensity of fluorescence is measured from a given area of a specimen, usually with a view to measuring the amount of fluorophore present and thereby estimating the amount of some substance present in that region of the specimen. For example, nuclear DNA may be measured by microfluorometry of nuclei subjected to Feulgen hydrolysis and stained with a fluorescent Schiff-type reagent.

A microfluorometer, the instrument required for microfluorometry, is essentially a fluorescence microscope with a stabilized light source, and a measuring device which measures the intensity of fluorescence from an area of the specimen defined by a diaphragm (see Fig. 8.1). This measured intensity is compared to the intensity to be measured from a standard containing a known amount of the fluorophore, and the amount of fluorophore present in the specimen is determined by proportion.

Although simple in principle, microfluorometry presents a number of traps for the unwary, against which due precautions must be taken.

Relationship between fluorescence measurement and fluorophore concentration

All microfluorometry is based on the concept of irradiating a specimen with nominally monochromatic light of known intensity, and measuring the intensity of light emitted from a given area of the specimen. The mathematical relationship between the intensity of the fluorescence and the amount of fluorescent substance present is discussed in the companion volume (Rost, 1991) and by Böhm & Sprenger (1968), Fukuda, Böhm & Fujita (1978) and Oostveldt *et al.* (1978). The following is a pragmatic description of the approximate relationships used in practice.

The intensity of the measured fluorescence (I_f) is determined by the amount of light absorbed by the specimen (I_a), the quantum efficiency of the fluorophore (ϕ), and optical instrumental factors, represented by a constant (r). These are related by the equation:

$$I_f = rI_a\phi$$

The amount of light absorbed (I_a) is determined by the amount of light incident on the specimen (I_0) and the extinction coefficient (ϵ; a constant which expresses the ability of the fluorophore to absorb light), concentration (c) and thickness (d) of the specimen, according to the Beer–Lambert exponential law:

$$\log I_a = \epsilon cd \times \log I_0$$

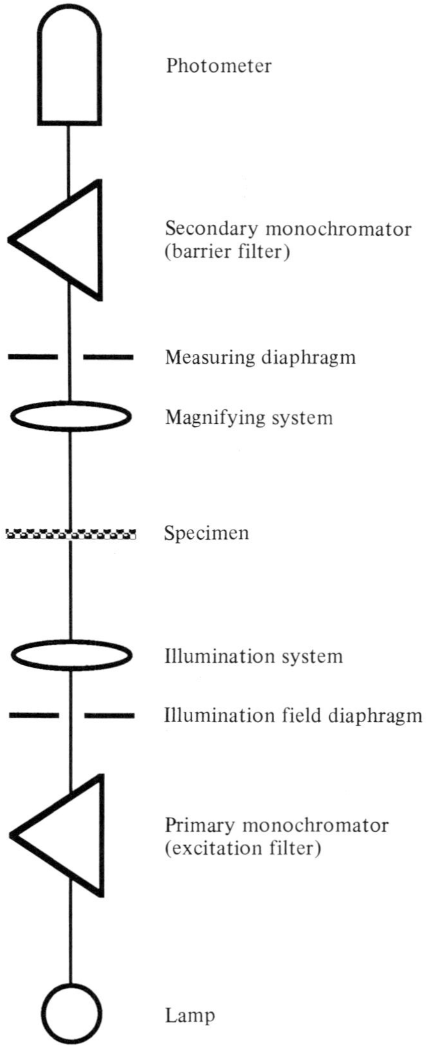

Fig. 8.1. Block diagram of a microfluorometer.

It is convenient to denote I_f/I_0 by a symbol, *F*, representing the relative fluorescence intensity, i.e. the fluorescence intensity for a constant light input. Under usual conditions, provided that the concentration of the fluorophore is not excessive, the exponential relationship can be simplified to:

$$F = 2{\cdot}3\ \epsilon cd$$

In other words, the emitted fluorescence is directly proportional to the concentration of the fluorophore. If the concentration of the fluorophore is high, the relationship becomes non-linear, and a closer approximation is:

$$F = 2{\cdot}3\ \epsilon cd\ (1 - 1{\cdot}15\ \epsilon cd)$$

This approximation is valid to within 4% provided that ϵcd is less than 0·17 (Gains & Dawson, 1979). The amount of substance present tends to be underestimated by microfluorometry. According to my calculations, the amount of the error is roughly equal to the extinction (defined as the negative logarithm of the proportion of light absorbed); so that if the result is to be obtained within 1% accuracy, the extinction should not exceed 0·01, while for a 10% accuracy the extinction should not exceed 0·1. It has been suggested that for practical purposes, the relationship can be taken as linear provided that the extinction of the fluorophore does not exceed 0·2 (Böhm & Sprenger, 1968).

Aperture defined volume

If, using epi-illumination, the objective is dipped into a fluid, or applied to the surface of a translucent solid, the region from which measurement is made is a volume, forming a cone defined by the aperture of the objective. This is referred to as *aperture defined volume*. This can be applied in the measurement of uranyl glass standards (Jongsma, Hijmans & Ploem, 1971) and fluids in Terasaki trays (Deelder, Tanke & Ploem, 1978).

Comparison with microdensitometry

It is natural to compare quantification by microfluorometry with quantification by microdensitometry, particularly as very often both techniques could in principle be used for the same basic purpose, as in the Feulgen procedure. There are two main methods available for microdensitometry: the scanning method and the two-wavelength method. Details are given by Piller (1977), Wied (1966) and Wied & Bahr (1970). The main points of comparison of quantification by microfluorometry and by microdensitometry are summarized in Table 8.1.

Theoretically, a major difference between microfluorometry and microdensitometry is that microfluorometry does not (generally speaking) suffer from distributional error. This in turn leads to one of the advantages which microfluorometry has over scanning microdensitometry: speed. Once the required area of the specimen has been selected, a measurement may be made by microfluorometry almost instantly; whereas by scanning microdensitometry, each area to be measured requires many point measurements. Absorption measurement by the two-wavelength method can also be made instanta-

Table 8.1. *Comparison of microfluorometry and microdensitometry*

Characteristic	Microfluorometry	Microdensitometry
Sensitivity	High	Low
Accuracy at		
low concentrations	Normal	Low
high concentrations	Low	Normal to $E \approx 0{\cdot}4$
Opaque specimens	Possible	Impossible
Optical specificity	At 2 wavelengths (excitation and emission)	At 1 wavelength (absorption)
Decay time resolution	Possible	Impossible
Type of measurement	Relative	Absolute
Speed of measurement	Fast	Slow
Distributional error	Absent[a]	May be present

Notes:
[a] Except at extremely high concentrations extending above the linear region.

neously, the two measurements at different wavelengths being made simultaneously using a dichromatic mirror and two detectors.

In general, microfluorometry is advantageous compared to microdensitometry for assaying low to very low concentrations. Sometimes both methods are available, as in quantification of DNA stained by the Feulgen method with Pararosaniline. Direct comparison of microdensitometry and microfluorometry on the same specimens may assist in calibration as well as validation of the overall technique. Such comparisons of microfluorometry with microdensitometry were made in respect of quantification of nuclear DNA by Böhm, Sprenger & Sandritter (1970, 1973), Schnedl *et al.* (1977), Geber & Hasibeder (1980), Tanke & van Ingen (1980), and Broekaert *et al.* (1986); and in respect of quantitative immunocytochemistry by Smith, Redick & Baron (1983).

Instrumentation

A microfluorometer is essentially a fluorescence microscope with a stable, uniform light source and a photometric device for measuring the intensity of fluorescence from a defined area of the specimen (see Fig. 8.1). Usually, a separate illumination system is required for phase-contrast or other examination with light of a longer wavelength, so that the specimen can be set in the measuring field before the excitation light is allowed to irradiate the specimen; this precaution reduces fading.

Equipment for microfluorometry has been available commercially for quite some years (see Figs. 8.2–8.5). Leitz, in particular, appears to have paid attention to the development of equipment for microfluorometry. Independent companies also supply some equipment which can be used in conjunction with a microscope. Microfluorometers can be based on a normal or inverted microscope. An inverted microscope with epi-illumination (Figs. 8.4 and 8.5) is particularly suitable for experimental arrangements requiring manipulation of the specimen or for studying fluid, tissue culture cells and other sediments. Such adaptation is now particularly easy with the Leitz Fluovert inverted microscope (Fig. 8.4), which has a normal trinocular body tube at the top of the microscope, identical with that of Leitz upright microscopes (see Chapter 4).

Photomultiplier tubes

Photomultiplier tubes depend upon the photoelectric effect, whereby a photon incident upon a metallic surface may release one or more electrons from the surface (see Chapter 2). Photomultiplier tubes have a metallic cathode coated with a compound of one or more rare earths. Incident photons knock out electrons from the cathode, which are attracted towards the first of a series of positively charged elements called dynodes. As

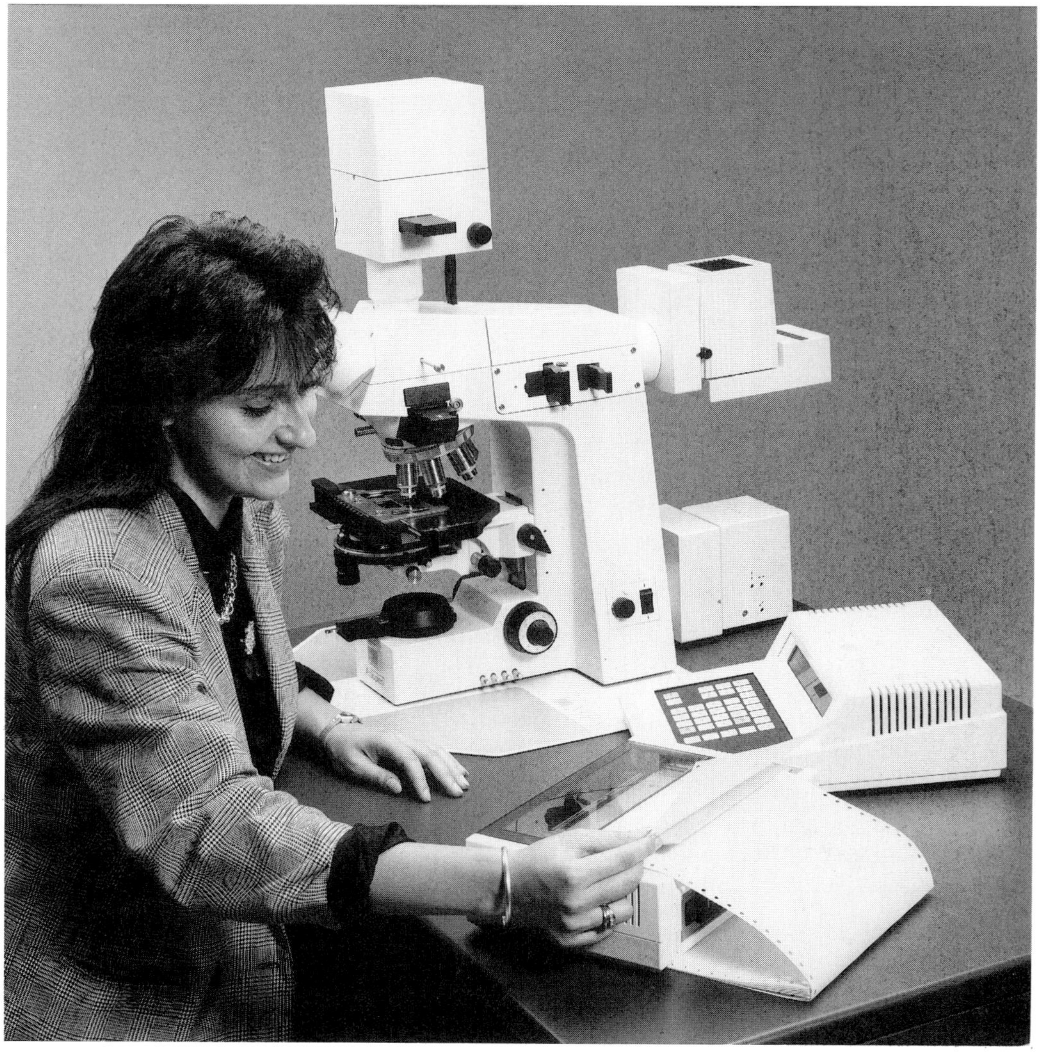

Fig. 8.2. Carl Zeiss (Oberkochen) MPM 200 photometer for microfluorometry on the Axioplan microscope. Photograph courtesy of Carl Zeiss Pty Ltd, Sydney.

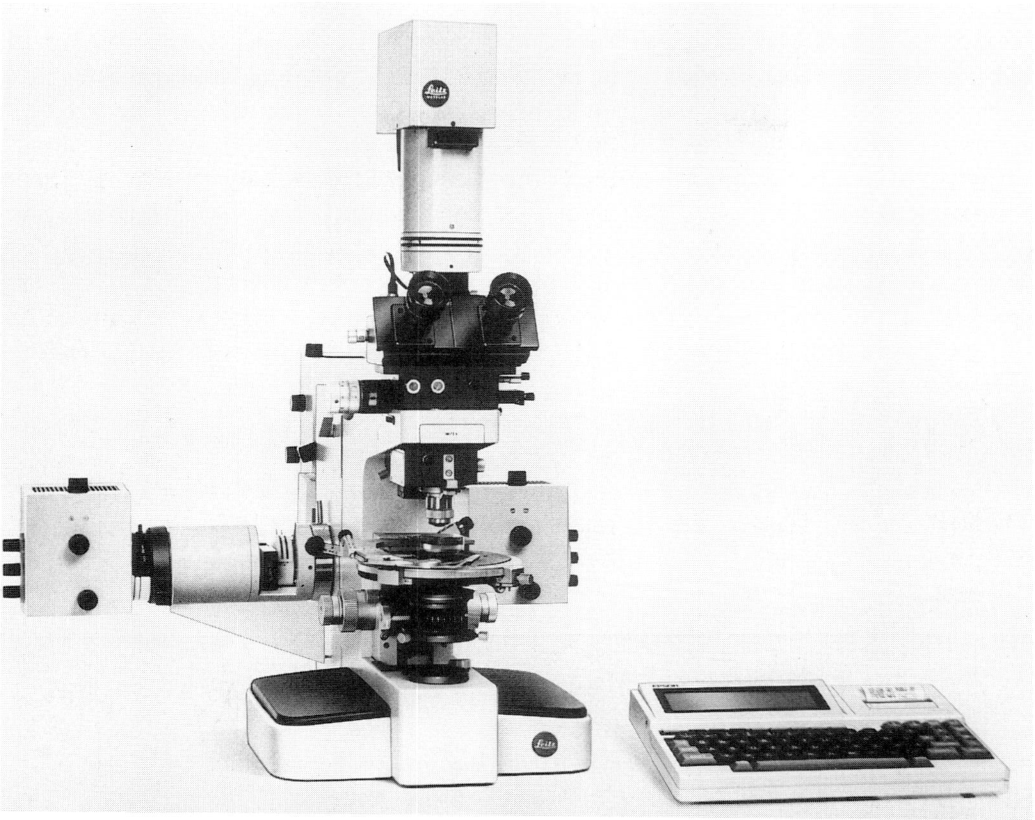

Fig. 8.3. Leitz MPV Compact 2 system on Orthoplan microscope. Photograph courtesy of Dr W. J. Patzelt of Ernst Leitz Wetzlar GmbH.

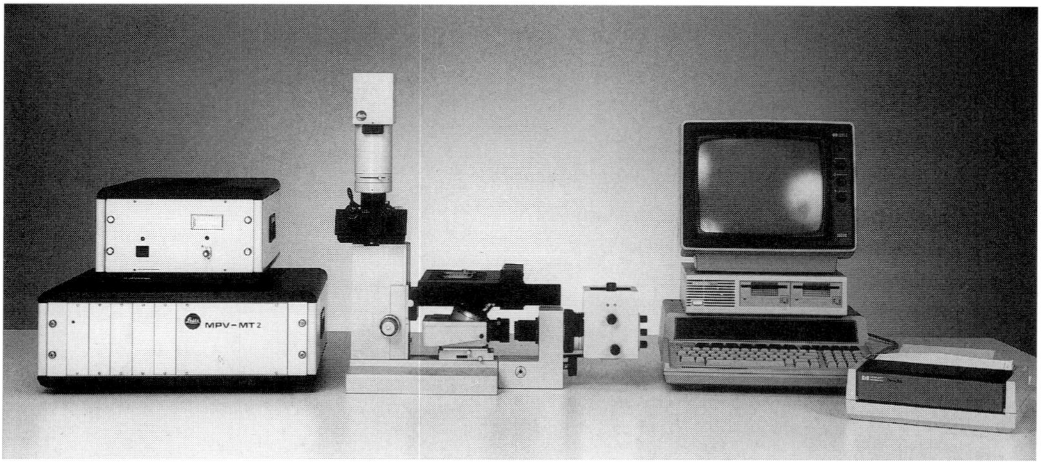

Fig. 8.4. Leitz Fluovert inverted microscope with fluorescence epi-illuminator and MPV-MT2 photometer. Photograph courtesy of Dr W. J. Patzelt of Ernst Leitz Wetzlar GmbH.

a photon releases an electron from the cathode, the electron is accelerated towards the first dynode because of the dynode's positive potential (typically about 100 V with respect to the cathode). The electron, on hitting the first dynode, will have attained a higher energy level than it had on leaving the cathode, because of having been accelerated through a potential of 100 V or so. The electron will therefore have enough energy, on hitting the dynode, to knock out several electrons. These are now attracted to a second dynode, held at a higher positive potential (another 70–100 V; now about 200 V with respect to the cathode). In consequence, an amplified current will flow towards the second dynode. The process is repeated at several dynodes, so that, although only a single electron may be released from the cathode by one photon, the result is a substantial pulse of current at the final anode. The increase in sensitivity is enormous. The average value of amplification of current per stage is about 4·5. Hence, a typical photomultiplier with 9 dynodes has a gain of about $(4{\cdot}5)^9$, i.e. approximately one million.

With no light incident on the photomultiplier, there is still a small amount of emission from the cathode, mainly thermionic (i.e. associated with the temperature of the cathode). This results in a small output, the so-called dark current.

The output from the photomultiplier can be measured in either of three basic ways: direct current (DC) measurement, alternating current (AC) measurement in conjunction with a light chopper, and photon counting.

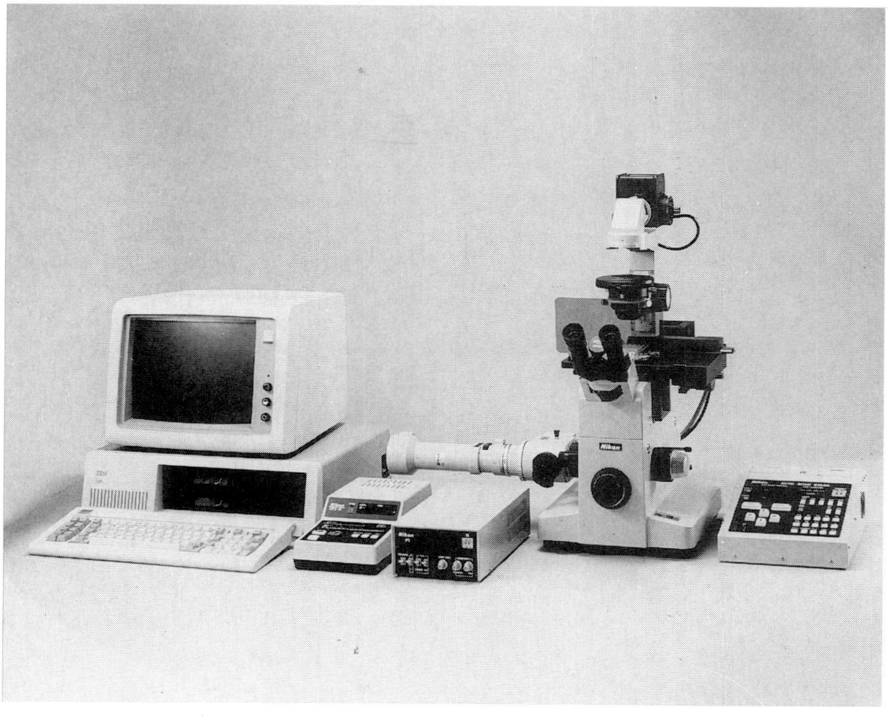

Fig. 8.5. Nikon P1 photometer on Diaphot TMD inverted microscope. Photograph courtesy of FSE Pty Ltd, Sydney.

DC current measurement measures the aggregate of the pulses due to the individual photons. This current is of the order of microamperes. Appropriate instrumentation includes high-impedance chart recorders (e.g. Gillis *et al.*, 1966) and digital voltmeters, usually placed in parallel with a load resistor of a megohm or so. DC current measurement in general is the simplest method, but its accuracy is critically dependent on the stability of the high-voltage supply for the photomultiplier.

A mechanical chopper in the excitation light path reduces the effect of stray light in the system. The photomultiplier signal is amplified by an AC amplifier with a filter to favour the frequency of the chopper. Constant stray light, not modulated by the chopper, is therefore ignored by the system. Also, an AC amplifier tends to be easier to construct and cheaper than a DC amplifier.

Photon counting is the best method (in my opinion) for measurement of very low light levels, and finds its major application in astronomy. In this method, the pulses associated with each photon are amplified, passed through a discriminator which eliminates the weakest pulses (which are mainly noise from the photomultiplier tube), and then counted during preset intervals. Besides giving a direct digital output, photon counting has relative immunity to variations in the high-voltage supply to the photomultiplier (which affects the height of the pulses but not their number), the temperature of the tube, and the position of the light spot on the photocathode. Photon counting circuits for microfluorometry were described by Pearse & Rost (1969), Rost & Pearse (1971), Cova, Prenna & Mazzini (1974), Wittig, Rohrer & Zetzsch (1984) and Candy (1985a,b).

Lasers

A laser is essentially a device for generating very pure monochromatic light, ideal for fluorescence excitation. The light may be generated either continuously or in short pulses. For microfluorometry, brief pulses have the advantage of enabling measurements to be taken before there is time for photodecomposition to occur, and if the pulse is short enough (or at least ends sharply) the decay time of fluorescence can be determined. A laser is an optical oscillator which emits light. Like other oscillators, it consists of an amplifier, a positive feedback mechanism to render the amplifier unstable and make it oscillate, a resonator to determine the frequency of oscillation, and a power source. A laser has three main components (Fig. 8.6): the *active medium*, the *pumping source*, and the *resonator*.

The active medium is a solid, liquid, or gas which acts as an amplifier. It is a fluorescent substance such as a ruby crystal or a solution of a fluorescent dye. Amplification takes place by the phenomenon of stimulated emission. If a quantum of energy with frequency identical with that which would be emitted spontaneously impinges upon an excited atom, both quanta are emitted simultaneously and in phase with each other. Each of the two quanta is then capable of stimulating emission by other excited atoms, producing four quanta in phase, and so on. For amplification to occur, the medium has to be kept in a state of population inversion, in which the number of atoms in an excited state is greater than the number in a lower energy level.

The pumping source excites the active medium and is powerful enough to achieve this population inversion; power for the laser is provided by the pump. Solid-state (e.g. ruby) and liquid lasers are excited by light sources such as flash lamps or another laser; gas lasers are pumped by electrical discharges.

The optical resonator consists of a pair of mirrors facing each other and enclosing a space. When the active medium is placed inside the resonator, the system acts as an oscillator. The mirrors of the resonator provide positive feedback to the amplifier, to produce oscillation. One mirror is made semi-transparent to provide an output path for the laser light. The result is the output of a beam of light, which is characterized by high intensity, monochromaticity, and coherence. In some designs, the laser can be tuned over a range of wavelengths by changing the resonant wavelength of the resonator. This can be achieved by replacing one mirror with a diffraction grating; turning the grating alters the effective frequency of the grating and hence the resonant wavelength.

For further information on lasers, see Andrews (1986), Herrman & Wilhelmi (1987), and Letokhov (1986). Laser dyes are described by Rost (1990), Maeda (1984) and Drexhage (1973).

Sources of inaccuracies in microfluorometry

Measurements by microfluorometry are subject to errors of two main types, related respectively to the *specimen* and to the *apparatus*.

It is important to distinguish between *precision* and *accuracy*. A measurement may be very precise, e.g. to four decimal places, without being particularly accurate (say only to two decimal places). The precision of a measurement depends only on the readout device, whereas accuracy is determined by the experimental conditions as a whole. The specificity of the method also has to be considered.

Optical specificity

In microdensitometry, some degree of optical specificity is obtained by employing monochromatic light of a wavelength corresponding to the absorption maximum of the desired substance. In microfluorometry, two wavelengths are involved (excitation

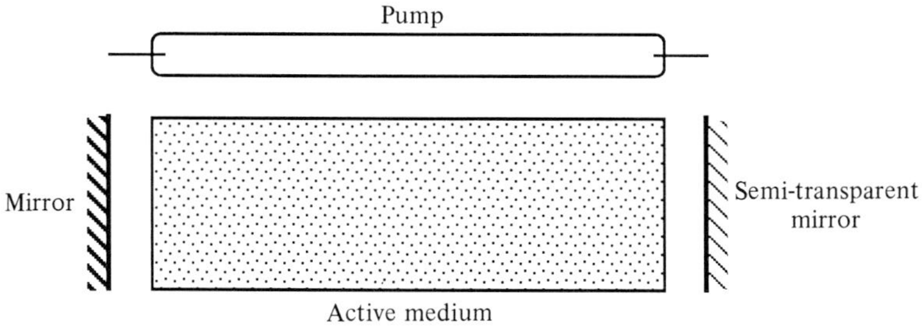

Fig. 8.6. Diagram of a laser, showing the active medium, pumping source (here a gas-discharge tube), and resonator (here two plane parallel mirrors facing each other; one is partially transparent to pass the output beam).

and emission) and in principle optical specificity can be maximized by appropriate choice of wavelengths for excitation and for emission measurements. In addition, the technique automatically differentiates fluorescent from non-fluorescent substances, and the fluorescence decay times can be used to differentiate between different fluorophores.

Sources of inaccuracies, related to the specimen

Fluorescence fading (photobleaching) almost always occurs during irradiation (see Chapter 2). After fading has taken place, some recovery may occur if the specimen is kept in the dark, preferably at low temperature; this can be studied quantitatively (see Fluorescence recovery after photobleaching, below). The rate of fading depends upon the individual fluorophore, its chemical environment, and the intensity and quantum energy of the irradiation.

In quantitative microfluorometry, the steps required to minimize inaccuracies due to fading include:

1. Selection, where possible, of a resistant fluorophore (e.g. for the Feulgen reaction, selection of a resistant fluorochrome).
2. Selection of mounting media to reduce the likelihood of photochemical reactions.
3. Reduction of the amount of irradiation, by:
   (a) Localization of the specimen by phase-contrast or another method involving light of a low quantum energy.
   (b) Choice of an appropriate excitation wavelength without extraneous light of shorter wavelength.
   (c) Minimization of the time lag between the commencement of irradiation and the completion of measurement, e.g. by using a photometric device triggered by the illumination shutter.
   (d) The use of an efficient detector system so that emitted light is not wasted.
4. Use of a brief flash of excitation, so that the measurement is made before significant fading can occur.

Concentration quenching occurs at high concentrations of fluorophore, where there is so much absorption of light from the excitation source that the intensity of excitation is significantly reduced in the inner region of the specimen; this results in a fall-off of observed fluorescence, known as concentration quenching.

Reabsorption of fluorescence is a further physical process to be considered. If the excitation (absorption) and emission curves of the fluorophore overlap, as they commonly do, some of the emitted light may be reabsorbed. The possible complications due to reabsorption have been considered by several workers, notably Rigler (1966).

The fluorochrome used for microfluorometry should be chosen with the requirements of quantification specifically in mind. The fluorochrome should, as far as possible, stain specifically, conform to the Beer–Lambert laws, be resistant to fading under irradia-

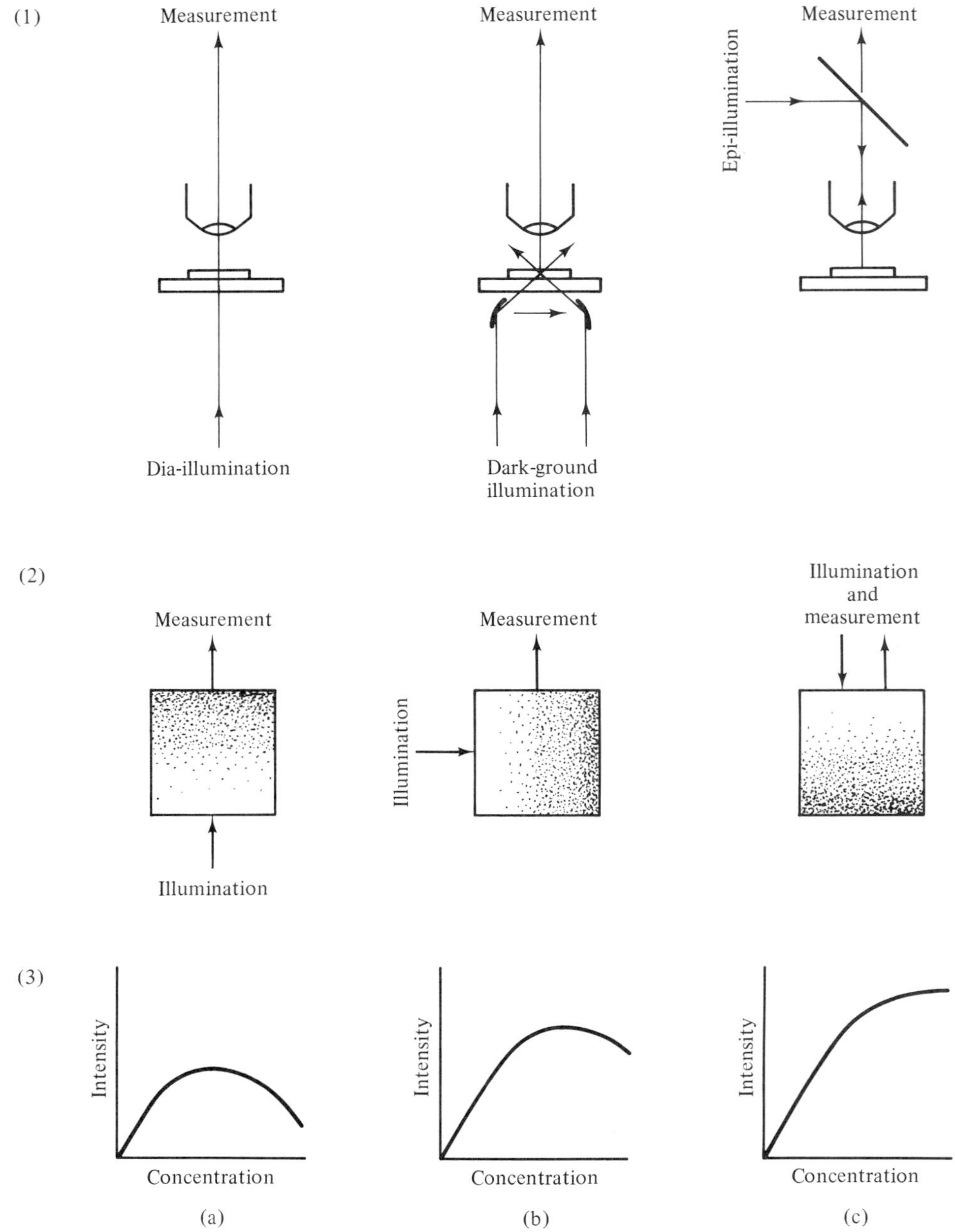

Fig. 8.7. The relationships between the concentration of a fluorophore and the intensity of the measured fluorescence as observed with three optical arrangements. The first row (1) shows optical systems in a microfluorometer: (a) dia-illumination (with a substage condenser), (b) darkground illumination, and (c) epi-illumination. The second row (2) shows the nearest corresponding arrangement with a cuvette in a chemical fluorometer. The third row (3) shows the relationship between concentration of fluorophore (horizontal axis) and measured fluorescence intensity (vertical axis). The advantage of epi-illumination over dia-illumination and darkground illumination is evident. Diagram from Rost (1991), modified from Rost (1980); after Udenfriend (1962) and Rigler (1966).

tion, be not metachromatic, have a high quantum efficiency (in the interest of sensitivity), have widely separated excitation and emission spectra (to reduce reabsorption), and be excited at a wavelength substantially longer than that giving rise to autofluorescence in the specimen.

Instrumental errors

Major instrumental sources of error include instability of the light source, non-homogenous illumination of the object (i.e. illumination with light of uneven intensity or wavelength distribution), and non-linearity of the photometric system.

The direction of illumination in relation to the direction of measurement affects the relationship between the concentration of the fluorophore and the measured fluorescence intensity (Fig. 8.7). It has already been indicated in Chapter 3 that three main possibilities exist for illumination: through a substage brightfield condenser, through a substage darkground condenser, and through the objective (epi-illumination). In relation to cuvette fluorometry, unless the specimen is illuminated from the same side as fluorescence is measured, at high fluorophore concentrations the apparent fluorescence intensity will be low due to reabsorption of emitted light by the specimen itself. Rigler (1966) discussed the applications of this principle to microspectrofluorometry, and concluded that epi-illumination was to be preferred (see also Benson & Knopp, 1984) because:

1. Error due to reabsorption of fluorescence is reduced (see Fig. 8.7).
2. Correct focussing of the objective automatically focusses the illumination system, thereby maintaining standard conditions from one measurement to the next.
3. With high-power, high-NA objectives such as are usually required for microfluorometry, brighter illumination is obtained.
4. Irradiation is confined to the area being examined.
5. The high optical correction of microscope objectives, far exceeding that of condenser systems, enables high-quality imaging of the field diaphragm into the object plane, homogeneous distribution of light intensity, and reduction of light-scatter.

Background fluorescence and stray light are usually due to a combination of the following: dark current and electrical noise in the measuring system; stray light of the emission wavelength, derived from the excitation source; autofluorescence of the optical system; and autofluorescence of the tissue and mountant. Tissue autofluorescence can be minimized by illumination at a wavelength corresponding to the excitation maximum of the specific fluorophore, which is thereby favoured in comparison with other fluorescent substances present unless they happen to have similar excitation maxima. It is usually necessary to subtract a 'blank' measurement of tissue autofluorescence from a control area.

Standardization

Unlike microdensitometry, in which an absolute determination of optical density is made, microfluorometry is strictly comparative and relies on a standard for its calibration. Microfluorometry therefore relies on the comparison of the fluorescence of the object to be measured with that of a standard preparation, under stable conditions. The requirements for accurate standards have been discussed by, among others, Ritzén (1967), Sernetz & Thaer (1970, 1973), and Jongsma *et al.* (1971). To ensure proper comparability of standard and test objects, the following criteria (largely due to Ritzén, 1967; see also Lichtensteiger, 1970) may be laid down in respect of the standard:

1. Its size should be similar to that of the objects being measured, so that conditions of microscopy can be identical.
2. The standard should have the same fluorescence excitation and emission spectra as the object, otherwise it will respond differently to changes in the excitation. This means that it should be made of the same chemical substance, in the same chemical environment.
3. The standard is required to be constant, not fading during the measurement process and preferably also constant during storage.
4. The standard must be reproducible.
5. The standard must contain an accurately known amount of fluorophore.
6. If the concentration of fluorophore is high, the standard must be thin enough to ensure sufficiently uniform excitation throughout the entire layer, and to minimize the reabsorption of fluorescence emitted.

Several types of fluorescence standards have been described; the best at present (for most purposes) is the capillary tube method of Sernetz & Thaer (1970, 1973). Other standards include uranyl glass, phosphor particles, fluorochrome solutions in microdroplets or a cuvette, and fluorochromed carbohydrate or latex beads.

Fluorescent (uranyl) glass

Fluorescent glass was one of the first standards used, as a plate in the object plane or between the illumination field diaphragm and condenser, or immediately below the objective (assuming epi-illumination). Fluorescent glass contains fluorescent inorganic ions, usually uranyl, or rare earth ions. Such glass does not fade significantly in the short term, even under strong irradiation. Because of their dimensions, plates cannot be measured under the same conditions of microscopy as the objects to be investigated; this problem is solved by using microspheres or thin fibres of fluorescent glass.

Phosphor crystals

Crystals of phosphors can be mounted on a slide and embedded in DPX or other non-fluorescent medium under a coverslip, and used as standard. They are small enough to be used under conditions of microscopy similar to those of the object to be measured, and are reasonably stable under irradiation. On the other hand, their luminescence characteristics are different from those of fluorochromes likely to be used in microfluorometry.

Fluorochrome solution

A cuvette with fluorochrome solution has been used instead of a glass plate, but this is generally less satisfactory and has been superseded by microdroplets and capillary tubes as described below. Using epi-illumination, the objective can be dipped into a dish of fluorochrome solution using the aperture-defined volume principle. Fluorochrome solutions are best used in capillary tubes or as microdroplets, as described below.

Sernetz–Thaer capillary tubes

The capillary tube method of Sernetz & Thaer (1970, 1973) is based on capillary tubes of quartz glass or glass with an inside diameter of 6–8 μm, an external diameter of about 30 μm, and about 50 mm in length. The internal diameter of the tube is best measured by interference microscopy. These tubes are filled with a fluorochrome solution of known concentration, and the fluorescence of a measured length of capillary tube measured.

The excited section with a length of about 50 μm is very small in comparison with the total capillary length. During fluorescence excitation, the fluorochrome solution of the excited section is continuously exchanged with non-irradiated solution, so that photodecomposition does not affect the measurement. In capillary tubes, the fluorescence intensities were found to remain constant over two months and many measurements, and may be expected to continue to do so provided the fluorescent standard solution itself does not change its properties responsible for fluorescence emission.

Microdroplets

The microdroplet method is based on the measurement of the fluorescence intensity of single droplets with diameters up to about 20 μm sprayed onto a layer of silicone oil. The fluorochrome content of single droplets is calculated from their measured lateral diameters and the known concentration of fluorochrome in the solution. This method meets all requirements in respect of the standard's position in the object plane and of the standard's similarity to microscopic objects. It also permits the relevant fluorochrome or substances stained with fluorochromes to be used as the standard. However, the method is less accurate than the use of capillary tubes, because volume determination in the microdroplet method requires the accurate measurement of a very small diameter, and the fluorochrome within the sphere is subject to photodecomposition. An individual droplet can be measured only once. The microdroplet method can be adapted to standardization of induced fluorescence (e.g. formaldehyde-induced fluorescence) by measuring out droplets of known volume and drying the droplet before exposure to vapour treatment.

Fluorochromed carbohydrate beads

Microscopical beads of dextran (Sephadex) or agarose (Sepharose), labelled with a fluorochrome such as FITC, have advantages similar to those of microdroplets. For a review, see van der Ploeg & Duijndam (1986).

Applications of microfluorometry

The main stimuli for the development of microfluorometry were the need for quantification of fluorescent cytochemical procedures (such as immunocytochemistry and the Feulgen method for DNA), and the fact that individual microfluorometric measurements can be made much more quickly than corresponding measurements by scanning densitometry and therefore microfluorometry greatly facilitated the repetitive measurements required for the study of substantial cell populations. Microfluorometry therefore came to be most widely used for the measurement of nuclear DNA. For the same reason, classical microfluorometry has now been substantially replaced by flow cytofluorometry. An overall review of applications of microfluorometry was given by Ruch (1970). A few examples of applications of microfluorometry are noted below.

Nucleic acids

In biology, microfluorometry has most commonly been applied to the measurement of nuclear DNA (for staining methods, see Volume II, Chapters 13 and 16). Reviews were written by Ruch (1973), Böhm & Sprenger (1968) and Böhm *et al.* (1973). For microfluorometry, the most valuable methods have been the Feulgen method, with conventional (Pararosaniline) or Acriflavine-Schiff (for current method, see Tanke & van Ingen, 1980), and various intercalating dyes (see Curtis & Cowden, 1983).

Immunofluorescence

The quantification of immunocytochemistry has been briefly reviewed by Sternberger (1986) and Ploem (1970). Quantitative immunocytochemistry has many difficulties, of which probably the least is the microfluorometry. One of the main problems is standardization, both of the immunological reaction and of its quantification. Recent studies applying quantitative immunofluorescence to cytochemistry have tended to use flow cytofluorometry.

Neurotransmitter amines: formaldehyde-induced fluorescence

The quantification of formaldehyde-induced fluorescence (FIF) of neurotransmitter amines was reviewed by Lichtensteiger (1970) and Jonsson (1971). Quantification is now commonly performed using scanning microfluorometry (see below). Serotonin (5-hydroxytryptamine, 5-HT) presents particular problems because its FIF fades very rapidly.

Probes

Microfluorometry can be applied to measurement of the amount of a probe taken up. Studies with those probes which significantly change their fluorescence spectra according to their microenvironment require measurements at at least two different wavelengths. For example, measurement of intracellular pH or calcium ion concentration requires such determination. Kassotis *et al.* (1987) described a microfluorometer with provision for automatic rapid changing between two excitation wavelengths for this kind of measurement.

Opaque specimens

Microfluorometry with epi-illumination can be employed on opaque specimens, for which microdensitometry is impossible. This has proved to be particularly useful in studying the NADH fluorescence of the surface of intact organs.

Petrology
Quantitative studies on rocks (particularly coal) are reviewed in Volume II, Chapter 11, where their fluorescence is described.

MICROSPECTROFLUOROMETRY

Microspectrofluorometry (MSFy) is concerned with the measurement of fluorescence excitation and emission spectra from specific regions of specimens. The instrument required, called a microspectrofluorometer, differs from a microfluorometer in having two variable monochromators, one each in the excitation and emission optical pathways. Emission spectra are determined by measurements made at varying settings of the emission monochromator, the excitation remaining constant; excitation spectra are determined by measurements made at varying excitation wavelengths while emission is measured in a constant wavelength band.

The practical object of microspectrofluorometry is usually the identification of a substance by its fluorescence characteristics; for example, the technique has been applied to the identification of neurotransmitter amines rendered fluorescent by formaldehyde. On the other hand, spectral measurements on a substance of known fluorescence characteristics, used as a probe, give information about the local conditions within the specimen, as in the studies of nucleic acids by microspectrofluorometry of the metachromasia of Acridine Orange staining of DNA and ribonucleic acid (RNA) (Rigler, 1966).

Previous overall reviews of microspectrofluorometry include mine (Rost, 1974) and those of Nordén (1953) and Rigler (1966). The majority of reviews are devoted to instrumentation, techniques and applications (e.g. Wreford & Smith, 1982; Rost & Pearse, 1974). General principles of spectrofluorometry are discussed by Slayter (1970).

Microspectrofluorometry is a powerful technique for the investigation of fluorescence in regions of microscopic size such as single cells. Microspectrofluorometry cannot at present give quite as specific and reliable identification as biochemical analysis, but it can be used for analysis in respect of an accurately defined anatomical region, which can be as small as a single cell or part of a cell, even down to a single fluorescing granule (Marques & Rost, 1973). Compared to measurement of absorption spectra, microspectrofluorometry offers greater sensitivity and greater specificity, the latter because both excitation and emission wavelengths can be selected.

Instrumentation for microspectrofluorometry

For the purposes of this section, I define a microspectrofluorometer (MSF) as a fluorescence microscope equipped for the measurement of excitation and/or emission spectra from a defined region of the specimen (see, e.g., Figs. 8.8 and 8.9). Microspectrofluorometers therefore differ from microfluorometers (described above) in having a light source of continuously variable wavelength, and/or means of measuring emission at continuously variable wavelengths. This usually implies the presence of at least one prism or grating monochromator: however, excitation spectra can be obtained with a tuneable laser, and this method is likely to be increasingly applied.

Principles involved in microspectrofluorometric instrumentation have been reviewed by Nordén (1953), Rost (1974), and Wreford & Smith (1982). The basic principles of microfluorometers also apply.

The instrumental requirements of microspectrofluorometry are essentially those of a microfluorometer with the substitution of monochromators (giving variable wavelengths) for the excitation and emission filters, and a reference channel to monitor the output from the excitation monochromator. During measurements of excitation

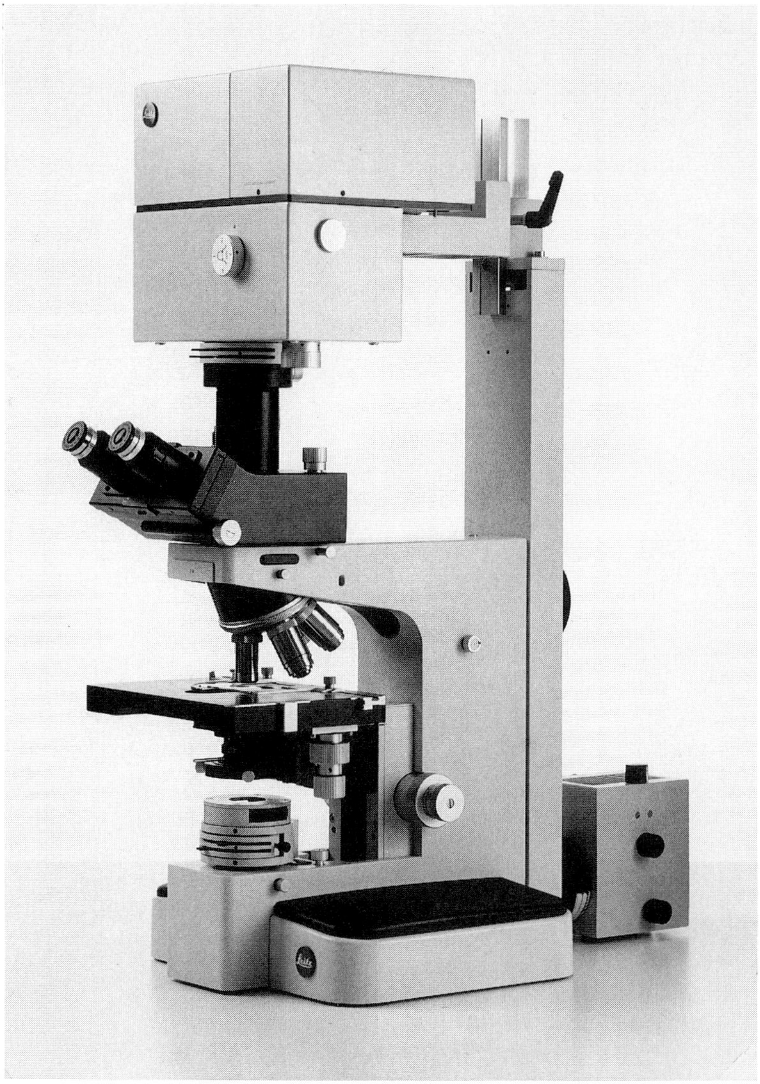

Fig. 8.8. The Leitz MPV-3 microspectrofluorometer. Photograph courtesy of Dr W. J. Patzelt, Ernst Leitz Wetzlar GmbH.

spectra, it is usual to employ a xenon arc lamp (which has a relatively continuous spectrum) and a conventional monochromator on the excitation side; on the emission side, the monochromating device may be a barrier filter, a broad- or narrow-band bandpass filter, or a monochromator. The purpose of the reference channel is twofold: to permit the calculation of corrected excitation spectra, and to compensate for fluctuations in the intensity of the lamp. Emission spectra can be scanned with a video camera.

A variable-frequency (tuneable) laser is a modern alternative to the combination of a continuous-spectrum light source and a monochromator. Its main advantage is much greater spectral purity. A laser of fixed wavelength can be used for excitation during the measurement of emission spectra.

Because excitation spectra commonly extend into the ultraviolet (UV), it is usually necessary to employ special UV-transmitting optics for the illumination system (see Chapter 3), including the coverslip over the specimen (assuming epi-illumination).

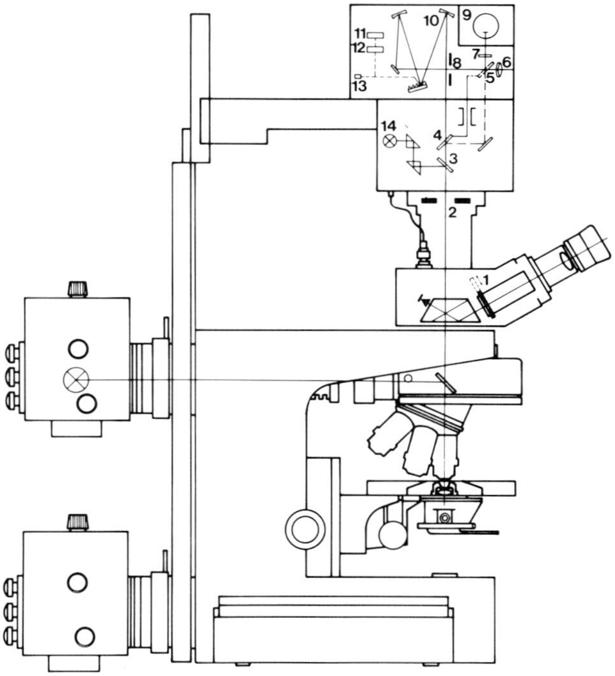

Fig. 8.9. Schematic diagram of a Leitz 'Orthoplan' microscope, with Ploem-Opak 2 epi-illuminator and MPV-SP photometer. The photometer system includes a grating monochromator for measurement of emission spectra. Key: 1, light flap; 2, entrance slit; 3, beam splitter for pilot lamp (14); 4, swing-in mirror for measurement without the monochromator (mirror 5 is swung out); 5, swing-in mirror for spectral measurement with the monochromator (mirror 4 is swung out); 6, magnifying lens for adjustment of slits (2 and 3); 7, filter turret; 8, exit slit; 9, photomultiplier; 10, monochromator; 11, incremental angle transducers; 12, drive for monochromator; 13, adjustment device; 14, pilot lamp. Diagram courtesy of Dr W. J. Patzelt, Ernst Leitz Wetzlar GmbH.

Measurement of spectra

The excitation spectrum is obtained by recording changes in the strength of the total fluorescence emission while varying the excitation wavelength. The uncorrected excitation spectrum so obtained must be corrected for the varying output of the excitation system at different wavelengths. In principle, assuming that the quantum efficiency of fluorescence is independent of excitation wavelength, the corrected excitation spectrum should be identical with the absorption spectrum of the same compound.

The emission spectrum is obtained by recording the strength of emission at each wavelength, while maintaining the excitation constant. The excitation monochromator is normally set to the peak of the apparent (uncorrected) excitation spectrum; alternatively, the excitation monochromator can be replaced with a filter. The uncorrected emission spectrum so obtained is severely distorted, and must be corrected for several variable factors, namely: varying transmission of the emission system at different wavelengths, variable spectral bandwidth of the emission monochromator, and variable sensitivity of the photometer at different wavelengths.

To enable comparison of results obtained in different laboratories, or even on the same instrument under different conditions, the results obtained must be corrected for all systematic errors, and the results plotted in a standard manner (Chapman *et al.*, 1963). After correction, excitation spectra should be presented as a plot of the relative number of photons emitted as a function of the quantum energy (or wavenumber or frequency, which are linearly proportional to quantum energy) of the photons used for excitation. Emission spectra should be presented as a plot of the relative number of photons emitted as a function of quantum energy (or wavenumber or frequency). A scale linear in photon quantum energy (or wavenumber or frequency) is chosen to represent more accurately the relations between absorption and excitation spectra and between emission and excitation spectra (see Chapter 2), and also to avoid compression of details at the short-wavelength end of the spectrum.

Applications of microspectrofluorometry

In biology, microspectrofluorometry has so far been most widely applied to the identification of neurotransmitter amines and related substances, and to characterizing intracellular probes. A previous brief review is that of Rost & Pearse (1974). Microspectrofluorometry has much to offer for forensic studies (Pabst, 1980), since it offers a non-destructive characterization of microscopically defined areas of specimens which can be quite tiny.

Induced fluorescence

Microspectrofluorometry has been widely used for the identification of arylethylamines ('biogenic amines') by spectral analysis of fluorescence induced with formaldehyde or glyoxylic acid (see Volume II, Chapter 12). In general, neurotransmitter amines can be divided into broad groups according to the emission maximum of their formaldehyde-induced fluorescence: serotonin (Em 520 nm), catecholamines (Em 490 nm) and *m*-tyramines (Em 415 nm). Some further differentiation is possible based on the excitation spectra. The application of microspectrofluorometry to the differentiation of neurotransmitter amines was reviewed by Wreford & Smith (1982). Recent examples include the studies of Pohle, Ott & Müller-Welde (1984) and Scheuermann, Stilman & De Groot-Lasseel (1988).

Fluorochromes and probes

The most obvious application of microspectrofluorometry in the study of fluorochromes is the determination of excitation and emission spectra, to provide data for a logical choice of filters and other instrumental conditions for fluorescence microscopy. Instrumentation for this purpose can be quite simple, since precise determination of peaks is not required and UV excitation spectra need be measured only down to 350 nm.

Microspectrofluorometry of fluorochromes taken up by or injected into living cells can confirm the localization of the dye in particular regions, and if more than one dye is used, determine the proportion of each present. In the earliest application of microspectrofluorometry, Nordén (1953) confirmed the uptake of 3,4-(*a*)-benzpyrene by living cells, as part of a study on carcinogenesis. Identification and quantification of metabolites of introduced compounds is possible.

Microspectrofluorometry of intracellular probes quantifies spectral shifts due to the environment of the fluorochrome. Metachromasia of fluorochromes such as Acridine Orange gives information about the substrate to which the dye is bound. Intracellular pH and free calcium can be determined by microspectrofluorometry of spectral shifts of indicator fluorochromes.

Autofluorescence

Autofluorescence is due to the presence of a fluorescent substance or substances in the specimen (described further in Volume II, Chapters 9, 10 and 11). Microspectrofluorometry has been applied to studies of the autofluorescence of elastic and collagenous fibres, photoreceptor pigments, and amyloid. In coal petrology, microspectrofluorometry assists in quantifying the diagenesis of coals. Microspectrofluorometry has been used for characterization of oil droplets ingested by marine plankton following contamination of the sea by oil from ships (Parker, 1969).

SCANNING, VIDEO INTENSIFICATION AND IMAGE PROCESSING

This section deals with a variety of techniques in which the fluorescent specimen is scanned, measurements of the fluorescence intensity being taken during the scanning, enabling an image to be built up as a video frame. Scanning of a fluorescence image may be useful where morphological features and fluorescence properties can usefully be correlated. The scanning may be carried out mechanically or with a television camera.

The final image is commonly held in digital form, and can be subjected to various forms of image processing by computer.

Scanning has in principle a number of virtues. Firstly, the picture produced on a monitor screen is much brighter than the view in the microscope, obviating the necessity for a darkened room and dark-adapted eyes, and the greater sensitivity allows reduction of the intensity of irradiation, thereby reducing fading and damage to living tissues. Secondly, the video screen can easily be viewed simultaneously by two or more observers. Thirdly, the video output can be recorded either continuously or intermittently (time-lapse), for future study. Time-lapse video recording can be used to study movements of fluorescent probes in living cells. Fourthly, digitization of the image permits image analysis, including quantification, exactly as in absorptiometry. The term fluorescence digital imaging microscopy (F-DIM) has been applied to this technique (Arndt-Jovin *et al.*, 1985). Taking any frame, or an average of several frames, a computer can analyse the data, present a modified version (e.g. with increased contrast) on the monitor screen, calculate areas or integrated intensity over a specified area, and transmit the picture directly to a graphics printer to produce a picture on paper. Fifthly, the monitor screen can be photographed; the exposure time is much shorter than direct photography of the fluorescence, and relatively constant (being determined by the brightness setting of the monitor). Sixthly, scanning with an excitation beam has the additional advantage that only one minute area (that of the scanning spot) is irradiated at any one time, and then only briefly, so that photobleaching of the specimen is reduced. Moreover, there is only minimal generation of autofluorescence in the optical system, thereby increasing the signal/noise ratio and the overall sensitivity (Ploem, 1986). Finally, use of a photomultiplier, video camera or charge-coupled device (CCD) expands the usable emission wavelengths to beyond the visible, to the UV and infrared.

The main disadvantages of scanning are complexity and expense of the apparatus, and possibly limited detail in the image (determined by the pixel size, see below).

In a scanning fluorescence microscope, the specimen can be scanned in any of three ways, as follows:

1. With the illumination confined to a focussed spot, preferably produced with a laser, and emitted light measured from the entire field (or at least an area large in relation to the excited spot), the specimen is scanned mechanically in raster fashion with respect to the excitation spot, or the spot is moved ('flying spot') so as to scan the specimen.
2. With the specimen uniformly illuminated, the emission is scanned, either by a video camera or by mechanical means whereby a pinhole aperture is placed in front of a photomultiplier tube and either the specimen is moved in raster fashion or mirrors shift the image of the object in corresponding fashion across the pinhole.
3. Both the area of illumination and the area from which measurement is made are confined to a single spot, the laser (or other point source) and a pinhole aperture in the emission measuring system being focussed on a single spot in the object plane, and this spot is scanned systematically over the specimen (or alternatively, the

spot is axial, and the specimen is moved mechanically). Because both illumination and detection systems are confined to points focussed on the same spot, this is known as *confocal microscopy*. A system using multiple confocal spots and in which the optical image is viewed directly is known as tandem scanning light microscopy.

During the process of scanning, the electrical output may be recorded either as an analogue signal, or digitally. In the latter case, each number recorded represents the emission intensity of a small region of the specimen known as a *pixel*. The dimensions of the pixels determine the resolution of the picture, and the total number of pixels in the scanned area determines the amount of detail in the final image. Since the spectral sensitivity of scanning fluorescence microscopy is limited only by the detector, studies are possible concerning otherwise invisible fluorescence in the UV and infrared. In the infrared, scanning fluorescence microscopy has been applied to Indocyanine Green (Moneta *et al.*, 1987), and may be useful for studying chlorophylls. There are numerous substances which are fluorescent in the UV, hitherto studied biochemically and which could perhaps now be studied by fluorescence microscopy.

Scanning with the excitation beam

An excitation-scanning fluorescence microscope is rather similar in principle to a scanning electron microscope; the specimen is scanned at the excitation wavelength by a narrow beam of light from a laser, and the emitted light is detected, quantified, and the resulting pixels built up into an image by a computer. Use of a narrow excitation beam irradiating only one small area at any one time, and then only briefly, reduces photobleaching of the specimen and the generation of autofluorescence in the optical system, thereby increasing the signal/noise ratio and the overall sensitivity.

Excitation-beam scanning was introduced by Mellors & Silver (1951), who used event-counting systems whose purpose was to count the number of cells with fluorescence above a present level. In the original apparatus of Mellors & Silver (1951), the specimen was scanned by a 'flying spot' for excitation, generated by a spiral series of apertures in a revolving disc in the excitation pathway. Modern systems employ a laser beam, scanned in a raster over an area of the specimen using oscillating mirrors driven electrically.

Applications

Cancer diagnosis is an obvious field of application for scanning-fluorescence microscopy. The concept of automated scanning in a fluorescence microscope of a diagnostic specimen for detecting neoplastic (cancer) cells appears to have been originated by Mellors & Silver (1951). I believe that commercial systems are now available or are projected for the near future.

Catecholamines in nerve fibres tend to be concentrated in varicosities along the length of nerve fibres, and the fibres themselves are irregularly distributed in the tissue. For quantification, rather than measurements from arbitrarily defined circular or rectangular areas including both nerves and background tissue, it has been found preferable to scan large areas and to use image analysis to determine the mean catecholamine concentration in the fluorescent areas assumed to correspond to the actual nerves (see Schipper, Tilders & Ploem, 1980a; Schipper *et al.*, 1980b; Larsson, Goldstein & Dahlström, 1984).

Video fluorescence microscopy

Scanning the emission while exciting the complete field is most easily accomplished by examining the image with some form of closed-circuit television (CCTV) system. This technique is now generally known as video intensification microscopy (VIM), so named by Willingham & Pastan (1978). However, since video intensification can be applied to modalities of microscopy other than fluorescence, perhaps it might be generally more appropriate to refer to video fluorescence microscopy (VFM). Video microscopy in general is described by Inoué (1986), who also gives a useful glossary of technical terms; see also Weiss *et al.* (1989), Bright & Taylor (1986), Arndt-Jovin *et al.* (1985), Inoué (1988), Jarvis (1988) and Shotton (1988). Instrumentation for video fluorescence microscopy has been reviewed by Arndt-Jovin *et al.* (1985).

Quantification

Quantitative video intensification microscopy (QVIM) is a logical extension of VIM. Benson, Knopp & Longmuir (1980) reported that their instrument could measure fluorescence intensities with an error of 1·5% of full scale in 65 536 different positions in a microscope field; a video frame freeze acquisition time of 33 ms allowed time-dependent changes of that order of time or slower to be followed. Methodology for QVIM has been discussed by Sisken, Barrows & Grasch (1986) and Smith *et al.* (1986).

Applications

Qualitative applications

The major application of VIM has been to dynamic studies of living cells, where the particular advantages of VIM are most felt. These applications include studies of uptake of fluorescent or fluorescently labelled substances by cultured cells, capillary angiography in humans, and lateral transport in membranes. For static studies, VIM is particularly useful where convenience in dealing with a large number of specimens is more important than achieving the highest image quality. In the future, VIM is likely to find valuable application for studying fluorescence of substances which fluoresce in the infrared or the UV.

Quantitative applications

QVIM has been applied to measurements of nuclear DNA, intracellular oxygen, and to kinetic studies involving photobleaching (see below).

Fluorescence photobleaching recovery

After fading has taken place, some recovery may occur if the specimen is kept in the dark, preferably at low temperature (Ritzén, 1966a; Dowson, 1984). At very low temperatures, recovery of fluorescence in crystalline material may be complete, that of biological material less so (Tiffe & Hundeshagen, 1982).

Fluorescence photobleaching recovery (FPR; or fluorescence redistribution after photobleaching, FRAP) is a technique for measuring two-dimensional lateral mobility of fluorescent particles, for example the mobility of fluorescently labelled molecules on a cell surface. In principle, a small spot on the cell surface, about 10 μm in diameter, is irreversibly photobleached by brief exposure to an intense laser beam, and the subsequent recovery of the fluorescence is monitored by the same, but greatly attenuated, laser beam. Recovery of fluorescence in this case is due to inwards migration of unbleached molecules from the surrounding cell surface. To avoid damage to other components of the system, such as the nucleus, photobleaching is carried out at a wavelength which is not absorbed by them and therefore does not affect them. Inwards movement of unbleached molecules may be due to diffusion or flow. The kinetics of the recovery are related to the diffusion coefficient or flow velocity of the labelled molecules, and the degree to which the fluorescence regains the prebleach value gives the mobile fraction of the measured probe molecules. Use of parallel lines for the photobleached field allows Fourier image analysis, simplifies calculations of diffusion and gives the possibility of detecting diffusional anisotropy. Multipoint measurement (using QVIM) within a bleached area enables immediate quantification of systematic flow and higher spatial resolution than the pattern method. Use of a polarized light beam can show rotation of unbleached fluorophores into the plane of polarization. Applications and technique were reviewed by Koppel, Primakoff & Myles (1986), Smith *et al.* (1986) and Jacobson *et al.* (1983). Methodology is described in detail by Zs.-Nagy *et al.* (1984). The above description of fluorescence photobleaching recovery is based on summaries by Axelrod *et al.* (1976), Jacobson *et al.* (1983) and Peters (1983).

Confocal fluorescence microscopy

In a confocal fluorescence microscope (CFM; see Figs. 8.10 and 8.11), both the area of illumination and the area from which measurement is made are confined to a single spot; the laser (or other point source) and a pinhole in the emission measuring system are focussed on the same spot in the object plane, and the spot is scanned systematically over the specimen. For a given NA, the resolution of a confocal microscope can be at least 1·4 times greater than that of a conventional microscope. In fluorescence microscopy, there is a further gain in resolution because the resolution of the scanning system is dependent on the excitation wavelength, rather than on the emission wavelength as in a conventional fluorescence microscope. A confocal fluorescence microscope possesses a unique depth discriminating property whereby out-of-focus planes are imaged with less intensity than the in-focus plane. In consequence, background fluorescence may be sharply reduced and it is possible to obtain three-dimensional data by building up a series of optical sections or by stereoscopic imaging.

There are two main types of confocal microscope. In the original version, known as a tandem scanning reflected light microscope (TSRLM) (Petráň *et al.*, 1968, 1985a,b), the scanning is performed rapidly with multiple beams, and the image is viewed directly. In the other type, usually known as a laser scanning confocal microscope, scanning is performed with a single beam, fluorescence is measured and the image is built up as a video frame.

In the tandem scanning confocal fluorescence microscope, the illumination takes the form of multiple narrow beams from an array of apertures focussed by the objective into in the focussed-on plane of the specimen. Fluorescence from the illuminated patches of the specimen is imaged onto an identical array of apertures in the intermediate image plane of the microscope. Both arrays of apertures are scanned in

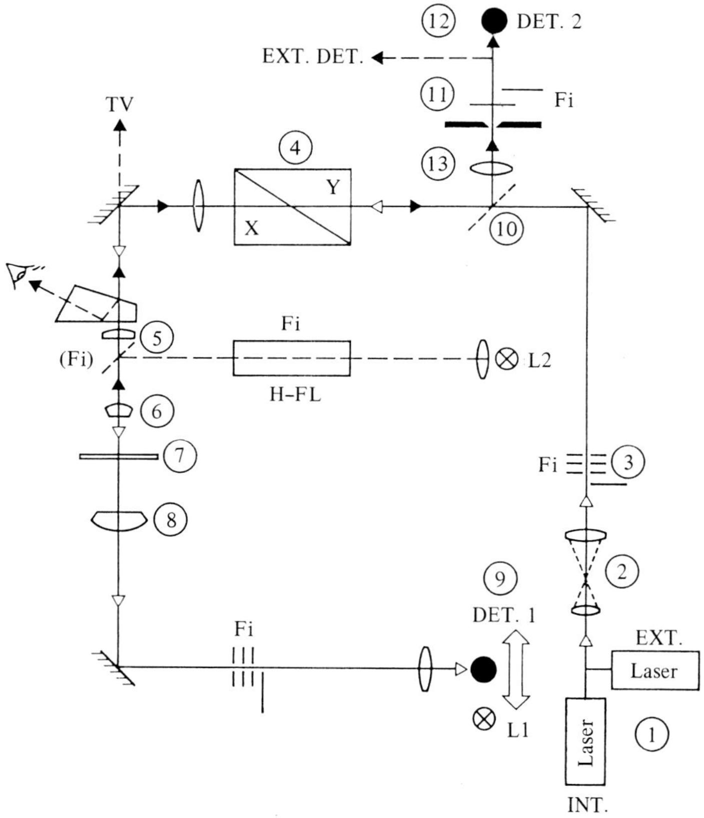

Fig. 8.10. Zeiss LSM laser scanning microscope (schematic). 1, lasers; 2, beam expander; 3, laser light attenuators; 4, *X–Y* deflection unit; 5, tube lens; 6, objective; 7, specimen; 8, condenser; 9, detector for transmitted light; 10, beamsplitter; 11, barrier filters; 12, detector for fluorescence and reflected light; 13, confocal spatial filter; DET. 1, DET. 2, detectors; EXT. DET, external detector; Fi, filter; H-FL, filter slider with matched excitation and emission filter set; L1, L2, lamps; TV, video camera. Diagram courtesy of Carl Zeiss Pty Ltd, Sydney.

tandem – hence the name tandem scanning. All extraneous light from out-of-focus layers or scattered from optical surfaces in the microscope is intercepted by the solid portions of the aperture array in the intermediate image plane or by light traps in the microscope. The TSRLM is used like a conventional light microscope: the observer looks through the eyepiece at a steady image. Focussing up and down, a mental image can be constructed of three-dimensional structures in the usual way. Further details of the fluorescence mode of the TSRLM are given by Boyde (1985, 1987, 1988, 1989) and Boyde *et al.* (1990).

In single-beam confocal fluorescence microscopes, scanning is carried out either by oscillating mirrors which simultaneously scan images of the illumination and observation apertures over the specimen, or else the illumination and observation apertures are axial and the specimen is scanned mechanically in *X*, *Y* and optionally *Z* axes. Such instruments have been described by Cox (1984), Wijnaendts van Resandt *et al.* (1985), van der Voort *et al.* (1987), and Takamatsu & Fujita (1988). A confocal module for a scanning fluorescence microscope was described by Shack *et al.* (1987). Confocal fluorescence microscopes are now available commercially from Biorad, Leitz and Zeiss (Oberkochen).

The Zeiss LSM series (Laser Scan Microscopes; Fig. 8.11) are based on a large microscope stand whose special feature is a wide distance between the objective axis and the stand, so that large objects can be accommodated. Several models already exist, for various applications in biology, medicine, and materials sciences. Both confocal and line scan operating modes are provided. For fluorescence, standard excitation wavelengths are 488 nm and 514 nm (argon laser) and 633 nm (helium–neon laser), supplemented by a conventional fluorescence epi-illumination system with 50 W mercury arc lamp.

Image processing

The video signal for a picture can be digitized, i.e. converted to an array of numbers each representing the light intensity at a particular spot (i.e. the brightness of a particular pixel). The image is usually digitized in a matrix of 512×512 pixels. The picture can then be stored in computer memory, recorded as computer data (e.g. on magnetic tape or a floppy disc) and displayed on a computer monitor or printed with a digital printer (laser printer), and subjected to analysis by computer. The major limitations of current technology are difficulties in discriminating the profiles of individual objects in the field, particularly with biological material, and sizing of particles from their profiles. For reviews of this technique applied to microscopy, see Arndt-Jovin *et al.* (1985) and Jarvis (1988). For more general reviews, see Watkinson (1988), Green (1983), Talmi (1983), and Rosenfeld & Kak (1982).

Digital enhancement techniques involve computer processing of a digitized image, in order to optimize the image in respect of specified criteria. Given a rapid digitizer and a sufficiently fast computer, manipulation is possible in real time to permit substantially continuous observation. Possible changes to the image include expansion or contraction of the contrast range by manipulation of the grey scale, the sharpening of edges,

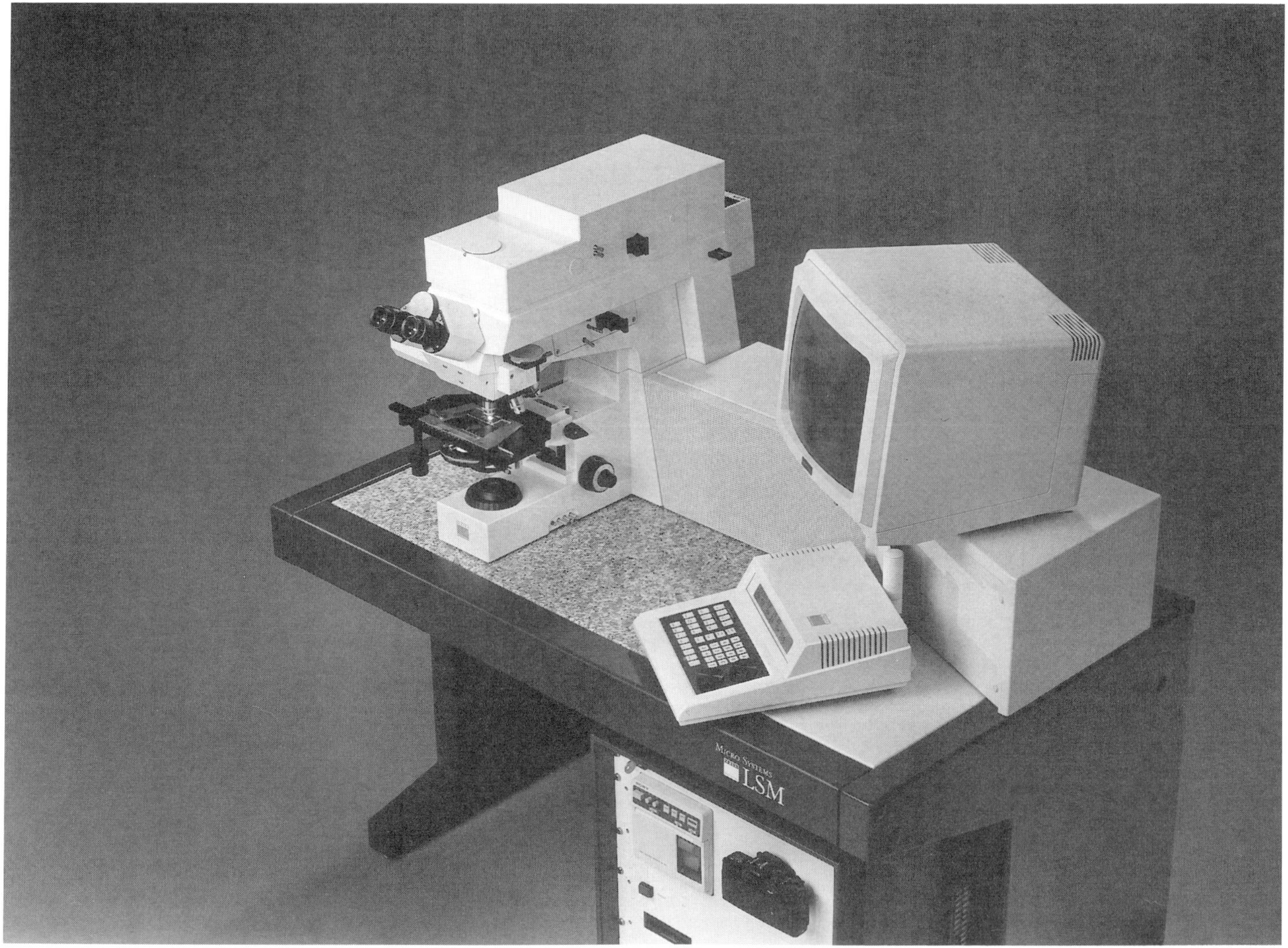

Fig. 8.11. Zeiss LSM laser scanning microscope. Photograph courtesy of Carl Zeiss Pty Ltd, Sydney.

reduction of random video noise by averaging several images, and subtracting fixed patterns by displaying the difference between images to show changes.

Analogue devices can be used for image processing, and may be faster in operation and cheaper to assemble.

Image analysis of the fluorescence image enables the correlation of morphological and quantitative data. Quantitative fluorescence image analysis was described by Takamatsu, Kitamura & Fujita (1986). Image analysis of microscope images in general was recently reviewed by Jarvis (1988) and Serra (1986); see also van der Ploeg *et al.* (1977), and the general reviews listed above.

TIME-RESOLVED FLUORESCENCE

It has already been indicated in Chapter 2 that, if fluorescence is excited by a sharp pulse of light, fluorescence emission takes place over a short period of time following excitation. The decay time of fluorescence (τ) varies from one substance to another and this characteristic can be used in the identification of unknown fluorophores and also to permit the quantification of a fluorophore in the presence of a second fluorophore with similar or overlapping excitation and emission spectra but with a different decay time. Time-resolved fluorometry is now widely used for chemical analysis: the technique was well explained by Wahl (1975). The application of this technique to fluorescence microscopy has the same advantages as the corresponding applications of fluorometry and spectrofluorometry, namely the possibility of characterizing a substance present in a defined region of a specimen (Brand & Gohlke, 1972).

Starting with a microfluorometer, measurement of fluorescence decay time requires the addition of standard instrumentation for the measurement of decay time. One such instrument (Morgan, 1987) is commercially available (Philips Scientific & Industrial).

Phosphorescence

Phosphorescence (see Chapter 2) can be distinguished from fluorescence by the longer delay time of the luminescence. Phosphorescence can be distinguished in the presence of fluorescence by time-resolution, using flash excitation and observation after the fluorescence lifetime. Because of the relatively long times involved, this can be arranged by a system of rotating mechanical shutters, one each in the excitation and emission pathways, so arranged that the emission pathway is blocked while the specimen is illuminated and vice versa. I have seen only one phosphorescence microscope, that devised by Professor R. Barer at Sheffield. Parker (1969) described a microspectrophosphorometer (mentioned above), in which a fluorescence microscope and a conventional spectrophosphorometer were combined.

MICROPOLARIMETRY OF FLUORESCENCE

A microfluorometer adapted for measurement of fluorescence polarization was described by MacInnes & Uretz (1966). In that instrument, the incident light (using dia-illumination) is unpolarized. Another micropolarimeter was described by dos Remedios *et al.* (1972a,b) for rapid measurement of polarization of tryptophan fluorescence from single muscle fibres during relaxation, contraction, and rigour.

FLOW CYTOFLUOROMETRY

Flow cytofluorometry is a particular form of flow cytometry, in which fluorescence of cells is measured. Cells or cellular components in aqueous suspension are made to pass through the focal point of a fluorescence microscope or equivalent device, and the fluorescence emission is measured as each cell passes. Cell size can be determined simultaneously by measurement of light-scatter or electrical resistance. By use of double-staining and simultaneous measurement of emission at two wavelengths, it is possible to determine two fluorometric parameters simultaneously. By flow cytometry, it is possible to measure a large number of cells in a short time, greatly facilitating statistical studies or characterization of a minor component of cell populations. Some flow cytometers can sort the cells according to predetermined parameters of the measurements, for subsequent visual examination, biochemical analysis or other procedures.

Recent reviews of flow cytometry include those of Shapiro (1988) and Steinkamp (1984). *Quantitative fluorescence microscopy* (Rost, 1991) has a chapter on this topic by H. J. Tanke.

Appendix 1
German–English vocabulary

Translations given are those relevant to fluorescence microscopy; some words have other meanings in different contexts. For technical words and expressions not given here, try Bradbury *et al.* (1989), Bindmann (1975) or Walther (1967).

This vocabulary was prepared initially for my own benefit, to help with understanding of the German literature. I was at that time unaware of the specialist dictionary of Bindmann (1975), and that of the Nomenclature Committee of the Royal Microscopical Society (Bradbury *et al.*, 1989) had not yet appeared. Initially a list of words was selected from the index to Beyer (1977), and in most cases checked with the context. I also made substantial use of the dictionary of Schoenewald (1952) and to a lesser extent that of De Vries (1952). German manufacturers' brochures were searched for technical terms, and additional words were added when found in German papers.

Abbildungsfehler, image aberration
Abbildungsgleichung, correction (of image aberration)
Abbildungsmasstab, scale
Abbildungstiefe, depth of focus
Abbildungsvermögen, resolving power
Abgleichlänge, adjustment length (of objective or ocular)
Achse, axis
Akusto-optische, acousto-optical
Anlaufglas, sharp cut-off glass (for barrier filter)
Anpassung, adaptation (of the eye, to darkness)
Anregung, excitation
Antikörper, antibodies
Apertur, aperture
Aperturblende, aperture diaphragm
Auflicht, epi-illumination, incident-light
Auflicht-Fluoreszenz, epi-fluorescence (fluorescence with epi-illumination)
Auflösung, resolution
Auflösungsvermögen, resolving power (resolution)
aufrechtes Bild, erect image
ausbleichen, to be bleached, fade
Ausstrahlung, radiation
Ausstrich, smear
Ausstrichpräparat, smear preparation
Austrittspupille, exit pupil

Beleuchtung, illumination
Bild, image
Bildbrennpunkt, rear focal point
Bildfeldwölbung, field curvature
Bildhelligkeit, image brightness
Bildpunkt, point of image
Bildschirmaufzeichnung, photography from video monitor
Bildwinkel, angle of view (in eyepiece)
Blende, diaphragm (iris)
Bogenlampe, arc lamp
Brechung, refraction
Brechungsexponent, refractive index
Brechungsindex, refractive index
Brechzahl, refractive index
Breitband, broad-band
Breitbandanregung, broad-band excitation
Breitbandfilter, broad-band filter
Brennpunkt, focal point, focus
Brennweite, focal length

Dauerpräparat, permanent preparation
Deckglas, coverglass, coverslip
dichroisch, dichroic
Dichroismus, dichroism
dichromatisch, dichromatic
Difluorezenz, difluorescence
Doppelbrechung, birefringence
Dunkelfeld, darkfield, darkground
Dünnschnittpräparat, thin section (preparation)
Durchflussfluoreszenzcytophotometrie, flow cytofluorometry
Durchlässigkeit, transmission
Durchlicht, transmitted light, dia-illumination

Ebene, plane
Eigenfluoreszenz, autofluorescence
Einschlussmittel, mounting medium
Eintrittspupille, entry pupil
Emissionsspektrum, emission spectrum
Empfindlichkeit, sensitivity
Erregerfilter, excitation filter
Erregerlicht, exciting light, light used for excitation
Erregung, excitation
Expositionszeit, exposure time
Exzitation, excitation

Farbe, colour, stain, dye, pigment
Färbabweichung, chromatic abberration
Färbeflüssigkeit, staining solution
Färbeindex, colour index
färben, to stain, dye, colour
Farbenbild, spectrum
Farbenbrechung, optical dispersion, refraction of colours

Farbglas, coloured glass (filter)
Farbglashalter, filter-glass carrier
farbig, coloured, stained
farblos, colourless
Farblösung, staining solution
Farbmethode, staining method
Farbstoff, dye, stain, pigment
Farbstoffspeicherung, (vital) (intracellular) dye deposit
Farbteiler, beamsplitter
Farbtemperatur, colour temperature
Farbton, shade or tone of colour
Färbung, staining, dyeing, colouring, etc.
Feldblende, field diaphragm
Feldzahl, field number
Fernsehmikroskopie, television microscopy
Filtersatz, filter set
Filtrierung, filtration
Fluoreszenz, fluorescence
Fluoreszenzabfall, fluorescent decay
Fluoreszenz-Antikörpertechnik, fluorescent antibody technique (immunofluorescence)
Fluoreszenzeinrichtung, fluorescence apparatus
Fluoreszenzfährigkeit, ability to fluoresce
Fluoreszenzlöschung, (fluorescence) quenching
Fluorochromierung, fluorochromy
Fluorochrommarkiert, fluorochrome-labelled
Fluorometrie, fluorometry
Focusebene, focal plane
Formdoppelbrechung, form birefringence

Gegenfarbe, counterstain
Gitter, grating (diffraction grating)
Glühlampe, incandescent (tungsten filament) lamp

Halbwertsbreite, half-bandwidth
Halogen-Glühlampe, tungsten-halogen lamp
Heiztisch, heating stage
Hellfeld, brightfield
Hilfspiegel, (collector mirror in lamphouse)
Hochleistung, high intensity
Höchstdrucklampe, high-pressure (arc) lamp
Huygenssches Okular, Huygenian eyepiece

Immersionsöl, immersion oil
Immunfluoreszenz, immunofluorescence
Inhomogenitätsfehler, distribution error
Irisblende, iris diaphragm

justieren, to adjust

Kantenfilter, sharp cut-off filter (barrier filter)
Kardioidkondensor, cardioid condenser
Kellnersches Okular, Kellner eyepiece

Klappkondensor, swing-out condenser
Kleinbildformat, minature format (= 35 mm film)
Kleinbildkamera, minature (35 mm) camera
Kohlenbogenlampe, carbon arc lamp
Köhlersche Beleuchtung, Köhler illumination
Kollektor, collector
Koma, coma
Kondensor, condenser
Konfocal, confocal
Kreuztisch, mechanical stage
Kühltisch, cooling (cold) stage
Kurzpassfilter, short-pass filter

Lambert–Beersches Gesetz, Beer–Lambert Law
Lampenhaus, lamphouse
Längsaberration, axial aberration
Leuchtdichte, optical density
Leuchtfeldblende, (illumination) field diaphragm
lichtelektrisch, photoelectric
Lichtempfindlichkeit der Filme, film sensitivity (speed)
Lichtquelle, light source
Lichtquellenbild, image of light source
lichtschwach, dim
Lochblende, pinhole diaphragm

Mattscheibe, ground-glass screen
Messeinrichtung, measuring device
Messfeld, measuring field
Metachromasie, metachromasia

Nachvergrösserungssystem, subsidiary magnification system
Nachweis, demonstration
Niederdruck, low-pressure (of a gas discharge lamp)
Niedervolt-Glühlampe, low-voltage lamp

Objekt, object (specimen)
Objektbrennpunkt, front focal point
Objektiv, objective
Objektivrevolver, nosepiece
Öffnungsfehler, spherical aberration
Okular, eyepiece, ocular
Orthoskopisches Okular, orthoscopic eyepiece

Primärfilter, excitation (primary) filter
Primärfluoreszenz, primary fluorescence, autofluorescence

Quecksilber-Höchstdrucklampe, high-pressure mercury arc lamp
Queraberration, lateral aberration
Querschnitt, transverse section

Raumfilter, spatial filter
Revolver, Objektive-, nosepiece
Rotsperrfilter, red-stopping filter ('minus red' filter)

Sammellinse, converging lens
Schärfentiefe, depth of field
Schieber, slide
Schliff, polish
Schliffpräparat, polished specimen
Schmalband, narrow-band
Schmalbandanregung, narrow-band excitation
Schnitt, section
Schnittdicke, section thickness
Schwarzweiss, black-and-white
Sekundärfluorescence, secondary fluorescence (as distinct from primary or autofluorescence)
Sichtfeld, visible field
Siliconöl, silicone oil
Spektrum, spectrum
Sperrfilter, barrier filter
Spiegelkondensor, mirror condenser (for darkground)
sphärische Aberration, spherical aberration
Spiegel, mirror
Stativ, stand
Stokes'sche Regel, Stokes Law
Strahl, ray, light
Strahldichte, radiation strength (intensity)
Strahlengang, ray path, light path
Strahlenquelle, light source, source of rays
Strahlteiler, beamsplitter
Strahlungsfluss, ray path
Streulicht, stray light
Stutze, adapter

Teilerspiegel, half-silvered mirror, beamsplitter
Tisch, stage
Tubuslinse, tube lens

Umgekehrtes Mikroskop, inverted microscope
Umlenksspiegel, diagonal mirror
Uranglas, uranium glass

Vergrösserung, magnification, enlargement
verkehrtes Bild, inverted image
Verlaufinterferenzfilter, wedge interference filter
Vielschichteninterferenzfilter, multi-layer interference filter
Vitalfärbung, vital staining

Wellenlänge, wavelength
Winkel, angle

Xenon-Höchstdrucklampe, high-pressure xenon arc lamp

Zubehör, accessory
zusätlich, supplementary
Zweischalenfehler, astigmatism
Zwischenbild, intermediate image
Zwischentubus, intermediate tube

Appendix 2
Checklist of features of fluorescence microscopes

Some years ago I was asked to make a comparison between a particular make of fluorescence microscope and other currently available models. For the survey which I carried out, I made a checklist of features which might be found. The following is an updated version. Not all features mentioned will be relevant to every application.

Microscope stand

Upright/inverted
Suitable for general use as a microscope?
Firmly constructed?
Adequate stage?
Square stage?
Rotating stage? Full 360° rotation?
Convenient in use?
Integral construction for both dia- and epi-illumination?
If not, is an epi-illuminator available, which can attach easily and firmly?

Optics – general

Width of field (field number of eyepieces)
Simultaneous or alternate other modes:
- Transmitted light
- Phase-contrast (positive, negative)
- Interference-contrast
- Darkground
- Polarization
- Reflected light

Ease and speed of conversion from fluorescence to other modes?

Light sources

Mercury arc: 50 W, 100 W, 200 W
Xenon arc: 75 W, 150 W, 450 W
Tungsten-halogen: 100 W, 150 W, 250 W
Laser

Lamphouse for arc lamp

Controls adequately insulated from heat?
Mirror present behind arc?

Collector lens achromatic – focus not shifting with change of wavelength
Collector lenses of different kinds to suit different arc lamps?
Collector lens unfrosted?

Mirror-house
Can the same light source be used alternatively for epi- or dia-illumination?
How many lamps can be mounted simultaneously?
How many lamps can be used simultaneously?
Can the excitation light be blocked off easily?

Filters
Ease of exchange between different filter combinations?
Interchangeable filter blocks with matched set of excitation filter, dichromatic mirror and barrier filter
Slots for additional barrier and/or excitation filters
How many filter combinations can be accommodated simultaneously in slide or turrets?
Range of appropriate filter combinations for different purposes
- Narrow-band excitation filters
- Broad-band excitation filters

Suitable barrier filters
- Non-fluorescent barrier filters

Background (field without specimen) really dark, fairly dark, not dark

Illumination
Epi-illumination by dichromatic mirrors?
Dia-illumination by substage condenser?
Darkground condenser
- Evenness of illumination?
- Maximum diameter of illuminated area?
- Available with oil-immersion objectives?

Brightfield condenser
- Achromatic aplanatic
- Numerical aperture
- Coverage of field of low-power objective

Objectives
Special objectives for fluorescence
Low autofluorescence
Infinity-corrected
High numerical aperture in relation to magnification
Wide exit pupil (entrance pupil for epi-illumination)
Immersion media: oil/water/glycerol/silicone/etc.
Immersion objectives: high/medium/low powers
Working distance of special objectives
Phase-contrast objectives also suitable for epi-illumination
Interference-contrast, with simultaneous epi-illumination

Oculars (eyepieces)
Lowest magnification?
Range of magnifications
Field numbers
High eyepoint for spectacle wearers

Body tube

- Gives 100% light to oculars
- Gives 100% light to phototube
- Gives part light to oculars, part to camera? (Not generally suitable for fluorescence)

Camera

- 35 mm format
- Other formats
- Exchangeable film magazines
- Illuminated graticule
- Efficient light transmission to film
- Simultaneous rotation of camera and graticule
- Automatic exposure
    - Automatic system compensates for reciprocity failure
    - Automatic exposure can easily be compensated for proportion of field fluorescing
    - Exposure time memory
- Manual exposure
    - Longest timed exposure which can be set?
- Automatic film advance
- Exposure compensation (specific, or by changing film speed setting)
    - Rapid changing by 1/3 or 1/2 stop increments
- Metering system
    - Sensitive enough to measure fluorescence
    - Integrated measurement from field
    - Centre-weighted measurement from field
    - Spot measurement
    - Spot need not be in centre of field

Fluorometry

- Provision for attachment of photometer
- Stable light source
- Monochromatic light source

Video

- Port for attachment of video camera
    - C-mount, ENG-mount
    - Independent of camera mounting

Appendix 3A
The laboratory microscope

This appendix is intended primarily to give practical information relevant to the routine operation of laboratory microscopes. Further details are to be found in standard texts of general microscopy such as those of James (1976) and Beyer (1977), in booklets supplied by the manufacturers of microscopes, and in monographs concerning special techniques (see below).

Principles of operation

The basic function of a microscope is of course for examination of a specimen and to reveal details too minute to be visible with the naked eye. For this purpose the instrument has an optical system consisting of an objective and an eyepiece whose essential characteristics are resolution and magnification. The resolution of a microscope can be defined as the ability to separate minute details that are very close together. This is usually measured as the least distance between two minute circular objects of high contrast at which they can still be seen to be separate. Resolving power depends primarily on the objective's numerical aperture, to which it is proportional. The magnification of a microscope is a function of both the objective and the eyepiece.

If the light rays from the specimen do not fully utilize the numerical aperture of the objective, its theoretical resolving power is not obtained. It is the function of the condenser to provide a cone of light which passes through the specimen and which has an angle sufficiently wide to fill the aperture of the objective. The condenser must be adjusted properly or else the quality of the image suffers.

Setting up and adjusting a microscope

When a microscope is properly adjusted it gives its best performance and is least likely to cause eyestrain and headache for the operator. It is convenient to provide a checklist near each microscope, especially to assist staff in training. The following is an example of a general checklist which may require modification to suit individual types of microscope. It assumes the type of stand in which the stage moves up and down for focussing rather than the older variety in which the body tube moves and carries the objective and the eyepiece.

It must be noted that there are two diaphragms to be adjusted, namely the field diaphragm and the aperture diaphragm. The former is usually situated on or within the base of the microscope and determines the diameter of the field of illumination. The aperture diaphragm is usually located underneath the condenser and regulates the angle of the cone of light passing through the specimen and entering the objective. Their respective actions can be demonstrated by placing on the stage a block of fluorescent glass or plastic, or a small beaker containing a solution of a fluorescent dye such as Fluorescein or Rhodamine B.

Checklist

1. *Power supply:* Plug in and switch on the power supply (there may be a separate switch on the microscope base). Adjust the rheostat until the lamp gives a suitable brightness (this may require readjustment later).

2. *Check the condenser:* Rack the substage condenser up fully. The top (swung in if necessary) should generally be only slightly below the level of the stage; if it does not reach this level, check whether the condenser has slipped down in its holder (it is usually held by a knurled screw). Open the condenser aperture diaphragm fully. Check whether a filter is present in the filter holder; if there is one of the frosted variety remove it temporarily or replace it with a clear one.
3. *Select low power conditions:* There may be a movable condenser top which must be swung out or an auxiliary lens which usually has positions marked L and H for low and high powers, respectively.
4. *Nosepiece:* Swing the lowest-power objective (usually $\times 4$ or $\times 10$) into place, ensuring that the nosepiece clicks into position.
5. *Specimen:* Lower the stage and place a slide on it. This usually involves opening a clip, inserting the slide, and allowing the clip to swing gently back to hold the slide. Do not allow the clip to spring back sharply because this will probably break or chip a corner of the slide. Ensure that the slide is sitting flat on the stage with its coverglass uppermost. Bring the specimen into the centre over the top lens of the condenser.
6. *Focus the specimen:* Rack the stage up with the coarse-focus knob while looking at the stage to make sure that the specimen does not strike the objective. Continue until the specimen is closer to the objective than the distance that you expect it to focus at (say about 10 mm for the usual $\times 4$ objective). Next look through the eyepieces and rotate the coarse-focus knob in the opposite direction until the specimen comes into focus. This procedure ensures that the objective does not hit the specimen, avoiding damage to both specimen and objective.
7. *Focus the condenser:* Close the field diaphragm on the base of the microscope until its image is smaller than the field of view. Focus its image into the same plane as the specimen by turning the condenser focussing knob while looking down the eyepiece. Sometimes colour fringes are visible which change with focussing up and down: these indicate poor colour correction in the condenser. Check that the diaphragm image is central in the field of view; if it is not, adjustment is usually by centring screws in the condenser mount. Open up the field diaphragm until its image (including coloured fringes, if any) is just clear of the field of view. Usually, if this adjustment is made with the lowest power objective, the diaphragm can be left in this position; however, sometimes it is helpful to adjust the diameter of the field diaphragm (to just beyond the field of view) with the objective actually in use, particularly when an oil-immersion lens is being used. This is particularly important for photomicrography (and for fluorescence by dia-illumination, so as not to fade the specimen outside the field of view).
8. *Check that the light source (lamp filament) is centred:* Some modern microscopes have no provision for centring the filament, others provide centring screws. An image of the lamp filament can sometimes be made visible by moving an auxiliary lens (if the H position gives source-focussed illumination) or by removing the eyepiece. Alternatively a white card can be held just below the condenser to show an image of the lamp. Some manufacturers provide a ground-glass to be placed on the base of the microscope under the condenser.
9. *Adjust the condenser aperture:* At this stage any frosted filter previously removed can be put back. Remove the eyepiece and look down the body tube. The rear of the objective should appear illuminated. Adjust the condenser aperture diaphragm until only about two-thirds to seven-eighths of the objective aperture is illuminated. Replace the eyepiece. In principle this adjustment must be repeated each time the objective is changed. In practice, except in critical conditions, the condenser aperture can be adjusted to suit the highest power dry objective ($\times 40$ usually) and left in this position for all the dry objectives; however, the diaphragm needs to be readjusted when an oil-immersion objective is used and again on changing back to dry objectives.
10. *The binocular head:* One or both eyepieces should be adjustable for focus. If there is only one adjustable eyepiece, close the eye over that; otherwise close either eye. With the eye which is open, focus the specimen sharply. Close it and open the other eye, and with it focus the specimen by adjusting the second eyepiece. The specimen should then be in focus for both eyes. Adjust the interocular distance as required until clear binocular vision is obtained.
11. *Oil immersion:* When it is known that the oil-immersion objective lens will be used and the highest

possible resolution is required, it may be helpful to oil the condenser before placing the slide on the stage in order to obtain its full aperture. This procedure is not required if the aperture of the condenser or objective is 0·9 or less. Put a few drops of oil on top of the condenser, place the slide on the stage, and very gradually raise the condenser until the oil touches the lower surface of the slide. Avoid air bubbles in the oil; if any occur, remove the slide, wipe off the oil, and start again. Focus the condenser as described above.

Find the area of interest in the specimen and focus it with the highest-power dry objective. Turn to the lowest-power objective (which is short and well clear of the oil on the slide, as well as adjacent to the oil-immersion objective), place one large drop of oil on the slide over the condenser and gently rotate the nosepiece to bring the oil-immersion objective into place. Some spring-loaded objectives can be locked in the up position, brought over the specimen and lowered gently into the immersion fluid. Avoid passing the ×40 objective over the oil because it will get oil on its front surface. Focus carefully with the fine focus knob. If the object was focussed for the high dry objective, the oil-immersion objective is usually almost in focus or requires focussing upwards slightly with the fine adjustment. Remove an eyepiece and adjust the condenser aperture diaphragm; replace the eyepiece. Make final adjustments of the aperture and field diaphragms to give the most suitable resolution, contrast, and depth of field.

When the observations are finished wipe the slide, the objective, and the condenser (if oiled) with appropriate solvent on a soft tissue. Special lens-cleansing tissue is best for the objective but otherwise a roll of really soft toilet tissue is most economical.

For best results the immersion oil recommended by the microscope manufacturer should be used. Some years ago a number of immersion oils were found to be carcinogenic and so old stocks should be discarded. In some cases glycerol can be used instead; its refractive index is 1·47 compared with 1·52 and its advantages are economy and ease of cleaning because it is miscible with water.

After oil has been put on the slide, returning to the high dry objective can present problems. In haematology and other fields that require scanning at medium power and frequent changes to high power it is advantageous to have two oil-immersion objectives, one with high and one with medium magnification (for instance ×100 and ×50).

Special techniques

Darkground illumination: Darkground (darkfield) microscopy is the simplest form of microscopy other than transmitted light microscopy. It requires only a stop in the condenser blocking out the central rays of light that would enter the objective while allowing the peripheral rays to illuminate the object. The aperture diaphragm of the condenser must be fully open. Only those rays reflected, refracted or diffracted by the object enter the objective and hence the object is seen as bright against a black background.

The required stop in the condenser can be built in, supplied as an accessory, or fitted to the revolver of a Zernike phase-contrast condenser. A simple stop can be made and fitted into the filter-holder, if there is one, underneath the condenser. This only requires an opaque disc of appropriate diameter on a transparent disc such as the filter itself or a piece of clear plastic. The required diameter of the stop disc can be found by trial and error or by measurement; in the latter case the aperture field diaphragm is closed down until its aperture is just greater than that of the objective: the diameter of the stop should be slightly larger than that of the diaphragm opening thus obtained (the diaphragm must of course be reopened for use). For temporary use, a small coin (such as a British new penny; Australian 5 cents) can be placed on the centre of the filter underneath the condenser.

Polarization: The mechanism is too complex to describe here; see Patzelt (1986), Ganse (1977) and James (1976). The additional equipment required is in principle only a pair of Polaroid™ discs. This can be obtained quite cheaply in sheet form and cut to size. One disc (the polarizer) is usually placed either over the field diaphragm on the base of the microscope or in the filter holder under the condenser. The other disc (the analyser) in a binocular microscope is best fitted in the body tube under the prisms of the binocular head, and in a monocular microscope is most conveniently fitted in the eyepiece. In a fluorescence microscope, it may be possible to use a barrier-filter slot. The presence of an epi-illuminator mirror much complicates the system (see Chapter 2). Polaroid cannot be used to polarize the beam used for fluorescence

excitation – the Polaroid will fade rapidly and lose its polarizing capacity. A calcite prism must be used.
Phase-contrast: Unstained biological objects (for example cells and other matter in urine) are very poorly visualized with an ordinary microscope due to lack of contrast. The contrast is usually increased by reducing the aperture of the condenser either by partially closing the condenser aperture diaphragm or by lowering the condenser. Both these techniques give an increase in contrast at the expense of a reduction in resolution.

A better method of examining such objects is a phase-contrast system in which differences in phase delay of the light due to differing refractive indices of the objects are converted into changes in the brightness of the image. Although phase-contrast systems used to be expensive, their relative cost has dropped considerably and there now seems to be no reason why every pathology laboratory should not have one. The expense is particularly reduced if phase-contrast equipment is bought as original apparatus for a new microscope, because it can be used for conventional microscopy as well. Equipment required for phase-contrast microscopy consists of a special condenser and a range of special objectives.

A phase-contrast condenser usually carries several phase rings on a revolving disc, while a blank position for ordinary microscopy has no ring. Some condensers have an additional position for darkfield microscopy. For each objective an appropriate ring must be selected; typically there are three, namely one each for low power, the ×40 objective, and the oil-immersion objective.

To set up the system, first the microscope must be adjusted as described above and then each phase ring must be centred in relation to its corresponding objective. This is best done with the aid of an auxiliary telescope, supplied to fit in place of an eyepiece, which gives a magnified view of the phase plate of the objective. However, it is possible to adjust the phase rings quite well by removing an eyepiece and looking down the tube.

Appendix 3B

Faults in the microscope and possible remedies

This appendix is intended as a guide to finding the source of faults in the fluorescence microscope. It is assumed that the specimen is actually fluorescent! Unless otherwise stated, the faults may occur with both epi- and dia-illumination.

The lamp will not light

Faulty electrical connection: test each electrical connection, commencing at the power point. Caution: the electricity is lethal! In Britain, Singapore, etc.: check the fuse in the plug
Blown fuse in the power unit: disconnect from the mains, check the fuse and replace if necessary
Unserviceable lamp: replace the lamp
Power unit not functioning: refer to manufacturer

Nothing is visible in the microscope

The lamp has not lit (see above)
The mirror house arrangement does not allow the light from the lamp to reach the specimen (is it set for dia- or epi-illumination?)
(Trinocular body) The beamsplitter knob is set to reflect all the light away from the eyepieces
The light is so much reduced that the image is invisible (see The image is too dim, below)
Wrong filter combination (excitation and/or barrier filters inappropriate for the object; or two excitation filters in optical path simultaneously)
Nosepiece positioned between objectives (no objective properly in place)
The barrier filter slide is improperly positioned
The specimen is not under the objective
The specimen is not in focus
(Epi-illumination) The shutter in the epi-illuminator is closed
(Epi-illumination) No filter block in place
(Epi-illumination) The filter block is improperly positioned
(Epi-illumination) The objective is too far from the specimen
(Dia-illumination) The condenser is not focussed
(Dia-illumination) The specimen is opaque
(Darkground condenser) The condenser is set too low (or too high, less likely)
Wrong filter combination for the subject

The image is too dim

Wrong filter combination
Wrong barrier filter (cutoff wavelength too long)
Wrong excitation filter (wavelength not suitable for fluorophore)
Two incompatible excitation filters present simultaneously

Neutral-density filter present
The condenser (or epi-illuminator) aperture diaphragm is partially closed
The objective aperture diaphragm is closed
(Trinocular body tube) Beamsplitter set to send only part of light to oculars (diverts portion of light up phototube)
The lamp is old, blackened inside, needs replacing
The lamp is not properly centred
Tungsten-halogen lamp: not at full current
The lamphouse collector lens requires focussing
The lamphouse reflector requires adjustment
The eyepiece magnification is too high
Intermediate-tube magnification-changing device: increases magnification and reduces light
Wrong collector lens in the lamphouse (not suitable for the lamp installed)
Inappropriate use or omission of auxiliary lens in illumination system
(Darkground condenser) Condenser not focussed correctly (too low or too high)
Wrong lamp in use (e.g. tungsten filament instead of arc lamp)
(Epi-illumination) The objective numerical aperture (NA) is too low
(Dia-illumination) The substage condenser is not focussed
(Dia-illumination) No oil, or insufficient oil on the condenser
(Dia-illumination) The specimen is too thick
(Dia-illumination) The objective magnification is too high
(Dia-illumination) Auxiliary lens for high/low magnification wrongly set
(Substage condenser) Top component of condenser swung out, fully or partially
(Substage condenser) Wrong kind of top (low NA)
(Substage condenser) Phase-contrast rings present

The illumination of the field is uneven

Lamp is incorrectly centred
Reflector in the lamphouse is incorrectly adjusted
Collector lens is not focussed correctly
Diffusing glass required in lamphouse
Darkground condenser not focussed correctly
Field diaphragm opening too small
Nosepiece improperly positioned, not at indexed position (objective not properly in place)
The excitation filter slide (if present) not positioned correctly
(Dia-illumination) Condenser not centred
(Dia-illumination) Bubble in the oil on the condenser

The lamp flickers too much

Lamp was not 'burnt-in' correctly
Bad lamp
Lamp is getting old
Problems in electrical supply

Only one side of the image is visible, or stronger than the other side, with a clear line of demarcation

The beamsplitter or sliding mirror of the trinocular body tube is at an intermediate position (pull it fully out or push it fully in as appropriate)
The barrier filter slide is incorrectly positioned as above

Excitation light coming through to eye

No barrier filter
No excitation filter

Mismatched filter set (wrong barrier filter or wrong excitation filter or both)
(Red background) Red-stop filter omitted
Filters positioned for dia-illumination, lamp for epi-illumination or vice versa
Excitation filter broken (light comes through break)
(Darkground condenser) NA of objective too great for condenser
(Oil-immersion objective with iris) Iris too widely open

Frequent breakage of excitation filters
Heat filter is not the one nearest to the lamp
Filter tightly clamped in housing
Inadequate ventilation
Cold draughts of air

Substage condenser cannot be focussed high enough
Specimen slide is too thick (exceeds 1·2 mm)
Condenser has slipped down in its mount
Long-working-distance condenser required

Image is of poor contrast
Unsuitable excitation filter (excitation wavelength unsuitable for fluorophore, or not specific enough)
Improper filter combination (barrier and excitation filters do not match)
Try a longer-wavelength barrier filter
Barrier filter with poor transmission characteristics
Fluorescent barrier filter (bad design)
Red-stop (BG38) filter omitted from primary filter set
Field diaphragm too wide
Masking by neighbouring artefacts (try reducing the field diaphragm)
Wrong type of immersion fluid (oil? water? glycerol? silicone? air?)
Fluorescent immersion fluid
Bubbles in immersion fluid
Fluorescent foreign matter in optical system
Dirt on objective
Objective fluorescences (unsuitable objective)
Specimen in fluorescent mountant
Excitation filter broken (light comes through break)
(Epi-illumination) Dirt on mirror surface in illuminator
(Dia-illumination with darkfield condenser only) NA of objective greater than that of darkfield condenser
The substage light-trap is not in place
The specimen's fluorescence has faded
Normal substage illuminator lamp left on

Image blurred
Not in focus
Dirt on optical system, probably front surface of objective
Dry objective has immersion fluid or mountant on front surface
Immersion objective used dry, or with wrong fluid
Bubble in immersion fluid
Barrier filter faulty (cement may have come loose)
Condensation on ocular (from observer's breath)
Ocular and objective do not match (e.g. of different make)

Coverglass of incorrect thickness
Correction collar on objective incorrectly set for coverglass thickness
Image too dim in the camera: see Chapter 6

Spots, flecks or blurs which remain in the field when specimen is moved
Dirty lens surface. Try rotating the ocular or objective: corresponding rotation of the spots identifies the dirty component

Specimen fades too quickly
Illumination is too bright
Improper excitation (broad-band rather than narrow-band; too much ultraviolet)
Inappropriate mountant, or pH of mountant wrong
(Dia-illumination) Field diaphragm too wide (fades a big area)
Aperture diaphragm should be partially closed

Objectives not parfocal
Objective not fully screwed into nosepiece
Spring-loaded oil-immersion objective locked in up position
Spring-loaded objective stuck in partially retracted position
Mixed objective types
Objectives and eyepieces corrected for a tubelength different from that of the microscope

'If all else fails, try reading the instructions!'

Appendix 4A

Data on photographic films for fluorescence photomicrography

'Professional' and 'amateur' films

Some colour films are supplied in both 'professional' and 'amateur' versions. The reason is that the colour balance of film changes with time; the changes are minimized by storage at a low temperature. Most professional films are intended to be stored at or below 13 °C until shortly before use, at which time they must be allowed to warm to room temperature. If such storage is not possible, 'amateur' films are preferable. In any case, for photomicrography of fluorescence, exact colour balance is unlikely to be important, so that the additional expense and trouble of using professional-type films is hardly likely to be justified.

Currently available 35 mm films

This list includes all 35 mm films currently commercially available to me, i.e. in Australia in July 1990, excluding films unsuitable for fluorescence microscopy – e.g. artificial-light colour films, and black-and-white (B&W) films with a speed rating below 100/21°. I have checked the lists with local representatives, but accept no responsibility for errors or omissions. For simplicity, the tables show Kodak designations for developers; similar developers are made by other manufacturers. Equivalents, where available, are as follows:

Kodak	C-41	E-4	E-6	K-14
Agfa	AP-70		AP-44	
Fuji	CN-16		CR-56	
Ilford	XP-1 (for XP1 film only)			
Konica	CNK-4			

Currently available 35 mm films

Manufacturer, name of film	Type	ISO rating	'Amateur' (A) or 'professional' (P)
Colour positive (reversal, transparency) films, daylight type			
These all require E-6 or equivalent processing, except for some Kodak films, as indicated.			
AGFA-GEVAERT			
Agfachrome	50RS	50/18°	P
Agfachrome	CT-100	100/21°	A
Agfachrome	100RS	100/21°	P
Agfachrome	CT-200	200/24°	A
Agfachrome	200RS	200/24°	P
Agfachrome	1000RS	1000/31°	P
(CT and RS denote 'amateur' and 'professional' respectively. The Agfa processing kit is designated AP44.)			
FUJI			
Fujichrome	RF-50	50/17°	A
Fujichrome	RFP-50	50/17°	P
Fujichrome Velvia	RVP-50	50/17°	P
Fujichrome	RD-100	100/21°	A
Fujichrome	RDP-100	100/21°	P
Fujichrome	RH-400	400/27°	A
Fujichrome	RHP-400	400/27°	P
Fujichrome	RSP-1600	1600/33°	P
(The Fuji processing kit is designated CR-56. Velvia has a particularly fine grain. RFP, RVP, RDP, RHP are 'professional' films. RSP-1600 is intended for pull- or push-processing, at EI 800, 1600 or 3200.)			
KODAK			
Kodachrome	25	25/15°	A, P[a]
Kodachrome	64	64/19°	A, P[a]
Kodachrome	200	200/24°	P[a]
Ektachrome	EPR64	64/19°	P
Ektachrome	EPN100	100/21°	P
Ektachrome	HC	100/21°	A
Ektachrome	EPP100	100/21°	P
Ektachrome	EPD200	200/24°	P
Ektachrome	EL400	400/27°	A
Ektachrome	P800/1600	800/30°–1600/33°	P
Ektachrome	Infrared		R
R, Requires E-4 or EA-5 process. Commercial processing is available from a limited number of laboratories; locate source of processing before embarking on use of film.			
[a] Requires K-14 process.			
(Kodachrome has particularly fine grain, and is the only colour positive film so far to have stood the test of time in respect of archival qualities. Ektachrome P800/1600 is specifically designed for push-processing.)			
KONICA			
Konica Slide Film		100/21°	A

Currently available 35 mm films (cont.)

Manufacturer, name of film	Type	ISO rating	'Amateur' (A) or 'professional' (P)
Colour negative films			
Many of the films listed below are also sold under other brand names. All require C-41 or equivalent processing.			
AGFA-GEVAERT			
	XRG100	100/21°	A
	XRS100	100/21°	P
	XRG200	200/24°	A
	XRS200	200/24°	P
	XRG400	400/27°	A
	XRS400	400/27°	P
	XRS1000	1000/31°	P
(XRG and XRS denote 'amateur' and 'professional' films. The Agfa-Gevaert processing kit is designated AP 70; see Agfa-Gevaert Technical Data A20.)			
FUJI			
Fujicolor	Reala	100/21°	A
Fujicolor	Super HR-II 100	100/21°	A
Fujicolor	Super HG200	200/24°	A
Fujicolor	Super HG400	400/27°	A
Fujicolor	Super HR-II 1600	1600/33°	A
Fujicolor	NSP	160/23°	P
Fujicolor	NLP	160/23°	P
(Fujicolor NSP and NLP are professional-type films; NLP is intended for long exposures. The Fuji developer is designated CN-16.)			
KODAK			
Kodacolor Gold	100	100/21°	A
Kodacolor Gold	200	200/24°	A
Kodak Veracolor	III	160/23°	P
Kodak Veracolor	400	400/27°	P
Kodak Ektacolor Gold	160	160/23°	P
Kodacolor Gold	400	400/27°	A
Kodacolor Gold	1600	1600/33°	A
Kodacolor	VR1000	1000/31°	A
Ektar	25	25/15°	P
Ektar	125	125/22°	P
Ektar	1000	1000/31°	P
Kodak Ektapress Gold	100	100/21°	F
Kodak Ektapress Gold	400	400/27°	F
Kodak Ektapress Gold	1600	1600/33°	F

F, Kodak Ektapress Gold films are professional films intended for photojournalists and others who use large amounts of film and are unable to provide refrigerated storage.

(Ektar films have particularly fine grain; they are 'professional' films, requiring storage at 10 °C. Kodacolor Gold is an 'amateur' type film, intended for general use, with a wider exposure latitude than the Ektar series. Veracolor and Ektacolor are 'professional' films.)

Currently available 35 mm films (cont.)

Manufacturer, name of film	Type	ISO rating	'Amateur' (A) or 'professional' (P)
KONICA			
Konica	SRG100	100/21°	A
Konica	SRG200	200/24°	A
Konica	SRG400	400/27°	A
Konica	SRG3200	3200/36°	A
(All these are 'amateur' films. The Konica developer is designated CNK-4.)			
Chromogenic B&W film			
ILFORD			
	XP1-400	50/18°–1600/33°	
(Requires XP-1 or C-41 process. Can also be reversal-processed by E-6 process, in which case rate at ISO 100.)			
Conventional B&W negative films			
(Films rated below ISO 100/21° not included.)			
AGFA-GEVAERT			
Agfapan	100	100/21°	
Agfapan	400	400/27°	
FUJI			
Neopan	100	100/21°	P
Neopan	400	400/27°	P
Neopan	1600	1600/30°	P
ILFORD			
	FP4	125/22°	A, P
	HP5	400/27°	A, P
KODAK			
Plus-X pan		125/22°	A, P
Tri-X pan		400/27°	A, P
T-max	TMX100	100/21°	P
T-max	TMY400	400/27°	P
T-max	TMZ3200	3200/33°	P
Recording Film	2475	3200/33°	
High-Speed Infrared film		c. 50/20°	

Polaroid black-and-white films

'Relative speed' takes into account both the ISO speed rating and the additional magnification required to fill the full area of the film, as compared to the 35 mm format. The image on the 73 mm × 95 mm format is 7·7 times dimmer than that on the 35 mm (24 mm × 36 mm) format.

Type and name	Image dimensions (mm × mm)	Image diagonal (mm)	Speed, ISO	Speed relative
Colour transparency				
Polachrome CS	24 × 36	43	40	40
B&W ultrahighspeed print				
Type 612	73 × 95	120	20 000	2500
B&W high-speed print				
Type 87	70 × 73	101	3000	550
Type 667	73 × 95	120	3000	400
Type 107	73 × 95	120	3000	400
Type 57	90 × 114	145	3000	250
Type 53	90 × 114	145	800	70
Type 553	90 × 118	150	800	64

Appendix 4B
Reciprocity failure

Depending upon the kind of camera, one of three kinds of information is needed about any particular film. Given a manually controlled camera, if the exposure meter indicates a particular time of exposure at the film's rated speed, the corresponding corrected exposure time is required. Automatic cameras without correction for reciprocity failure require fudging of the ISO film setting. The problem involves: (1) as above, except that only an approximation is required; and (2) obtaining a value for the film's speed (exposure index) at the expected length of exposure. Cameras with an automatic shutter and a choice of settings for correction of reciprocity failure require a value of the reciprocity exponent (p) so that the appropriate reciprocity setting can be chosen.

Manufacturers' reciprocity data usually, however, specify at various exposure times the increased exposure required, expressed in stops, on the assumption that the photographer will open up the lens aperture accordingly. However, the data can also be interpreted as an indication of the loss of speed rating – e.g. if a 2-stop increase in exposure is recommended, that means that the film is 4 times less sensitive, and that the exposure specified is the corrected exposure corresponding to an uncorrected exposure 4 times shorter. From these data and Table A4B.2 it is possible to calculate p and thence the corrected time (T_c) corresponding to any measured, uncorrected, exposure (T_m) of the same magnitude (p unfortunately is not necessarily constant).

Table A4B.1. *Data on reciprocity failure of various films*
Data, calculated by me from data supplied by the manufacturers or from my experience, of the speed of various films at long exposures, and of the reciprocity exponent (p) to be set on automatic cameras with automatic reciprocity correction.

The values given are approximate, and only intended as a guide. The user should experiment to determine the best setting for the conditions involved. If a film is not listed, the user should try the values given for a similar film. The mean value for p is about 1·37, which is also the value chosen by Wild.

This table can be used in several ways, as follows.

1. With automatic cameras with compensation settings for reciprocity failure, set the film speed (exposure index, EI) value shown for 1 s exposure, and the nearest available reciprocity setting to the p value given in the table. For example, with the Olympus PM-10 camera, to use Fujichrome P1600 film, set the film speed to 1600 and reciprocity setting to no. 3; to use Kodachrome 64, set the film speed to 50 and reciprocity setting to no. 6.
2. With automatic cameras with no reciprocity correction, estimate the likely length of exposure and set the exposure index accordingly. For example, for Kodak T-max 400 film at about 10 s exposure, set the film speed to 250.
3. With manual cameras, fitted with an exposure meter, measure the exposure (T_m) at the exposure index shown for 1 s, find the p value for the film from the table, and calculate the corrected exposure (T_c) using Table A4B.2.

Film type	Exposure Index (EI)				Reciprocity exponent
	nominal	at exposure time of			
	(ISO)	1 s	10 s	100 s	p
Agfachrome 200	200	200	125	64	1·43
Agfachrome 1000RS	1000	1000	640	400	1·24
Agfachrome XRG & XRS					1·83
Fujichrome 50 (RFP)	50	50	32	—[a]	1·25
Fujichrome 100 (RDP)	100	100	64	—[a]	1·24
Fujichrome 400 (RHP)	400	400	320	200	1·24
Fujichrome P1600	1600	1600	1000	—	1·38
Fujicolor Reala	100	80	50	25	1·34
Fujicolor HG 200	200	160	100	50	1·29
Fujicolor Super HG 400	400	320	160	80	1·42
Fujicolor HRII 1600	1600	640	320	160	1·42
Ilford Pan F	50	50	32	16	1·36
Ilford HP5	400	400	250	—	1·36
Ilford XP1-400	400	320	200	100	1·45
Kodachrome 25	25	20	12	6	1·66
Kodachrome 64	64	50	32	25[a]	1·44
Kodachrome 200	200	200	160[a]	125[a]	1·09[b]
Kodak Ektar 1000	1000	1000	800[a]	—[a]	1·28[b]
Kodak Plus-X	125	64	40	25	1·27
Kodak Tri-X	400	200	125	80	1·31
Kodak T-max 100	100	80	64	50	1·11
Kodak T-max 400	400	320	250	125	1·21
Polaroid B&W	3000	3000	1600	800	1·44
Polacolor 108	80	80	50	25	1·36

Notes:

[a] Not recommended by the manufacturer at these exposure times.

[b] Based on my experience of a small number of rolls of film.

Black and white films

Ilford HP5 film has reciprocity characteristics described in Fig. A4.1. For automatic reciprocity correction, the reciprocity exponent (p) is about 1·36.

XP1-400 film requires no correction for reciprocity failure for exposures in the range 1/1000000 to 1/2 s. Exposures longer than 1/2 s require correction. Ilford's published data give corrections for measured exposures up to 20 s (Fig. A4.2). The speed rating to be set on automatic cameras is given in the table below. Values were calculated by the author from manufacturer's data. The latter, for exposures of 10 s or more, fit nicely to the equation below, where DIN can be taken as equal to the logarithmic component of the ISO rating:

$$\text{DIN} = 27 - 3 \log (\text{actual exposure in s})$$

assuming a basic rating of 400/27° ISO. For automatic reciprocity correction the reciprocity exponent (p) is about 1·45.

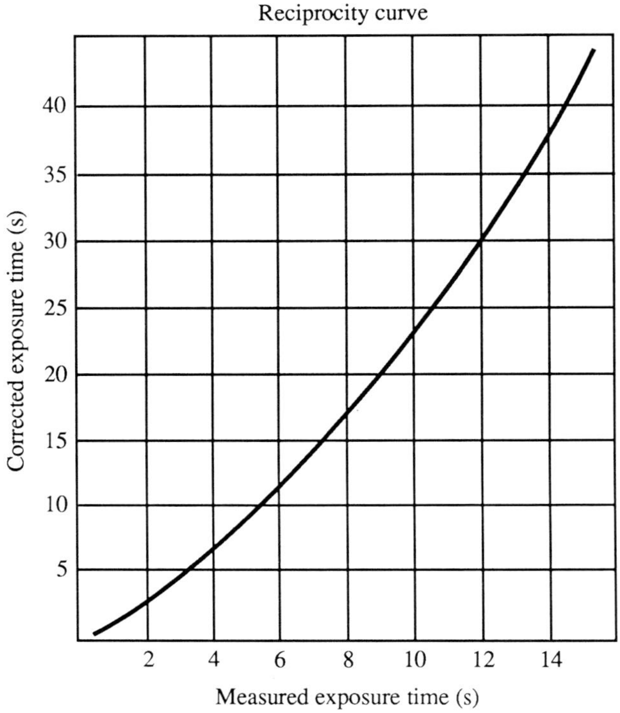

Fig. A4.1. Reciprocity failure of Ilford HP5 (manufacturer's data).

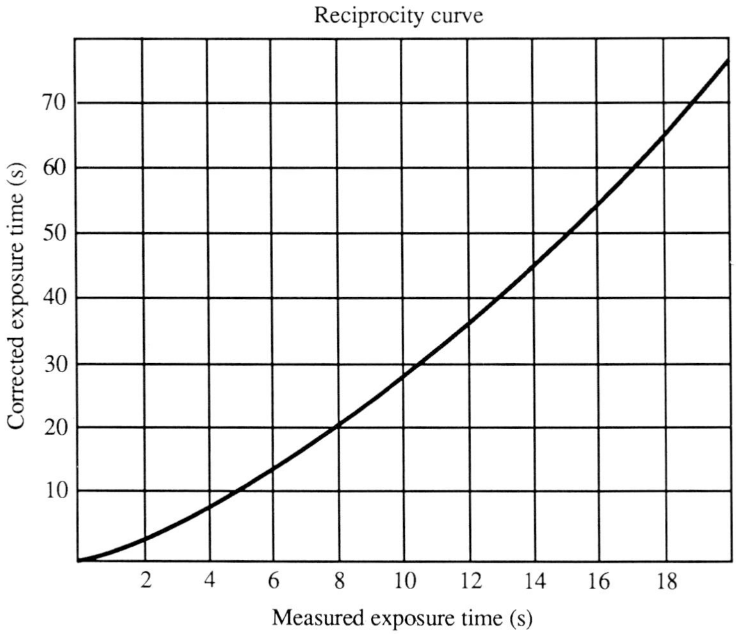

Fig. A4.2. Reciprocity failure of Ilford XP1-400 film (manufacturer's data).

Speed rating of Ilford XP1-400 film at various exposure times

Exposure time (actual) s	Speed rating ASA	Speed rating DIN
1/2	400	27
1	320	26
2	270	25
4	200	24
10	200	24
15	160	23
50	120	22
100	100	21

Colour films

Agfa: Agfachrome 200 has characteristics described by Fig. A4.3.

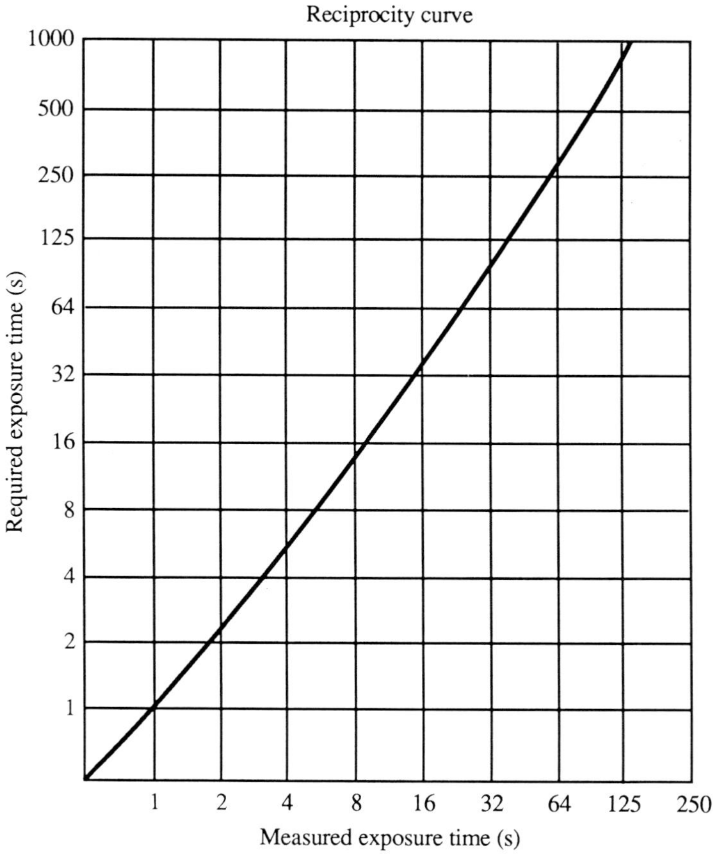

Fig. A4.3. Reciprocity failure of Agfachrome 200 (manufacturer's data).

Automatic correction for reciprocity failure

Automatic correction for reciprocity failure uses the formula $T_c = (T_m)^p$, where T_c is the corrected time, T_m is the uncorrected time, and p is a constant dependent upon the film. The following table gives the corrected time (T_c) for a given uncorrected time (T_m) and value of T_m.

The table can be applied in two ways. Given an estimated exposure time T_m, uncorrected, and a value for p, the corrected exposure time (T_c) can be obtained from the body of the table. Alternatively, given a camera system with built-in compensation for varying reciprocity characteristics, an appropriate setting can be selected for a new film type if a corrected exposure time is known for a given uncorrected time. For example: for the film in use, the manufacturer states that for an exposure time of 100 s, an increase of 2 stops, exposure is required, i.e. the film is 4 times less sensitive, so 100 s is the corrected exposure time (T_c) for an uncorrected exposure time (T_m) of 25 s. In the table above, looking along the line where $T_m = 25$ s, the nearest T_c to 100 s is 106 s, with a p of 1·45. Using the Olympus PM-10 camera, an appropriate setting would be number 5 ($p = 1{\cdot}44$).

Table A4B.2. *Corrected exposure time (*T_c*) at different values of measured time (*T_m*) and reciprocity exponent (*p*)*

$p=$	1·05	1·1	1·15	1·2	1·25	1·3	1·35	1·4	1·45	1·5
$T_m = 1$	1·0	1·0	1·0	1·0	1·0	1·0	1·0	1·0	1·0	1·0
2	2·1	2·1	2·2	2·3	2·3	2·5	2·5	2·6	2·7	2·8
3	3·2	3·3	3·5	3·7	3·9	4·2	4·4	4·7	4·9	5·2
4	4·3	4·6	4·9	5·3	5·7	6·0	6·5	7·0	7·5	8·0
6	6·6	7·1	7·9	8·6	9·4	10	11	12	13	14
8	8·9	9·8	11	12	13	15	17	18	20	23
10	11	13	14	16	18	20	22	25	28	32
12	13	15	17	19	22	25	28	32	36	41
16	18	21	24	28	32	37	42	49	56	64
20	23	26	31	36	42	49	57	66	77	89
32	38	45	54	64	76	91	108	128	152	181
40	48	57	69	83	100	120	145	174	210	252
64	79	97	119	147	181	223	274	337	416	512
80	99	123	154	192	239	297	370	461	574	715
100	126	158	200	251	316	398	501	631	794	1000
128	163	208	265	338	431	549	699	891	1136	1448
160	206	265	342	441	569	733	945	1218	1570	2023
200	260	339	442	577	752	980	1277	1665	2170	2828
256	338	446	588	776	1024	1351	1782	2352	3104	4096

Corrections applied by commercial cameras

The Olympus PM-10 camera has values of p corresponding to the settings of the reciprocity correction switch as follows (calculated by me from published data supplied by the manufacturer).

Setting	0	1	2	3	4	5	6	7
p	1·00	1·09	1·18	1·27	1·36	1·44	1·53	1·62

The Wild Photoautomat MPS 45 employs a reciprocity exponent of 1·37; the figures given below were calculated by me from data supplied by the manufacturer. The instrument may be used optionally with or without the automatic correction.

Uncorrected time (s)	Correction factor	Corrected time (s)
1	1·00	1·0
2	1·25	2·5
4	1·7	6·7
8	2·2	17
16	2·8	44
32	3·6	115
64	4·7	300
100	5·5	550
128	6·0	770

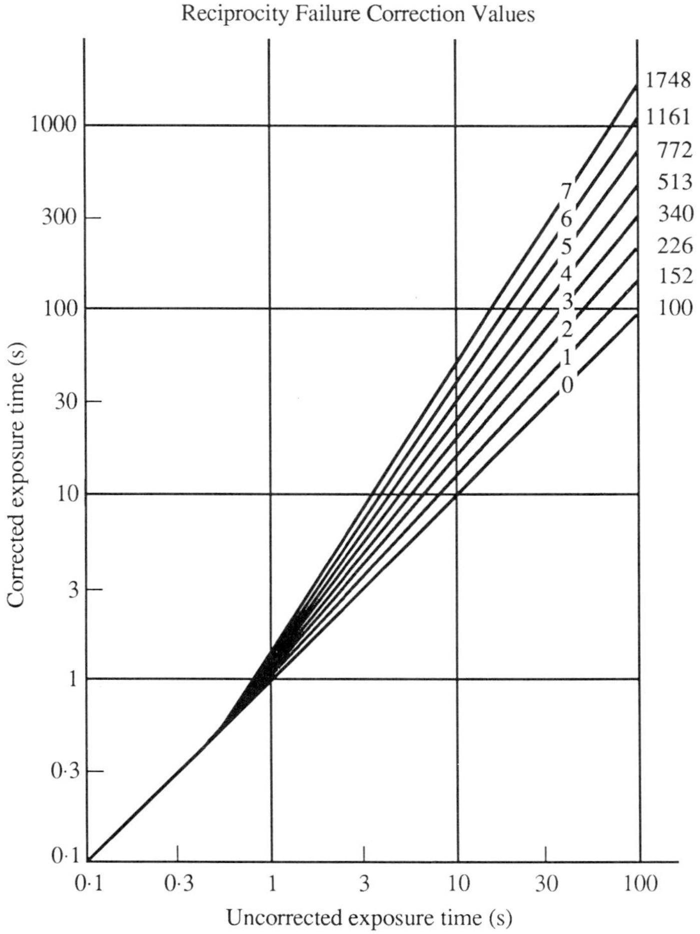

Fig. A4.4. Automatic corrections for reciprocity failure made by Olympus PM-10AD photomicrographic camera. There are seven different corrections which can be applied (1–7) and also a non-correction setting (0). Note the logarithmic scales.

Appendix 4C

Relative brightness of image on film in various formats

The table below indicates the relative brightness of the image theoretically expected when the image is projected onto film of various formats; it is assumed that the object is in each case projected onto the entire frame, i.e. the magnification is proportional to the size of the diagonal of the film. The formats are illustrated actual size in Fig. A4.5. It is assumed also that all other factors are kept constant. The brightness on standard 35 mm format is arbitrarily taken as 1·00. The dimensions given are those usually obtained for the picture, and may differ slightly from the nominal dimensions.

Format	Dimensions (mm × mm)	Diagonal (mm × mm)	Relative brightness
4 in. × 5 in.	100 × 125	160	0·07
Quarter-plate	83 × 108	136	0·10
¼ plate Polaroid	73 × 95	120	0·13
6 × 9	56 × 88	104	0·17
Polaroid	70 × 73	101	0·18
6 × 7	56 × 69	90	0·23
6 × 6	56 × 56	79	0·30
6 × 4·5	56 × 43	75	0·33
Superslide	38 × 38	54	0·65
35 mm	24 × 36	43	1·00
126	28 × 28	39	1·20
'Robot'	24 × 24	34	1·60
Half frame	17 × 24	29	2·20
110[a]	13 × 17	21	4·00
Kodak disc[a]	8 × 10	13	11·00

Notes:
[a] These two smallest sizes are included for completeness: it seems unlikely that either would give adequate image quality for photomicrography.

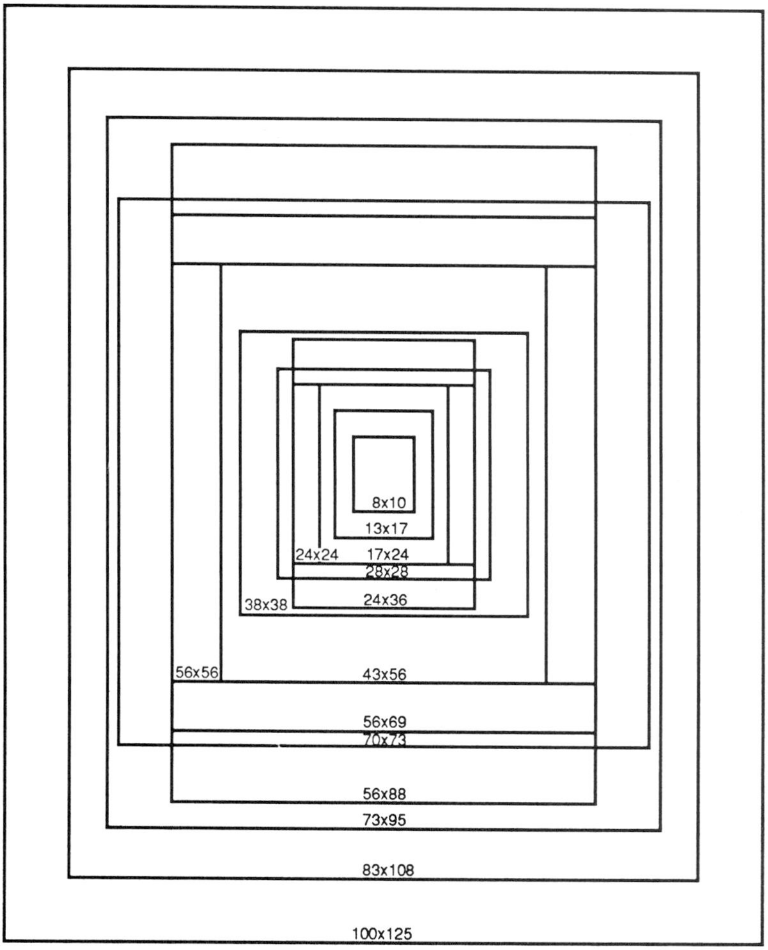

Fig. A4.5. Diagram showing various camera formats, actual size.

Appendix 4D

Faults in photomicrographs and possible remedies

See also Appendix 3B, which deals with difficulties with the microscope.

Pictures unsharp all over

Focussing procedure incorrect (operator error)
Focussing apparatus inadequate (try a focussing telescope)
Stage slipping down during exposure
Objectives of inadequate quality
Vibration in camera or microscope during exposure
Other causes of unsharp image – see Appendix 3B

Pictures unsharp at corners

Objective inadequately corrected
Wrong projective

Colour pictures seem wrong colour

Colour shift due to long exposure (reciprocity failure)
'Artificial light' colour film
Wrong barrier filter
Improper processing of film
(Prints) inappropriate filters used for printing

Pictures consistently underexposed

Film speed set wrongly on control box
Overcorrection for proportion of field fluorescent
No correction, or insufficient correction, for reciprocity failure
Meter needs recalibrating
Subjective appearance (decrease speed setting to suit)
Exposure correction setting set for underexposure

Pictures consistently overexposed

Film speed set wrongly on control box
Correction not applied for proportion of field fluorescent
Excessive correction for reciprocity failure
Meter needs recalibrating
Subjective appearance (change speed setting to suit)
Exposure correction setting set for overexposure

Dark corners

Field diaphragm too small
Uneven illumination of field – see Appendix 3B

Bright background or poor contrast

Inappropriate barrier filter, leaking excitation light
Autofluorescence in optical system
(See also Appendix 3B)

References

These references are arranged according to current Anglo-American cataloguing rules (Gorman, M. & Winkler, P.W. (Eds.) (1978) *Anglo-American cataloguing rules*, 2nd edn. 619 pp. The Library Association, London). For names beginning with van, van der, etc., or with von, see under the capitalized surname; for names beginning with Van, see under Van. Names beginning with de appear under de.

Chapter 1: Principles of fluorescence microscopy

Axelrod, D., Thompson, N.L. & Burghardt, T.P. (1983) Total internal reflection fluorescent microscopy. *Journal of Microscopy* **129**, 19–38.

Bals, M.G. (1966) *Immunofluorescenta*, 3rd edn. Editura Academei Republicii Socialiste Romania Bucuresti, Bucharest.

Becker, E. (1985) *Fluorescence microscopy. Instruments, methods, applications.* 71 pp. Ernst Leitz Wetzlar GmbH, Wetzlar.

Beutner, E.H., Nisengard, R.J. & Albini, B. (Eds.) (1983) Defined immunofluorescence and related cytochemical methods. *Annals of the New York Academy of Sciences* **420**, 432 pp.

Birk, G. (1984) *Instrumentation and techniques for fluorescence microscopy.* 111 pp. Wild Leitz (Australia), Sydney.

Coons, A.H., Creech, H.H. & Jones, R.N. (1941) Immunological properties of an antibody containing a fluorescent group. *Proceedings of the Society for Experimental Biology (NY)* **47**, 200–202.

de Lerma, B. (1958) Die Anwendung von Fluoreszenzlicht in der Histochemie. *Handbuch der Histochemie* **1** (1), 78–159. G. Fischer Verlag, Stuttgart.

Gaub, H.E., Brian, A.A., Watts, T.H. & McConnell, H.M. (1986) Total internal reflection fluorimetry on single T-lymphocytes in contact with supported lipid bilayers. In: Taylor, D.L., Waggoner, A.S., Murphy, R.F., Lanni, F. & Birge, R.R. (Eds.) *Applications of fluorescence in the biomedical sciences*, Ch. 23, pp. 499–504. Alan R. Liss, New York.

Goldman, M. (1968) *Fluorescent antibody methods.* 303 pp. Academic Press, New York & London.

Haitinger, M. (1938) *Fluoreszenz-Mikroskopie. Ihre Anwendung in der Histologie und Chemie.* Akademische Verlagsgesellschaft M.B.H., Leipzig.

Haitinger, M. (1959) *Fluoreszenz-Mikroskopie.* 2nd edn, ed. Eisenbrand, J. & Werth, G. 168 pp. Akademische Verlagsgesellschaft Geest & Portig, Leipzig.

Hansen, P.A. (1964) *Fluorescent compounds used in protein tracing. Absorption and emission data.* 65 pp. University of Maryland, Bethesda, MD.

Hijmans, W. & Schaeffer, M. (1975) Fifth international conference on immunofluorescence and related staining techniques. *Annals of the New York Academy of Sciences* **254**, 1–627.

Holborow, E.J. (Ed.) (1970) *Standardization in immunofluorescence* (Proceedings of conferences, Florence 1967, London 1968). 282 pp. Blackwell, Oxford.

Kasten, F. (1981) Methods for fluorescence microscopy. In: Clark, G. (Ed.) *Staining procedures*, 4th edn, Ch. 3, pp. 39–103. Williams & Wilkins, Baltimore & London.

Kawamura, A. (1969) *Fluorescent antibody techniques and their applications.* 262 pp. University of Tokyo Press, Tokyo; University Park Press, Baltimore, MD, & Manchester.

Kawamura, A. & Aoyama, Y. (Eds.) (1983) *Immunofluorescence in medical science.* 262 pp. University of Tokyo Press, Tokyo; Springer-Verlag, Berlin, Heidelberg & New York.

Knapp, W., Holubar, K. & Wick, G. (1978) *Immunofluorescence and related staining techniques* (Proceedings of the sixth conference, Vienna, 1978). 363 pp. Elsevier, Amsterdam & New York; North-Holland Biomedical Press, Amsterdam.

Koch, K.-F. (1972) *Fluorescence microscopy. Instruments, methods, applications.* 48 pp. Ernst Leitz, Wetzlar [published in both German and English].

Köhler, A. (1904) Mikrophotographische Untersuchungen mit ultraviolettem Licht. *Zeitschrift für wissenschaftliche Mikroskopie* **21**, 129–165 & 273–304.

Lanni, F. (1986) Standing-wave fluorescence microscopy. In: Taylor, D.L., Waggoner, A.S., Murphy, R.F., Lanni, F. & Birge, R.R. (Eds.) *Applications of fluorescence in the biomedical sciences*, Ch. 24, pp. 505–524. Alan R. Liss, New York.

Nairn, R.C. (Ed.) (1976) *Fluorescent protein tracing*, 4th edn. 648 pp. Churchill Livingstone, Edinburgh & New York.

O'Brien, T.P. & McCully, M.E. (1981) *The study of plant structure: principles and selected methods.* 352 pp. Termacarphi, Melbourne.

Ploem, J.S. & Tanke, H.J. (1987) *Introduction to fluorescence microscopy.* 56 pp. Oxford University Press/ Royal Microscopical Society, Oxford.

Price, G.R. & Schwartz, S. (1956) Fluorescence microscopy. In: Oster, G. & Pollister, A.W. (Eds.) *Physical techniques in biological research*, Vol. **III**, Ch. 3, pp. 91–148. Academic Press, New York & London.

Rost, F.W.D. (1980) Fluorescence microscopy. In: Pearse, A.G.E. *Histochemistry theoretical and applied*, 4th. edn., Vol. **1**, Ch. 10, pp. 346–378. Churchill Livingstone, Edinburgh, London & New York.

Rost, F.W.D. (1991) *Quantitative fluorescence microscopy.* 240 pp. Cambridge University Press, Cambridge.

Sani, G. (1964) *Fluorescence microscopy in the cytodiagnosis of cancer.* 195 pp. Thomas, Springfield, IL.

Taylor, D.L., Waggoner, A.S., Murphy, R.F., Lanni, F. & Birge, R.R. (Eds.) (1986) *Applications of fluorescence in the biomedical sciences.* Alan R. Liss, New York.

Thaer, A.A. & Sernetz, M. (Eds.) (1973) *Fluorescence techniques in cell biology* (Proceedings of the conference on Quantitative fluorescence techniques as applied to cell biology, Battelle Seattle Research Centre, 1972). Springer-Verlag, Heidelberg, New York & Berlin.

Wick, G., Traill, K.N. & Schauenstein, K. (Eds.) (1982) *Immunofluorescence technology.* 442 pp. Elsevier Biomedical Press, Amsterdam, New York & Oxford.

Zelenin, A.V. (1967) *Luminisceutnaja citohimija nukleinovyh kislot* [Fluorescence cytochemistry of nucleic acids]. Nauka, Moscow.

Chapter 2: Fluorescence: physics and chemistry

Allen, C.W. (1976) *Astrophysical quantities*, 3rd edn. 310 pp. Athlone Press, London.

Barenboim, G.M., Domanskii, A.N. & Turoverov, K.K. (1969) *Luminescence of biopolymers and cells.* Plenum Press, New York & London.

Björn, L.O. & Björn, G.S. (1986) Studies on energy dissipation in phycobilisomes using the Kennard–Stepanov relation between absorption and fluorescence emission spectra. *Photochemistry and Photobiology* **44**, 535–542.

Chapman, J.H., Förster, T., Kortüm, G., Parker, C.A., Lippert, E., Melhuish, W.H. & Nebbia, G. (1963) Proposal for standardization of methods of reporting fluorescence emission spectra. *Applied Spectroscopy* **17**, 171 (reprinted in Udenfriend, S. (1969) *Fluorescence assay in biology and medicine,* Vol. **II**, pp. 592–593. Academic Press, New York & London).

Chen, R.F. (1967) Some characteristics of the fluorescence of quinine. *Analytical Biochemistry* **19**, 374–387.

Chen, R.F. (1973) *Extrinsic and intrinsic fluorescence of proteins.* In: Guilbault, G.G. (1973) (Ed.) *Practical fluorescence: theory, methods and techniques*, Ch. 12, pp. 467–541. Marcel Dekker, New York.

Drexhage, K.H. (1976) Fluorescence efficiency of laser dyes. *Journal of Research of the National Bureau of Standards. A. Physics and Chemistry* **80A**, 421–428.

Fischbeck, H.J. & Fischbeck, K.H. (1982) *Formulas, facts and constants*. 251 pp. Springer-Verlag, Berlin, Heidelberg & New York.

Fletcher, A.N. (1968) Fluorescence emission band shift with wavelength of excitation. *Journal of Physical Chemistry* **72**, 2742–2749.

Goodwin, R.H. (1953) Fluorescent substances in plants. *Annual Review of Plant Physiology* **4**, 283–304.

Grzywacz, J. (1967) Difference between prompt and delayed fluorescence spectra. *Nature* **213**, 385–386.

Guilbault, G.G. (1973) (Ed.) *Practical fluorescence: theory, methods and techniques*. 664 pp. Marcel Dekker, New York.

Guilbault, G.G. (1985) Principles of luminescence spectroscopy. Luminescent determination of clinically and agriculturally important samples. *Pure and Applied Chemistry* **57**, 495–514.

Haugland, R.P. (1989) *Handbook of fluorescent probes and research chemicals*. 235 pp. Molecular Probes, Eugene, OR.

Hercules, D.M. (Ed.) (1966) *Fluorescence and phosphorescence analysis*. 258 pp. Interscience, New York.

Johnson, G.D., Davidson, R.S., McNamee, K.C., Russell, G., Goodwin, D. & Holborow, E.J. (1982) Fading of immunofluorescence during microscopy: a study of the phenomenon and its remedy. *Journal of Immunological Methods* **55**, 231–242.

Krasnovsky, A.A. (1982) Delayed fluorescence and phosphorescence of plant pigments. *Photochemistry and Photobiology* **36**, 733–741.

Lakowicz, J.R. (1983) *Principles of fluorescence spectroscopy*. 496 pp. Plenum Press, New York.

Mielenz, K.D. (1976) Comments on spectrometry nomenclature. *Analytical Chemistry* **48**, 1093–1094.

Picciolo, G.L. & Kaplan, D.S. (1984) Reduction of fading of fluorescent reaction product for microphotometric quantitation. *Advances in Applied Microbiology* **30**, 197–234.

Pringsheim, P. (1949) *Fluorescence and phosphorescence*. 794 pp. John Wiley, New York.

Rost, F.W.D. (1991) *Quantitative fluorescence microscopy*. 240 pp. Cambridge University Press, Cambridge.

Seliger, H.H. & McElroy, W.D. (1965) *Light: physical and biological action*. 417 pp. Academic Press, New York.

Slayter, E.M. (1970) *Optical methods in biology*. 757 pp. Wiley-Interscience, New York, London, Sydney & Toronto.

Stepanov, B.I. (1957) A universal relation between the absorption and luminescence spectra of complex molecules. *Doklady Akademii Nauk SSR* **112**, 839–842 (English translation *Soviet Physics Doklady* **2**, 81–84).

Udenfriend, S. (1962) *Fluorescence assay in biology and medicine*. 505 pp. Academic Press, New York & London.

Udenfriend, S. (1969) *Fluorescence assay in biology and medicine*, Vol. **II**. 660 pp. Academic Press, New York & London.

Udenfriend, S., Guroff, G. & Zaltzman-Nirenberg, P. (1967) Use of fluorescence as a probe into mechanisms for concentrative transport of amino acids in living cells. In: Guilbault, G.G. (Ed.) *Fluorescence: theory, instrumentation, practice*, Ch. 9, pp. 359–442. Edward Arnold, London; Marcel Dekker, New York.

Wehry, E.L. (1973) Effects of molecular structure and molecular environment on fluorescence. In: Guilbault, G.G. (Ed.) *Practical fluorescence: theory, methods and techniques*, Ch. 3, pp. 79–136. Marcel Dekker, New York.

Wotherspoon, N., Oster, G.K. & Oster, G. (1972) The determination of fluorescence and phosphorescence. In: Weissberger, A. & Rossiter, B.W. (Eds.) *Physical methods in chemistry*, Part IIIB, Ch. 4, pp. 429–484. (In: Weissberger, A. (Ed.) *Techniques in chemistry*, Vol. **I**.) Wiley-Interscience, New York.

Chapter 3: The optical sytem of the fluorescence microscope

Abramowitz, M. (1985) *Microscope basics and beyond*, Vol. **1**, 26 pp. Olympus Corporation, Lake Success, NY.

Bals, M.G. & Velculescu, V.G. (1966) New high-efficiency light source for immunofluorescence microscopy. *Nature* **210**, 1073–1074.

Beyer, H. (Ed.) (1977) *Handbuch der Mikroskopie*, 2nd edn. 495 pp. VEB Verlag Technik, Berlin.

Bradbury, S. (1984) *An introduction to the optical microscope*. 85 pp. Oxford University Press/Royal Microscopical Society, Oxford.

Cripps, D.J., Hawgood, R.S. & Magnus, I.A. (1966) Iodine tungsten fluorescence microscopy for porphyrin fluorescence. *Archives of Dermatology* **93**, 129–134.

Determann, H. & Lepusch, F. (1973) *Das Mikroskop und seine Anwendung* (English edition: *The microscope and its application*, also French and Spanish editions). Ernst Leitz, Wetzlar.

Hartley, W.G. (1974) Microscope illuminating systems for transmitted light. *Proceedings of the Royal Microscopical Society* **3**, 167–179.

James, J. (1976) *Light microscope techniques in biology and medicine*. 336 pp. Martinus Nijhoff, Amsterdam.

Koch, K.-F. (1972) *Fluorescence microscopy: instruments, methods, applications*. 48 pp. Ernst Leitz, Wetzlar.

Kraft, W. (1973) Fluorescence microscopy and instrument requirements. *Leitz Scientific and Technical Information* **2**, 97–109.

Mellors, R.C. & Silver, R. (1951) A microfluorometric scanning method for the detection of cancer cells in smears of exfoliated cells. *Cancer* **5**, 458–468.

Ploem, J.S (1965) Die Möglichkeit der Auflichtfluoreszenzmethoden bei Untersuchungen von Zellen in Durchströmungskammern und Leightonröhren. *Acta histochemica* Supplementum **VII**, pp. 339–343.

Ploem, J.S. (1967a) The use of a vertical illuminator with interchangeable dichroic mirrors for fluorescence microscopy with incident light. *Zeitschrift für wissenschaftliche Mikroskopie* **68**, 129–142.

Ploem, J.S. (1967b) Enkele methoden voor toxiciteitsonderzoek met behulp van weefselkweekcellen. 169 pp. Doctoral thesis (Geneeskunde), University of Amsterdam.

Rigler, R. (1966) Microfluorometric characterisation of intracellular nucleic acids and nucleoproteins by Acridine Orange. *Acta physiologica scandinavica* **67**, Supplementum **267**. 122 pp.

Rost, F.W.D. (1980) Fluorescence microscopy. In: Pearse, A.G.E., *Histochemistry, theoretical and applied*, 4th edn, Vol. **1**, Ch.10, pp. 346–378. Churchill Livingstone, London.

Rost, F.W.D. (1991) *Quantitative fluorescence microscopy*. 240 pp. Cambridge University Press, Cambridge.

Schott (n.d.) *Schott color filter glass*. Jenaer Glaswerk Schott & Genossen, Mainz.

Taylor, C.E.D. (1970) Optical features of immunofluorescence: 1. In: Holborow, E.J. (Ed.) *Standardization in immunofluorescence*, pp. 15–22. Blackwell Scientific Publications, Oxford & Edinburgh.

Zieler, H.W. (1972) *The optical performance of the light microscope*, Part 1. Microscope Publications Ltd, London & Chicago.

Zieler, H.W. (1973) *The optical performance of the light microscope*, Part 2. Microscope Publications Ltd, London & Chicago.

Zeiss (1967) *Optical systems for the microscope*. Carl Zeiss, Oberkochen.

Chapter 4: Commercial fluorescence microscopes

Beck, R. (1984) The technological development of inverted microscopes. *Leitz Scientific and Technical Information* **8**, 179–189.

Becker, E. (1985) *Fluorescence microscopy. Instruments, methods, applications*. 71 pp. Ernst Leitz Wetzlar GmbH, Wetzlar.

Gundlach, H. & Wehland, J. (1988) Immunofluorescence microscopy with the new Zeiss photomicroscope Axiophot. *Zeiss Information* **29** (99E), 36–39.

Chapter 5: Setting up and using the fluorescence microscope

Albert, E.N. (1973) A combined fluorescent and electron microscopic stain for elastic tissue. *Histochemical Journal* **5**, 157–167.

Bagge, U. (1981) A method for combined fluorescence and non-fluorescence incident-light vital microscopy. *Microvascular Research* **22**, 123–124.

Barer, R. (1968) Maximum-intensity fluorescence microscopy. *Nature* **217**, 672.

Beyer, H. (Ed.) (1977) *Handbuch der Mikroskopie,* 2nd edn, 495 pp. VEB Verlag Technik, Berlin.

Biggiogera, M. & Pellicciari, C. (1988) Fluorescein mercuric acetate as a dye for fluorescence and electron microscope detection of -SH groups in mouse sperm protamines: a critical evaluation. *Acta histochemica* **83**, 1–10.

Bretherick, L. (Ed.) (1981) *Hazards in the chemical laboratory*, 3rd edn. 567 pp. The Royal Society of Chemistry, London.

Dhéré, C. & Biermacher, O. (1936) La feuille de Géranium vivante émet un rayonnement de fluorescence qui s'étend l'infrarouge jusqu'à λ 830 mμ. *Comptes Rendus de l'Académie des Sciences de Paris* **203**, 412–414.

dos Remedios, C.G., Millikan, R.G.C. & Morales, M.F. (1972) Polarization of tryptophan fluorescence from single striated muscle fibres. *Journal of General Physiology* **59**, 103–120.

Egensperger, R. & Holländer, H. (1988) Electron microscopic visualization of fluorescent microspheres used as a neuronal tracer. *Journal of Neuroscience Methods* **23**, 181–186.

Elkin, L. & Park, R.B. (1975) Chloroplast fluorescence of C_4 plants. I. Detection with infrared colour film. *Planta* **127**, 243–250.

Eisert, W.G. & Beisker, W. (1980) Epi-illumination optical design for fluorescence polarization measurements in flow systems. *Biophysical Journal* **31**, 97–112.

Ferrer, J.M., Garrigues, M.G. & Stockert, J.C. (1983) Mercurochrome, a fluorescent and electron opaque dye. *Microscopica acta* **87**, 293–300.

Furness, J.B., Costa, M. & Wilson, A.J. (1977) Water-soluble fluorophores, produced by reaction with aldehyde solutions, for the histochemical localisation of catechol- and indolethylamines. *Histochemistry* **52**, 159–170.

Ganse, H. (1977) Polarisations Mikroskopie. In: Beyer, H. (Ed.) *Handbuch der Mikroskopie,* 2nd edn, pp. 193–223. VEB Verlag Technik, Berlin.

Hartmann, H., Hund, A., Moll, S.H. & Thaer, A. (1978) Attachment for combined scanning electron and light microscopical examinations. *Beiträge zur elektronenmikroskopischen Direktabbildung von Oberflachen* **11**, 381.

Horisberger, M. & Vonlanthen, M. (1979) Fluorescent colloidal gold: a cytochemical marker for fluorescent and electron microscopy. *Histochemistry* **64**, 115–118.

James, J. (1976) *Light microscopic techniques in biology and medicine*. 336 pp. Martinus Nijhoff, Amsterdam.

Kieda, C., Roche, A.C., Delmotte, F. & Monsigny, M. (1979) Lymphocyte membrane lectins. Direct visualization by the use of fluoresceinyl-glycosylated cytochemical markers. *FEBS Letters* **99**, 329–332.

Lenga, R.E. (Ed.) (1988) *The Sigma-Aldrich library of chemical safety data*, 2nd edn. 4098 pp. Sigma-Aldrich Corporation, Milwaukee, WI.

Maranto, A.R. (1981) A technique enabling the electron microscopic examination of lucifer yellow-injected neurons. *Neuroscience Abstracts* **7**, 418.

Missmahl, H.P. (1966) Birefringence and dichroism of dyes and their significance in the detection of oriented structures. In: Wied, G.L. (Ed.) *Introduction to quantitative cytochemistry*, pp. 539–547. Academic Press, New York & London.

Molday, R.S., Dreyer, W.J., Rembaum, A. & Yen, S.P.S. (1975) New immunolatex spheres: visual markers of antigens on lymphocytes for scanning electron microscopy. *Journal of Cell Biology* **64**, 75–78.

Moneta, G., Brülisauer, M., Jäger, K. & Bollinger, A. (1987) Infrared fluorescence microscopy of skin capillaries with indocyanine green. *International Journal of Microcirculation: Clinical and Experimental* **6**, 25–34.

Moravec, J., Hatt, P.Y., Opie, L.H. & Rost, F.W.D. (1972) The application of the cytophotometer to the study of metabolic transitions of isolated rat heart. *Cardiology* **57**, 61–66.

Novello, A.F. & Brauer, M.M. (1986) Mercurochrome: a test for chromatin organization at the electron microscope level. *Cellular and Molecular Biology* **32**, 493–497.

Oster, G. (1955) Birefringence and dichroism. In: Oster, G. & Pollister, A.W. (Eds.) *Physical techniques in biological research*, Vol. **1**. Academic Press, New York & London.

Pabst, H. (1974) A new method of cleaning front lenses of objectives. *Leitz Scientific and Technical Information* **2**, 235.

Patzelt, W.J. (1986) *Polarized light microscopy*, 3rd edn. 102 pp. Ernst Leitz Wetzlar GmbH, Wetzlar.

Ploem, J.S. (1981) A new instrument permitting simultaneous scanning electron microscopy and fluorescence microscopy of the same specimen by integrating a LM in a SEM vacuum chamber. *Cytometry* **2**, 121.

Ploem, J.S. & Thaer, A. (1981) A light microscope integrated in the vacuum chamber of a SEM. *Proceedings of the Royal Microscopical Society* **16**, 253.

Puchtler, H., Meloan, S.N. & Paschal, L.D. (1980) Infrared fluorescence microscopy of stained tissues: principles and technic. *Histochemistry* **68**, 211–230.

Rost, F.W.D. (1972) Fluorescence microscopy. In: Pearse, A.G.E., *Histochemistry, theoretical and applied*, 3rd edn, Vol. **2**, Ch. 29, pp. 1171–1206. Churchill Livingstone, Edinburgh & London.

Rost, F.W.D. (1980) Fluorescence microscopy. In: Pearse, A.G.E., *Histochemistry, theoretical and applied*, 4th edn, Vol. **1**, Ch. 10, pp. 346–378. Churchill Livingstone, Edinburgh & London.

Thorball, N. (1982) The electron microscopy of fluorescent dextrans (FITCDextrans) in thin sections of tissue. *Experientia* **38**, 876.

Wouters, C.H. & Koerten, H.K. (1982) Combined light microscope and scanning electron microscope, a new instrument for cell biology. *Cell Biology International Reports* **6**, 955–959.

Wouters, C.H. & Ploem, J.S. (1987) Light and scanning electron microscopy in a combined instrument. In: Hayat, M.A. (Ed.) *Correlative microscopy in biology*, pp. 23–57. Academic Press, Orlando, FL, & London.

Chapter 6: Photography of fluorescence

Anon. (1988) *Conservation of photographs*. Kodak Publication F-40. Eastman Kodak Company, Rochester, NY.

Anon. (1989) *Kodak scientific imaging products*. Kodak Publication L-10. 78 pp. Eastman Kodak Company, Rochester, NY.

Bergquist, N.R. (1973) The pulsed dye laser as a light source for the fluorescent antibody technique. *Scandinavian Journal of Immunology* **2**, 37–44.

Blaker, A.A. (1977) *Handbook for scientific photography*. 379 pp. Freeman, San Francisco.

Blazek, M.C. (1983) How I photograph fluorescent minerals. *Jewellery Making Gems & Minerals* **543**, 72–74.

Bouleau, C. (1980) *The painter's secret geometry: a study of composition in art*, 2nd edn, 268 pp. Hacker Art Books, New York. (Translated by J. Griffin from *Charpentes, la géométrie secrete des peintres*. Editions du Seuil.)

Catala, R.L.A. (1964) *Carnaval sous la mer* (English edition: *Carnival under the sea)*. 141 pp. R. Sicard, Paris.

Cowen, B.C. (1969) A factual survey of fully automatic cameras currently available for use with the microscope. *Proceedings of the Royal Microscopical Society* **4**, 71–73.

Cripps, D.J., Hawgood, R.S. & Magnus, I.A. (1966) Iodine tungsten fluorescence microscopy for porphyrin fluorescence. *Archives of Dermatology* **93**, 129–134.

Croy, O.R. (1968) *Creative photomicrography*. 151 pp. Focal Press, London.

Dragesco, J. (1988) Reciprocity failure of photographic emulsions before and after hypersensitization by forming gas: a densitometric study. *Journal of the British Astronomical Association* **98**, 249–254.

Eastman Kodak Co. (1972) *Ultraviolet and fluorescence photography*. Publication M27. Rochester, NY.

Editors of Time-Life Books (1972) *Caring for photographs*. Time-Life International.

Elkin, L. & Park, R.B. (1975) Chloroplast fluorescence of C_4 plants. I. Detection with infrared colour film. *Planta* **127**, 243–250.

Elliot, J.G., Carpenter, J.A. & Hamdy, M.K. (1974) Technique for preparing high-quality photomicrographs by fluorescence microscopy. *Applied Microbiology* **28**, 1063–1065.

Gillett, K. & Yaldwyn, J. (1969) *Australian seashores in colour*. 112 pp. Reed, Sydney.

Jacobson, R.E., Ray, S.F. & Attridge, G.G. (Eds.) (1988) *The manual of photography*, 8th edn. 394 pp. Focal Press (Butterworths), London. (Formerly *The Ilford manual of photography*, first published 1890.)

James, J. (1976) *Light microscopic techniques in biology and medicine*. 336 pp. Martinus Nijhoff, Amsterdam.

James, T.H. (1972) Effects of moisture and oxygen on inherent and spectral sensitivity. *Journal of the Photographic Society* **20**, 182–186.

James, T.H. (1986) Chemical sensitization, spectral sensitization, and latent image formation in silver halide photography. *Advances in Photochemistry* **13**, 329–425.

Jonas, P. (1976) *Photographic composition simplified*. 96 pp. Amphoto, New York.

Koch, K.F. (1975) Fluorescence photomicrography. *Leitz Scientific and Technical Information* **6**, 149–156.

Lawson, D. (1972) *Photomicrography*. Academic Press, London & New York.

Loveland, R.P. (1970) *Photomicrography: a comprehensive treatise*. Vols. I and II. Wiley, New York, London, Sydney & Toronto.

Mertens, L.E. (1970) *In-water photography*. Wiley-Interscience, New York.

Myers, B. (1985) A less expensive interference filter for fluorescence photography. *Plastic and Reconstructive Surgery* **75**, 133–135.

Ploeg, M. van der, Bosman, F.T. & Duijn, P. van (1976) Preexposure of films in fluorescence photography. *Histochemical Journal* **8**, 201–204.

Quirk, S. (1990) Kodak print film review. *Southern Astronomy* **3**, 51–54.

Reynen, A.T.H. (1967) Grundlagen und Technik der Fluoreszenzmikrophotographie. *Acta histochemica* Supplementum **VII**, 367–370.

Rost, F.W.D. (1980) The laboratory microscope. *British Journal of Clinical Equipment* **5**, 146–150. (Reprinted in *Proceedings of the Royal Microscopical Society* (1981) **16**, 44–45.)

Rost, F.W.D. (1991) *Quantitative fluorescence microscopy*. 240 pp. Cambridge University Press, Cambridge.

Schuit, H.R.E. (1970) Photomicrographic recording of immunofluorescence. In: E.J. Holborow (Ed.) *Standardization in immunofluorescence*, pp. 159–162. Blackwell Scientific Publications, Oxford & Edinburgh.

Smith, A.G. (1986) Astronomical hypersensitization techniques applied to photomicrography. Kodak Technical Pan film 2415. *Journal of Microscopy* **144**, 39–44.

Smith, A.G. & Hoag, A.A. (1979) Advances in astronomical photography at low light levels. *Annual Review of Astronomy and Astrophysics* **17**, 43–71.

Young, M.R. (1969) Flash-fluorescence microscopy. *The Microscope* **17**, 15–18.

Chapter 7: The interpretation of the fluorescence microscopic image

Alexander, W.R.M. & Potter, J.L. (1963) Rhodamine-conjugated papain as counterstain in fluorescent microscopy. *Immunology* **6**, 450–452.

Berkley, K.J., Blomqvist, A., Pett, A. & Flink, R. (1980) Difference in the collaterization of neuronal projections from the dorsal column nuclei and lateral cervical nucleus to the thalamus and tectum in the cat: an anatomical study using two different double-labelling techniques. *Brain Research* **202**, 273–290.

de la Lande, I.S. & Waterson, J.G. (1968) Modification of autofluorescence in the formaldehyde-treated rabbit ear artery by Evans Blue. *Journal of Histochemistry and Cytochemistry* **16**, 281–282.

Hoffman, R. (1977) The modulation contrast microscope: principles and performance. *Journal of Microscopy* **110**, 205–222.

Hoffman, R. (1978) Applications of the modulation contrast microscope. *American Laboratory* **10** (4).

Holbrook, D.A. & Tishler, P.V. (1980) A simple procedure for distinguishing fluorescent bodies from certain artefacts in interphase Y-chromosome screening. *Cytogenetics and Cell Genetics* **2**, 59–60.

Horobin, R.W. (1982) *Histochemistry. An experimental outline of histochemistry and biophysical staining*. Fischer, Stuttgart; Butterworths, London.

Kuypers, H.G.J.M., Bentivoglio, M., Catsman-Berrevoets, C.E. & Bharos, A.T. (1980) Double retrograde labelling through divergent axon collaterals, using two fluorescent tracers with the same excitation wavelength which label different features of the cell. *Experimental Brain Research* **40**, 383–392.

Nairn, R.C. & Ploem, J.S. (1974) Moderne Immunofluoreszenzmikroskopie und ihre Anwendung in der klinischen Immunologie. *Leitz Mitteilungen für Wissenschaft und Technik* **6** (3), 91–95. (Reprinted in English as: Modern microscopy for immunofluorescence in clinical immunology. *Leitz Scientific & Technical Information* **2** (8), 91–95.)

O'Brien, T.P. & McCully, M.E. (1981) *The study of plant structure: principles and selected methods*. 352 pp. Termacarphi, Melbourne.

Ploeg, M. van der, Bonnet, J. & Duijn, P. van (1979) Pseudofluorescence of Giemsa-stained chromosomes. *Cytogenetics and Cell Genetics* **23**, 279–280.

Ritzén, M. (1967) Cytochemical identification and quantitation of biogenic amines. 63 pp. M.D. thesis, University of Stockholm.

Sander, G. (1970) Observer perception of fluorescence and autofluorescence. In: Holborow, E.J. (Ed.) *Standardization in immunofluorescence*, pp. 155–158. Blackwell Scientific Publications, Oxford & Edinburgh.

Schmued, L.C., Swanson, L.W. & Sawchenko, P.E. (1982) Some fluorescent counterstains for neuroanatomical studies. *Journal of Histochemistry and Cytochemistry* **30**, 123–128.

Sidgwick, J.B. (1979) *Amateur astronomer's handbook*, 4th edn, ed. Muirden, J. 568 pp. Pelham Books, London.

Tanke, H., Huikshoven, H. & de Jonge, S. (1984) Fluorescence microscopy of the fluorescent leprosy antibody absorption test (FLA-ABS). *International Journal of Leprosy* **52**, 93–96.

Tolansky, S. (1964) *Optical illusions*. 157 pp. Pergamon, Oxford.

Wolman, M. (1970) An optical artefact in the microscopic study of fluorescent materials. *Histochemie* **21**, 5–8.

Chapter 8: Quantitative and scanning fluorescence microscopy

Andrews, D.L. (1986) *Lasers in chemistry*. 160 pp. Springer-Verlag, Berlin, Heidelberg & New York.

Arndt-Jovin, D.J., Robert-Nicoud, M., Kaufman, S.J. & Jovin, T.M. (1985) Fluorescence digital imaging microscopy in cell biology. *Science* **230**, 247–256.

Axelrod, D., Koppel, D., Schlesinger, E., Elson, E. & Webb, W. (1976) Mobility measurement by analysis of fluorescence photobleaching recovery kinetics. *Biophysical Journal* **16**, 1005–1069.

Benson, D.M. & Knopp, J.A. (1984) Effect of tissue absorption and microscope optical parameters on the depth of penetration for fluorescence and reflectance measurements of tissue samples. *Photochemistry and Photobiology* **39**, 495–502.

Benson, D.M., Knopp, J.A. & Longmuir, I.S. (1980) Intracellular oxygen measurements of mouse liver cells using quantitative fluorescence video microscopy. *Biochimica et biophysica acta* **591**, 187–197.

Böhm, N. & Sprenger, E. (1968) Fluorescence cytophotometry: a valuable method for the quantitative determination of nuclear Feulgen-DNA. *Histochemie* **16**, 100–118.

Böhm, N., Sprenger, E. & Sandritter, W. (1970) Die Feulgen-Reaktion mit Acriflavin-SO_2. Ein Vergleich zwischen Absorptions- und Fluoreszenz Cytophotometrie anhand von Hydrolysekurven. *Acta histochemica* **35**, 324–329.

Böhm, N., Sprenger, E. & Sandritter, W. (1973) Absorbance and fluorescence cytophotometry of nuclear Feulgen DNA. A comparative study. In: Thaer, A.A. & Sernetz, M. (Eds.) *Fluorescence techniques in cell biology*, pp. 67–77. Springer-Verlag, Berlin, Heidelberg & New York.

Boyde, A. (1985) The tandem scanning reflected light microscpe (TSRLM) Part 2 – Pre-Micro '84 applications at UCL. *Proceedings of the Royal Microscopical Society* **20**, 131–139.

Boyde, A. (1987) Applications of tandem scanning reflected light microscopy and three-dimensional imaging. *Annals of the New York Academy of Sciences* **483**, 428–439.

Boyde, A. (1988) Confocal optical microscopy. *Microscopy and Analysis* Jan. 1988: 7–13.

Boyde, A. (1989) Fluorescence mode in the tandem scanning microscope. *Proceedings of the Royal Microscopical Society* **24**, 7.

Boyde, A., Jones, S.J., Taylor, M.L., Wolfe, L.A. & Watson, T.F. (1990) Fluorescence in the tandem scanning microscope. *Journal of Microscopy* **157**, 39–50.

Brand, L. & Gohlke, J.R. (1972) Fluorescent probes for structure. *Annual Review of Biochemistry* **41**, 843–868.

Bright, G.R. & Taylor, D.L. (1986) Imaging at low light level in fluorescence microscopy. In: Taylor, D.L., Waggoner, A.S., Lanni, F., Murphy, R.F. & Birge, R.R. (Eds) *Applications of fluorescence in the biomedical sciences*, pp. 257–288. Alan R. Liss, New York.

Broekaert, D., Ostveldt, P. van, Coucke, P., Bersaques, J. de, Gillis, E. & Reyniers, P. (1986) Nuclear differentiation and ultimate fate during epidermal keratinization. *Archives of Dermatological Research* **279**, 100–111.

Candy, B.H. (1985a) Photomultiplier characteristics and practice relevant to photon counting. *Review of Scientific Instruments* **56**, 183–193.

Candy, B.H. (1985b) Photon counting circuits. *Review of Scientific Instruments* **56**, 194–200.

Chapman, J.H., Förster, T., Kortüm, G., Parker, C.A., Lippert, E., Melhuish, W.H. & Nebbia, G. (1963) Proposal for standardization of methods of reporting fluorescence emission spectra. *Applied Spectroscopy* **17**, 171 (reprinted in Udenfriend, S. (1969) *Fluorescence assay in biology and medicine,* Vol. **II**, pp. 592–593. Academic Press, New York & London).

Cova, S., Prenna, G. & Mazzini, G. (1974) Digital microspectrofluorometry by multichannel scaling and single photon detection. *Histochemical Journal* **6**, 279–299.

Cox, I.J. (1984) Scanning optical fluorescence microscopy. *Journal of Microscopy* **133**, 149–154.

Curtis, S.K. & Cowden, R.R. (1983) Evaluation of five basic fluorochromes of potential use in microfluorometric studies of nucleic acids. *Histochemistry* **78**, 503–511.

Curtis, S.K. & Cowden, R.R. (1985) Microfluorometric estimates of proteins associated with murine hepatocyte and thymocyte nuclei, residual structures, and nuclear matrix derivatives. *Histochemistry* **82**, 331–339.

Deelder, A.M., Tanke, H.J. & Ploem, J.S. (1978) Automated quantitative immunofluorescence using the aperture defined microvolume (ADM) method. In: Knapp, W., Holubar, K. & Wick, G. (Eds.) *Immunofluorescence and related staining techniques*, Sixth international conference on immunofluorescence and related staining techniques, 1978, Vienna, pp. 31–44. Elsevier, Amsterdam & New York; North Holland Biomedical Press, Amsterdam.

dos Remedios, C.G., Millikan, R.G.C. & Morales, M.F. (1972a) Polarization of tryptophan fluorescence from single striated muscle fibres. *Journal of General Physiology* **59**, 103–120.

dos Remedios, C.G., Yount, R.G. & Morales, M.F. (1972b) Individual states in the cycle of muscle contraction. *Proceedings of the National Academy of Sciences, USA* **69**, 2542–2546.

Dowson, J.H. (1984) Partial recovery of fluorescence intensity after irradiation-induced fading of fluorescence, and its effects on techniques of quantitative fluorescence microscopy. *Experientia* **40**, 538–539.

Drexhage, K.H. (1973) Structure and properties of laser dyes. In: Shäfer, F.P. (Ed.) *Dye lasers* (Topics in applied physics, Vol. **1**), Ch.4, pp. 144–193. Springer-Verlag, Berlin, Heidelberg & New York.

Fukuda, M., Böhm, N. & Fujita, S. (1978) Cytophotometry and its biological application. *Progress in Histochemistry and Cytochemistry* **11** (1). 119 pp.

Gains, N. & Dawson, A.P. (1979) Relevance of the approximately hyperbolic relationship between fluorescence and concentration to the determination of quantum efficiencies. *Analyst* **104**, 481–490.

Geber, G. & Hasibeder, G. (1980) Cytophotometrische Bestimmung von DNA-Mengen: Vergleich einer neuen DAPI-Fluoreszenzmethode mit Feulgen-Absorptionsphotometrie. [Cytophotometric estimation of DNA contents: comparison of a new DAPI fluorescence method with Feulgen absorbance photometry.] *Microscopica acta* Supplementum **4**, 31–35.

Gillis, C.N., Schneider, F.H., Van Orden, L.S. & Giarman, N.J. (1966) Biochemical and microfluorometric studies of norepinephrine redistribution accompanying sympathetic nerve stimulation. *Journal of Pharmacology and Experimental Therapeutics* **151**, 46–54.

Green, W.B. (1983) *Digital image processing, a systems approach.* 192 pp. Van Nostrand-Reinhold, New York.

Herrman, J. & Wilhelmi, B. (1987) *Laser für ultrakurze Lichtimpulse*. Akademie-Verlag, Berlin. (English translation by Grossman, J. & Rudolph, W., *Lasers for ultrashort light pulses*. 302 pp. North-Holland Biomedical Press, Amsterdam.)

Inoué, S. (1986) *Video microscopy*. 584 pp. Plenum Press, New York & London.

Inoué, S. (1988) Progress in video microscopy. *Cell Motility and the Cytoskeleton* **10**, 13–17.

Jacobson, K., Elson, E., Koppel, D. & Webb, W. (1983) International workshop on the application of fluorescence photobleaching techniques to problems in cell biology. *Federation Proceedings* **42**, 72–79.

Jarvis, L.R. (1988) Microcomputer video image analysis. *Journal of Microscopy* **150**, 83–97.

Jongsma, A.P.M., Hijmans, W. & Ploem, J.S. (1971) Quantitative immunofluorescence. Standardization and calibration in microfluorometry. *Histochemie* **25**, 329–343.

Jonsson, G. (1971) Quantitation of fluorescence of biogenic amines. *Progress in Histochemistry and Cytochemistry* **2** (4). 36 pp.

Kassotis, J., Steinberg, S.F., Ross, S., Bilezikian, J.P. & Robinson, R.B. (1987) An inexpensive dual-excitation apparatus for fluorescence microscopy. *Pflügers Archiv* **409**, 47–51.

Koppel, D.E., Primakoff, P. & Myles, D.G. (1986) Fluorescence photobleaching analysis of cell surface regionalization. In: Taylor, D.L., Waggoner, A.S., Murphy, R.F., Lanni, F. & Birge, R.R. (Eds.) *Applications of fluorescence in the biomedical sciences*, pp. 477–497. Alan R. Liss, New York.

Larsson, P.-A., Goldstein, M. & Dahlström, A. (1984) A new methodological approach for studying axonal transport: cytofluorometric scanning of nerves. *Journal of Histochemistry and Cytochemistry* **32**, 7–16.

Letokhov, V.S. (Ed.) (1986) *Laser analytical spectrochemistry*. 430 pp. Adam Hilger, London.

Lichtensteiger, W. (1970) Katecholaminhaltige Neurone in der neuroendokrinen Steuerung. *Progress in Histochemistry and Cytochemistry* **1** (4). 92 pp.

MacInnes, J.W. & Uretz, R.B. (1966) Organization of DNA in dipteran polytene chromosomes as indicated by polarized fluorescence microscopy. *Science* **151**, 689–691.

Maeda, M. (1984) *Laser dyes: properties of organic compounds for dye lasers*. 335 pp. Academic Press, Tokyo, New York & London.

Marques, D. & Rost, F.W.D. (1973) Microspectrofluorometry of ultraviolet inducible fluorescence in supravitally-stained ascites tumour cells. *Histochemical Journal* **5**, 151–156.

Mellors, R.C. & Silver, R. (1951) A microfluorometric scanner for the differential detection of cells: application to exfoliative cytology. *Science* **114**, 356–360.

Miller, J.N. (1984) Recent developments in fluorescence and chemiluminescence analysis. *Analyst* **109**, 191–198.

Moneta, G., Brülisauer, M., Jäger, K. & Bollinger, A. (1987) Infrared fluorescence microscopy of skin capillaries with indocyanine green. *International Journal of Microcirculation: Clinical and Experimental* **6**, 25–34.

Morgan, C.G. (1987) Time resolved fluorescence microscopy. *Laboratory News*, April, 38 and 41.

Nordén, G. (1953) The rate of appearance, metabolism and disappearance of 3,4-benzpyrene in the epithelium of mouse skin after a single application in a volatile solvent. *Acta pathologica et microbiologica scandinavica* Supplementum **96**. 87 pp.

Oostveldt, P.M., Tanke, H.J., Ploem, J.S. & Boeken, G. (1978) Relation between fluorescent intensity and extinction in quantitative microfluorometry. *Acta histochemica* Supplementum **XX**, 59–63.

Pabst, H. (1980) Application of microspectrophotometry in forensic casework. *Microscopica acta* **S4**, 189.

Parker, C.A. (1969) Spectrophosphorimeter microscopy: an extension of fluorescence microscopy. *Analyst* **94**, 161–176.

Pearse, A.G.E. & Rost, F.W.D. (1969) A microspectrofluorimeter with epi-illumination and photon counting. *Journal of Microscopy* **89**, 321–328.

Peters, R. (1983) Fluorescence microphotolysis: diffusion measurements in single cells. *Naturwissenschaften* **70**, 294–302.

Petráň, M., Hadravsky, M., Benes, J., Kucera, R. & Boyde, A. (1985) The tandem scanning reflected light microscope. Part I – the principle, and its design. *Proceedings of the Royal Microscopical Society* **20**, 125–129.

Petráň, M., Hadravsky, M. & Boyde, A. (1985) The tandem scanning reflected light microscope. *Scanning* **7**, 97–108.

Petráň, M., Hadravsky, M., Egger, M.D. & Galambos, R. (1968) Tandem scanning reflected light microscope. *Journal of the Optical Society of America* **58**, 661–664.

Piller, H. (1977) *Microscope photometry*. 253 pp. Springer-Verlag, Berlin, Heidelberg & New York.

Ploeg, M. van der & Duijndam, W.A.L. (1986) Matrix models. Essential tools for microscopic cytochemical research. *Histochemistry* **84**, 283–300.

Ploeg, M. van der, Broek, K. van den, Smeulders, A.W.M., Vossepoel, A.M. & Duijn, P. van (1977) HIDACSYS: computer programs for interactive scanning cytophotometry. *Histochemistry* **54**, 273–288.

Ploem, J.S. (1970) Quantitative immunofluorescence. In: Holborow, E.J. (Ed.) *Standardization in immunofluorescence,* pp. 63–73. Blackwell Scientific Publications, Oxford & Edinburgh.

Ploem, J.S. (1986) New instrumentation for sensitive image analysis of fluorescence in cells and tissues. In: Lansing Taylor, D., Waggoner, A.S., Lanni, F., Murphy, R.F. & Birge, R.R. (Eds.) *Applications of fluorescence in the biomedical sciences*, pp. 289–300. Alan R. Liss, New York.

Ploem, J.S. & Tanke, H.J. (1987) *Introduction to fluorescence microscopy*. 56 pp. Oxford University Press/ Royal Microscopical Society, Oxford.

Pohle, W., Ott, T. & Müller-Welde, P. (1984) Identification of neurons of origin providing the dopaminergic innervation of the hippocampus. *Zeitschrift für Hirnforschung* **25**, 1–10.

Rigler, R. (1966) Microfluorometric characterisation of intracellular nucleic acids and nucleoproteins by Acridine Orange. *Acta physiologica scandinavica* **67**, Supplementum **267**. 122 pp.

Ritzén, M. (1966) Quantitative fluorescence microspectrophotometry of catecholamine formaldehyde products. *Experimental Cell Research* **44**, 505–520.

Ritzén, M. (1967) Cytochemical identification and quantitation of biogenic amines. 63 pp. MD thesis, University of Stockholm.

Rosenfeld, A. & Kak, A.C. (1982) *Digital picture processing*, Vols **1** and **2**, 2nd edn. Academic Press, New York.

Rost, F.W.D. (1974) Microspectrofluorometry. *Medical Biology* **52**, 73–81.

Rost, F.W.D. (1980) Quantitative histochemistry. In: Pearse, A.G.E. *Histochemistry theoretical and applied,* Ch. 11, pp. 379–417. Churchill Livingstone, Edinburgh, London & New York.

Rost, F.W.D. (1991) *Quantitative fluorescence microscopy*. 240 pp. Cambridge University Press, Cambridge.

Rost, F.W.D. & Pearse, A.G.E. (1971) An improved microspectrofluorimeter with automatic digital data logging: construction and operation. *Journal of Microscopy* **94**, 93–105.

Rost, F.W.D. & Pearse, A.G.E. (1974) Microfluorometry of primary and secondary fluorescence in biological tissue. *Histochemical Journal* **6**, 245–250.

Ruch, F. (1970) Principles and some applications of cytofluorometry. In: Wied, G.L. & Bahr, G.F. (Eds.) *Introduction to quantitative cytochemistry* II, pp. 431–455. Academic Press, New York.

Ruch, F. (1973) Quantitative determination of DNA and protein in single cells. In: Thaer, A.A. & Sernetz, M. (Eds.) *Fluorescence techniques in cell biology*, pp. 89–93. Springer-Verlag, Berlin, Heidelberg & New York.

Scheuermann, D.W., Stilman, C. & De Groodt-Lasseel, M.H.A. (1988) Microspectrofluorometric analysis of the formaldehyde-induced fluorophores of 5-hydroxytryptamine and dopamine in intrapulmonary neuroepithelial bodies after administration of L-5-hydroxytryptophan and L-DOPA. *Histochemistry* **88**, 219–226.

Schipper, J., Tilders, F.J.H. & Ploem, J.S. (1980a) A scanning microfluorimetric study on sympathetic nerve fibres: intraneuronal differences in noradrenaline turnover. *Brain Research* **190**, 459–472.

Schipper, J., Tilders, F.J.H., Groot Wasink, R., Boleij, H.F. & Ploem, J.S. (1980b) Microfluorometric scanning of sympathetic nerve fibres: quantification of neuronal and extraneuronal fluorescence with aid of histogram analysis. *Journal of Histochemistry and Cytochemistry* **28**, 124–132.

Schnedl, W., Roscher, U., Ploeg, M. van der & Dann, O. (1977) Cytofluorometric analysis of nuclei and chromosomes by DIPI staining. *Cytobiologie* **15**, 357–362.

Sernetz, M. & Thaer, A. (1970) A capillary fluorescence standard for microfluorometry. *Journal of Microscopy* **91**, 43–52.

Sernetz, M. & Thaer, A. (1973) Microcapillary fluorometry and standardization for microscope fluorometry. In: Thaer, A. & Sernetz, G. (Eds.) *Fluorescence techniques in cell biology*, pp. 41–49. Springer-Verlag, Berlin, Heidelberg & New York.

Serra, J. (1986) Morphological optics. *Journal of Microscopy* **145**, 1–22.

Shack, R.V., Bartels, P.H., Buchroeder, R.A., Shoemaker, R.L., Hillman, D.W. & Vukobratovich, D. (1987) Design for a fast fluorescence laser scanning microscope. *Analytical and Quantitative Cytology and Histology* **9**, 509–521.

Shapiro, H.M. (1988) *Practical flow cytometry*, 2nd edn, 353 pp. Alan R. Liss, New York.

Shotton, D.M. (1988) Review: video-enhanced light microscopy and its applications in cell biology. *Journal of Cell Science* **89**, 129–150.

Sisken, J.E., Barrows, G.H. & Grasch, S.D. (1986) The study of fluorescent probes by quantitative video intensification microscopy (QVIM). *Journal of Histochemistry and Cytochemistry* **34**, 61–66.

Slayter, E.M. (1970) Measurements of luminescence. In: *Optical methods in biology*, Ch. 25, pp. 552–574. Wiley-Interscience, New York.

Smith, L.C., Benson, D.M., Gotto, A.M. & Bryan, J. (1986) Digital imaging fluorescence microscopy. *Methods in Enzymology* **129**, 857–873.

Smith, M.T., Redick, J.A. & Baron, J. (1983) Quantitative immunohistochemistry: a comparison of microdensitometric analysis of unlabeled antibody peroxidase–antiperoxidase staining and of microfluorometric analysis of indirect fluorescent antibody staining for nicotinamide adenosine dinucleotide phosphate (NADPH)–cytochrome *c* (P-450) reductase in rat liver. *Journal of Histochemistry and Cytochemistry* **31**, 1183–1189.

Steinkamp, J.A. (1984) Flow cytometry. *Review of Scientific Instruments* **55**, 1375–1400.

Sternberger, L.A. (1986) *Immunocytochemistry*, 3rd edn. John Wiley & Sons, New York.

Takamatsu, T. & Fujita, S. (1988) Microscopic tomography by laser scanning microscopy and its three-dimensional reconstruction. *Journal of Microscopy* **149**, 167–174.

Takamatsu, T., Kitamura, T. & Fujita, S. (1986) Quantitative fluorescence image analysis. *Acta histochemica et cytochemica* **19**, 61–71.

Talmi, Y. (Ed.) (1983) *Multichannel image detectors*, Vol. **2**. ACS Symposium Series, No. 236. American Chemical Society, Washington, DC.

Tanke, H.J. & Ingen, E.M. van (1980) A reliable Feulgen-acriflavine-SO_2 staining procedure for quantitative DNA measurements. *Journal of Histochemistry and Cytochemistry* **28**, 1007–1013.

Thaer, A.A. & Sernetz, M. (1973) *Fluorescence techniques in cell biology*. 420 pp. Springer-Verlag, Berlin, Heidelberg & New York.

Tiffe, H.-W. & Hundeshagen, M. (1982) Investigation of fading and recovery of fluorescence intensity at 73–5 K. *Journal of Microscopy* **126**, 231–235.

Udenfriend, S. (1962) *Fluorescence assay in biology and medicine*. 505 pp. Academic Press, New York.

Voort, H.T.M. van der, Valkenburg, J.A.C., Spronsen, E.A. van, Woldringh, C.L. & Brakenhoff, G.J. (1987) Confocal microscopy in comparison with electron and conventional light microscopy. In: Hayat, M.A. (Ed.) *Correlative microscopy in biology*, pp. 60–81. Academic Press, Orlando, FL, & London.

Wahl, P. (1975) Nanosecond pulse fluorometry. In: Pain, R. & Smith, B. (Eds.) *New techniques in biophysics and cell biology*, pp. 233–285. John Wiley & Sons, Chichester.

Watkinson, J. (1988) *The art of digital audio*. 489 pp. Focal Press, London.

Weiss, D.G., Maile, W. & Wick, R.A. (1989) Video microscopy. In: Lacey, A.J. (Ed.) *Light microscopy in biology*, pp. 221–278. IRL Press at Oxford University Press, Oxford, New York & Tokyo.

Wehry, E.L. (1982) Molecular fluorescence, phosphorescence and chemiluminescence spectrometry. *Analytical Chemistry Review* **54**, 131–150.

Wied, G.L. (Ed.) (1966) *Introduction to quantitative cytochemistry*. 623 pp. Academic Press, New York & London.

Wied, G.L. & Bahr, G.F. (Eds.) (1970) *Introduction to quantitative cytochemistry II*. 551 pp. Academic Press, New York & London.

Wijnaendts van Resandt, R.W., Marsman, H.J.B., Kaplan, R., Davoust, J., Stelzer, E.H.K. & Stricker, R. (1985) Optical fluorescence microscopy in three dimensions: microtomoscopy. *Journal of Microscopy* **138**, 29–34.

Willingham, M.C. & Pastan, I. (1978) The visualization of fluorescent proteins in living cells by video intensification microscopy (VIM). *Cell* **13**, 501–507.

Wittig, B., Rohrer, F. & Zetzsch, C. (1984) Intelligent microcomputer interface for continuous registration and storage of spectra by photon counting. *Review of Scientific Instruments* **55**, 375–378.

Wreford, N.G.M. & Smith, G.C. (1982) Microspectrofluorometry in biogenic amine research. *Brain Research Bulletin* **9**, 87–96.

Zs.-Nagy, I., Ohta, M., Kitani, K. & Imahori, K. (1984) An automated method for measuring lateral motility of proteins in the plasma membrane of cells in compact tissues by means of fluorescence recovery after photobleaching. *Mikroskopie (Wien)* **41**, 12–25.

Appendix 1: German–English vocabulary

Beyer, H. (1977) (Ed.) *Handbuch der Mikroskopie*, 2nd edn. 495 pp. VEB Verlag Technik, Berlin.

Bindmann, W. (1975) *Fachwörterbuch Optik und optischer Gerätebau.* Deutsch-Englisch. 432 pp. VEB Verlag Technik, Berlin; Verlag für Augenheilkunde und Optik, Dresden.

Bradbury, S., Evennett, P.J., Haselmann, H. & Piller, H. (1989) *Dictionary of light microscopy.* 139 pp. Oxford University Press/Royal Microscopical Society, Oxford.

De Vries, L. (1952) *German–English medical dictionary.* 586 pp. McGraw-Hill, New York.

Schoenewald, F.S. (1952) *German–English medical dictionary.* 239 pp. H.K. Lewis, London.

Walther, R. (Ed.) (1967) *Polytechnical dictionary,* Vol. **2**, German–English. 1046 pp. Pergamon Press, Oxford; VEB Verlag Technik, Berlin.

Appendix 3A: The laboratory microscope

Ganse, H. (1977) Polarisations Mikroskopie. In: Beyer, H. (Ed.) *Handbuch der Mikroskopie,* 2nd edn, pp. 193–223. VEB Verlag Technik, Berlin.

James, J. (1976) *Light microscopic techniques in biology and medicine.* 336 pp. Martinus Nijhoff, Amsterdam.

Patzelt, W.J. (1986) *Polarized light microscopy*, 3rd edn. 102 pp. Ernst Leitz Wetzlar GmbH, Wetzlar.

Index to Volume I

Page numbers in italics refer to figures and tables. For synonyms of names of fluorochromes, consult Volume II, Appendix 7.